David Bohm: Causality and Chance, Letters to Three Women

Chris Talbot
Editor

David Bohm: Causality and Chance, Letters to Three Women

Springer

Editor
Chris Talbot
Oxford
UK

Birkbeck Library is acknowledged for the kind permission to publish transcriptions of the letters of David Bohm, whose originals are held in its archive

ISBN 978-3-319-55491-4 ISBN 978-3-319-55492-1 (eBook)
DOI 10.1007/978-3-319-55492-1

Library of Congress Control Number: 2017936037

Printed on acid-free paper

This Springer imprint is published by Springer Nature
The registered company is Springer International Publishing AG
The registered company address is: Gewerbestrasse 11, 6330 Cham, Switzerland

Preface

As a postgraduate theoretical physics student in the late 1960s and early 1970s, I became involved in left-wing politics as did many others of my generation. I was introduced to David Bohm's *Causality and Chance in Modern Physics* as one of the very few books available which combined Marxist philosophy and physics.[1] There were no references to Marxist works in Bohm's book, perhaps because he was constrained by the dominant Cold War ethos, but it was clearly in the dialectical materialist tradition. I found it extremely interesting but since very few people in left-wing circles had any knowledge of science, and certainly not quantum physics, I had no occasion to discuss it further. I also presumed, mistakenly, that Bohm's "hidden variable", later called "causal interpretation", approach to quantum mechanics had turned out to be a dead end. I did not think I could relate it to my own research. Not unlike today there were few job openings in theoretical physics, so with a young family to support I became a mathematics lecturer in a polytechnic, or a "new" university as they are now called, and concentrated on teaching and directed my research into engineering.

When I retired and was away from the pressure of teaching, administration and finding research funding, I returned to look at Bohm's work. I had read David Peat's colourful biography of Bohm and noted the many references in the period when Bohm was exiled in Brazil to personal letters to Hanna Loewy, Miriam Yevick and Melba Phillips. It seemed possible that these could give some insight into how Bohm had developed his ideas in *Causality and Chance*, so I inquired at Birkbeck College, University of London, library archives to find out what was available. I was delighted to find that there were photocopies of what eventually turned out to be 124 letters written by David Bohm in the first half of the 1950s. Quite a few were incomplete, Bohm's handwriting was often unclear, the photocopies were frequently poor and many were not in date order. Consequently I spent more than two years transcribing and editing the letters which are assembled here.

[1]There were English translations of Soviet books, but they were mostly uninspiring "official" publications.

In the meantime I read up on the Bohmian version of quantum mechanics, studying to the best of my ability the growing body of material from the Bohmian mechanics group, Bohm and Hileys' *The Undivided Universe*, James T. Cushing's *Quantum Mechanics*, and so on. I soon realized that I had been mistaken in thinking Bohm's approach to quantum mechanics had got nowhere, his "causal interpretation" was just as valid as the standard "Copenhagen" version of quantum theory taught in universities and deserved much more recognition than it had hitherto received. Reading the letters convinced me that the philosophy set out in "Causality and Chance" was also a major part of Bohm's work in that period, it was difficult to separate from the purely scientific work, and for reasons mainly related to the influence of Stalinism, has been largely ignored.

The letters are difficult to read through, containing a mixture of comments on scientific work, philosophy and politics together with details of personal experiences and problems. I decided the best way to draw out the importance of the letters was to write an introduction with various themes: philosophy, mathematics, the causal interpretation, probability and statistical mechanics, feminism and politics, attempting to explain Bohm's views and how they relate to scattered parts of the letters. I had to update my knowledge of theoretical physics to get to grips with the topics involved, and found I needed to gain a better understanding of Marxist philosophy. What passed for Marxist philosophy in the left-wing circles I had moved in was woefully inadequate, often highly disputative, and demonstrating only a nodding acquaintance with the actual ideas of Marx and Engels. I have taken the view that the theoretical outlook of Marxism, including the work on philosophy by Lenin and then the philosophical debates in the USSR in the 1920s, has an importance in its own right. I think the references I have given back that up. Historical studies, which I also reference, have shown the huge break from the 1920s to the repression that took place under Stalin's "revolution from above" in the 1930s onwards. Although there is still a widespread tendency, cultivated in the Cold War period, to identify Marxism with the Stalin period, it has no basis in historical fact.

In Brazil, Bohm was clearly fully committed to Stalinist politics, so I felt it necessary to try to understand something of the history of the USSR and the nature of Stalinism. Marxist philosophy was suppressed and distorted under Stalin's rule, and science also came under attack, most seriously in genetics with the well-known case of Lysenko but physics only narrowly escaped because of its importance in making an atomic bomb. Therefore I have added a chapter attempting to summarize, with references, what happened to science and philosophy in the USSR from the 1930s onwards. The letters show that Bohm's support for the USSR did not mean he restricted himself to Stalinist philosophical dogma. He had a sufficient knowledge of the Marxist classics to develop his own philosophy, especially in physics where he was at his most original and could base himself on a sound scientific understanding. In my opinion it was the suppression of genuine Marxism in the Soviet Union that contributed to Bohm's isolation in the Brazil period, and explains why *Causality and Chance* did not gain the interest it deserved on either side of the Iron Curtain.

Finally I felt that many readers might feel that I had concentrated on the science, philosophy and politics because these were the areas in which I was interested, at the expense of psychological issues. Therefore in the last chapter I have attempted to explain the huge pressures that Bohm was under in Brazil, and which go some way to explaining the emotional explosions that colour the letters.

I could only put together this book with the help of many people. I interviewed Basil Hiley two years ago and he kindly gave his support for editing the letters. Since then he commented favourably on a draft of my introduction and helped steer through copyright issues regarding the letters at Birkbeck College library. I have had considerable help from Olival Freire Jr. with whom I have corresponded and also met while he was in London. In discussion Olival generously shared his wide knowledge of Bohm's history and helped correct some of my mistakes in an earlier draft. I have also been helped by correspondence with other academics in the Bohmian area, the Finnish philosopher of mind Paavo Pylkkänen and the historian of science José Perillan of Vassar College, New York. Historical information was obtained from academics who remember David Bohm from the 1950s, Mario Bunge of McGill University and Silvan Schweber of Harvard. I must, of course, stress that the interpretation of the letters given in the introduction is entirely my own, as are any mistakes and omissions.

Working in the archives at Birkbeck was made possible by the helpfulness and interest in David Bohm of librarian Sue Godsell, now retired, and the current librarian Emma Illingworth who has maintained support and enthusiasm. Thanks to Huddersfield University for allowing me to continue with email and library access in retirement. I particularly need to thank Gratiela Neçsutu for meticulous proofing of my transcriptions of the letters. My editor Angela Lahee at Springer must be thanked for her interest in Bohm as well as her patience with my slow rate of work.

Finally I must thank my three children for their continued support and interest. I am especially indebted to my wife, Ann, for putting up with hours of discussion exploring the ideas of David Bohm and for her invaluable historical help as I attempted to unravel the complexities of the post-World War II world.

Oxford, UK Chris Talbot

Contents

Part I
Introduction

Chapter 1
Introduction

David Bohm (1917–1992) was a physicist, but his influence has gone far beyond the world of physics. He has become one of the most widely discussed intellectual figures of the twentieth century. While his work is still relevant to physicists and plays a part in contemporary areas of cutting edge research, Bohm's ideas have taken on a much wider cultural significance, reflecting his own wide-ranging genius. Growing up in the coal mining town of Wilkes-Barre, Pennsylvania, his outstanding talent eventually led him to be recruited into Robert Oppenheimer's theoretical physics group at Berkeley, where he joined the world's leading physics researchers at the beginning of World War II. He carried out brilliant investigations in a number of areas of front-line research, especially plasma physics, and was recognized as an equal to Richard Feynman in original thinking. At Berkeley he mixed with politically radical students in Oppenheimer's group, such as Joseph Weinberg and Rossi Lomanitz, and became a convinced supporter of Marxism and of the Soviet Union, joining the Communist Party for a brief period. Because of his political views, he was refused security clearance to work under Oppenheimer on the development of the atomic bomb at Los Alamos. However, his doctoral dissertation was regarded as so important that it was classified and used in the Manhattan Project.

At the beginning of 1947, Bohm was appointed to an Assistant Professorship at Princeton, where he delivered undergraduate lectures on quantum mechanics. He supervised research into plasma physics and other areas, and was able to hold discussions with Einstein on issues of fundamental physics. With the McCarthyite witch-hunt at its height during the Cold War, Bohm was summoned to appear before the House Un-American Activities Committee (HUAC) in 1949. He pleaded the Fifth Amendment and refused to testify. In response, in December 1950 he was indicted for contempt of Congress and arrested. He was driven to Trenton, the local capital, by a marshal, with whom he later recalled that, in his characteristic fashion, he had discussed Einstein's theories. He was eventually granted bail and released, but he was suspended from his post at Princeton. In May 1951 he appeared in court and was acquitted on all counts. His students had campaigned in his support and the physics

C. Talbot (ed.), *David Bohm: Causality and Chance, Letters to Three Women*,
DOI 10.1007/978-3-319-55492-1_1

department had praised his teaching and research, recommending that his contract be continued. However, in the anti-communist atmosphere of the time, Bohm was sacked by Princeton president Harold W. Dodds in June 1951. Unable to find work in the US and worried he could be imprisoned by a government he saw as increasingly fascistic, he took a job as Professor of Physics at the University of São Paulo, Brazil.[1] Most of the letters collected in this book were written during Bohm's stay in Brazil.

When he had spent a year at Princeton,[2] Bohm began lecturing on advanced quantum mechanics to graduate students. The notes from this course, based originally on courses given by Oppenheimer at Berkeley, became the basis for Bohm's highly regarded text-book *Quantum Theory*,[3] which was published in 1951, before Bohm left for Brazil. In the physics of quantum mechanics, there were, and still are, particularly difficult philosophical problems. It was the standard or "Copenhagen" approach to quantum mechanics that was being taught in universities in Bohm's day—and still is today, with very few exceptions.[4] Bohm's distinctive approach in his book was to develop the student's conceptual understanding, attempting to clarify the strange features of the theory rather than stressing the formal mathematical side, which most textbooks had done and still do. He brought out the intrinsic randomness of individual processes at the atomic level, and the so-called wave-particle duality. He investigated the measurement process which, some claimed, showed that the consciousness of the observer would determine the outcome of experiments, etc., etc. It is also evident, and this will become clearer when we look at the letters, that, without any explicit references, Bohm was attempting to develop a Marxist dialectical materialist approach to standard quantum mechanics.[5]

References

Bohm, D. (1951). *Quantum Theory*. Dover. (Reprint of New Jersey original, 1989).

Freire, O., Jr. (2015). *The quantum dissidents: rebuilding the foundations of quantum mechanics (1950–1990)*. Berlin Heidelberg: Springer.

Mullet, S. K. (2008). Little man: Four junior physicists and the red scare experience. Ph.D thesis, Harvard University.

Peat, D. (1996). *Infinite potential*. New York: Basic Books.

[1]For this earlier period see Peat (1996), Chaps. 1–7, Freire (2015), Chap. 2, and Mullet (2008).

[2]Peat (1996), p. 74.

[3]Bohm (1989).

[4]See Chap. 6 for a brief outline.

[5]See Chap. 6.

Chapter 2
The Causal Interpretation and Causality and Chance

In spite of the book's acclaim, as the letters show, Bohm was not satisfied that he had dealt adequately with the philosophical issues from the standpoint of Marxism. In a relatively short period at Princeton, in the year before leaving for Brazil, he developed his alternative "hidden variable", or "causal", as it was later called, approach to quantum mechanics. It is also known as the Bohm-de Broglie approach, since Bohm unwittingly repeated Louis de Broglie's so-called "pilot wave" theory of the 1920s. After criticism from other physicists, particularly Wolfgang Pauli, de Broglie had dropped this approach in 1927. Bohm had a thorough grasp of the standard theory and its weaknesses after writing his book and was therefore able to deal in detail with Pauli's criticisms.[1] He eventually published the two papers setting out the "causal" interpretation at the beginning of 1952, after he had arrived in Brazil.[2] Pauli, who was now the reviewer of Bohm's papers, had to admit they gave a consistent approach to quantum mechanics, despite his intense hostility to Bohm's philosophy. To this day, Bohm's version of quantum mechanics is just as valid as the standard theory. No experimental test that gives a result confirming the standard theory has yet been devised that does not also confirm Bohm's theory.

As we shall see from these letters, in Brazil, notwithstanding illness and depression, Bohm intensively developed scientific, philosophical and political views from his distinctive Marxist standpoint. In science, he not only worked on developing the "causal" approach so as to include spin and relativity, he also carried out concentrated work on probability theory and statistical mechanics. He had hoped to interest other physicists in the "causal" interpretation but as the letters show, he became increasingly discouraged at the possibility of achieving this. Physicists, even Communist Party members, wanted to see "results". Unless new physical phenomena, explicable only by Bohm's interpretation, were discovered, or developments were

[1] See Freire Jr. (2015), pp. 31–32 for more on this.

[2] Bohm (1952a,b).

C. Talbot (ed.), *David Bohm: Causality and Chance, Letters to Three Women*,
DOI 10.1007/978-3-319-55492-1_2

made by Bohm in new areas, such as particle physics, the physics community remained sceptical.[3]

It must be stressed that Bohm's ideas in this early period formed a closely integrated whole, with philosophy and politics being given as much, if not more, attention than science in the letters.[4] The politics was definitely Stalinist but was not intended for public discussion, and in any case, went through a drastic shake-up after Bohm's moving to Israel, with all the revelations about the USSR that became available. However, from 1952 onwards, Bohm clearly intended to put together a book on science and philosophy from a Marxist standpoint.[5] This was eventually published as *Causality and Chance in Modern Physics* in 1957.[6] As with *Quantum Theory*, there are no explicit references to Marxism in the book, but the letters help to shed much light on how the dialectical materialist ideas that went into it were developed, making it one of the few serious attempts to bring together the Marxist philosophical tradition and physics in the 20th century.[7]

It is worth recalling here how the "causal interpretation" is referred to in *Causality and Chance*. The new interpretation of quantum mechanics was not to be regarded as a finished or final theory. That would go entirely against the dialectical materialist conception, namely that scientific theories are not free from error but rather an "unending process in which the degree of truth in our knowledge is continually increasing".[8] In *Causality and Chance* Bohm explains that he had intended to show that "alternative interpretations of the quantum theory were in fact possible".[9] He even argued that his theory had "many aspects which seemed quite artificial and

[3]It was probably mainly this pragmatism and inherent conservatism that prevented Bohm from making a greater impact, though widespread anti-communism certainly also played a part. See Freire (2005) for a discussion on this issue.

[4]References to letters throughout this introduction are given in the form (X,**Y**, p. Z) where X is the chapter in Part 2, **Y** the letter number, and Z the page number.

[5]There are several references in the letters to Miriam Yevick in 1952 to the difficulties in writing a proposed book on philosophy. Then, in February 1953 (26, **96**, p. 317), he tells Miriam he is experimenting with a number of ideas, which he may publish first as articles. A little later, he writes to Melba Phillips about a book gradually taking shape (18, **42**, p. 163) and thanks her for efforts in trying to get a "paper" on causality published (18, **45**, p. 169). In April 1954, he tells Miriam that his book has been accepted by Routledge and Kegan Paul (30, **116**, p. 395), and in August (31, **120**, p. 414), he has a six-page summary of his ideas on probability (not found in the archives). Finally, in October 1955 (19, **52**, p. 180), he writes to Melba that the publishers have asked him to shorten the book, cutting out some technical material, and he also decides to cut out material on positivism. The revised version, with five chapters remaining, is probably the book in its present form.

[6]Bohm (1957).

[7]This is not to dismiss the work of Soviet physicists such as V.A. Fock (see Graham (1971), especially Chap. III), but they were working under even more disadvantageous circumstances than Bohm, as we will attempt to show in Chap. 12 on Soviet Physics and Philosophy.

[8]Bohm (1957), Chap. 5, especially Sect. 12.

[9]Bohm (1957), Chap. 4, Sect. 3.

unsatisfactory", and listed the criticisms of his new approach.[10] In particular, he made the remarkable point that:

> . . . our model in which wave and particle are regarded as basically different entities, which interact in a way that is not essential to their modes of being, does not seem very plausible. The fact that wave and particle are never found separately suggests instead that they are both different aspects of some fundamentally new kind of entity which is likely to be different from a simple wave or a simple particle, but which leads to these two limiting manifestations as approximations that are valid under appropriate conditions.[11]

This was written long before the seminal work of John Stewart Bell appeared, and the recognition of the phenomenon of quantum "entanglement", including the fact that at the quantum level, in either the standard or Bohm's interpretation, matter cannot be isolated in a localised, particle-like form. Bell would later use the term "beable". Bohm himself seems already to have intuited the existence of some such entity.

Whatever the limitations of Bohm's theory, it was and still remains a challenge to standard quantum mechanics. Even if one accepts that the standard interpretation does not necessarily imply that the observer's consciousness affects the outcome of quantum processes,[12] the idea that there is an ultimate purely random level in nature is at odds both with Bohm's version of quantum mechanics and the philosophy he put forward in *Causality and Chance*.

References

Bohm, D. (1952a). A suggested interpretation of the quantum theory in terms of hidden variables i. *Physical Review*, *85*(2), 166–179.

Bohm, D. (1952b). A suggested interpretation of the quantum theory in terms of hidden variables ii. *Physical Review*, *85*(2), 180–193.

Bohm, D. (1957). *Causality and chance in modern physics* (first ed., p. 1984). London: Routledge and Kegan Paul (Second edition with new preface by Bohm).

Freire, O., Jr. (2005). Science and exile: David Bohm, the hot times of the Cold war, and his struggle for a new interpretation of quantum mechanics. *Historical Studies on the Physical and Biological Sciences*, *36*(1). Also at http://arxiv.org/abs/physics/0508184.

Freire, O., Jr. (2015). *The quantum dissidents: Rebuilding the foundations of quantum mechanics (1950–1990)*. Berlin, Heidelberg: Springer.

Graham, L. R. (1971). *Science and philosophy in the Soviet Union*. London: Allen Lane.

Rosenblum, B., & Kuttner, F. (2011). *Quantum enigma: Physics encounters consciousness*. USA: Oxford University Press.

---

[10] As Bohm wrote in his second 1952 paper: "We should never expect to obtain a complete theory of this structure [the objectively real world of unlimited complexity]" (Bohm 1952b), and in 1953: "It is true that this model is somewhat crude, and that a deeper synthesis should be sought" (18, **43**, p. 165).

[11] Bohm (1957), Chap. 4, Sect. 5.

[12] Though many writers appear to think it does. See Rosenblum and Kuttner (2011) for a recent example.

Chapter 3
Bohm After Brazil

After Bohm moved to Israel in 1955, the few letters in the archives (all to Melba Phillips) show a distinctive change. In philosophy, he continued and deepened the study of Hegel he had begun in Brazil, especially under the influence of the Brazilian physicist and Communist Mario Schönberg. Whatever his misgivings about Marxism, however,[1] he went ahead with the publication of *Causality and Chance*.

More and more revelations about what was happening in the Soviet Union became available, which, while in Brazil, Bohm had insisted were only temporary problems, due to the backward conditions, to hostility from the west, and so on. But in Israel, as well as through Kruschev's revelations, he learnt directly about Russia and Eastern Europe from exiles, some of whom were long-standing Communists. The letters show he abandoned his support for Stalinist Communist Party politics, though continuing at that time to hold socialist ideals. Although he met and then married Saral, which put him on a more stable emotional footing, he appears to have gone through a considerable intellectual crisis.

Bohm appears to have spent much of the late 1950s and the 1960s pursuing "holistic" philosophy, studying the philosophy of Hegel[2] and A.N. Whitehead, conducting dialogues with the Indian speaker and writer Jiddu Krishnamurti and with a

---

[1]In (19, **54**, p. 182), in 1956 he wrote the following to Melba: "I have been studying Hegel (along with some other people here in Israel). It is true that Marx and Engels stood Hegel's ideas on their feet, by making them materialistic. Nevertheless, there was a tremendous wealth of ideas that they did not use, because the science of the time did not require them. But now, with the further development of science, these ideas applied to space, time and matter are surprisingly fruitful, as well as beautiful".

[2]An interesting summary of Bohm's views on Hegel in the 1960s is given by Paul Feyerabend, who discussed with Bohm when they were both at Bristol University (Radner and Winokur 1970), pp. 31–36 and 113–116). Feyerabend is well aware of the connections between Marxism and Hegel and teases the philosopher of science Imre Lakatos for pretending to be a Wittgensteinian and hiding his dialectical training in Hungary (by Georg Lukacs among others).

C. Talbot (ed.), *David Bohm: Causality and Chance, Letters to Three Women*,
DOI 10.1007/978-3-319-55492-1_3

research student, Donald Schumacher, on language, exchanging letters with Charles Biederman on art,[3] and so on.[4]

In terms of his work on physics in the 1960s period, he is mostly remembered for work with Yakir Aharonov, his research student in Israel and at Bristol, which led to the proposal of the Bohm-Aharonov effect.[5] But on the causal interpretation, we know from Basil Hiley that "in the first ten years I worked with David Bohm his '52 paper was not discussed at all", the main reason being that "first of all David Bohm was not that interested in it and secondly I didn't believe it."[6] Only when a student insisted that Hiley studied it did he become interested.[7] With the involvement of a number of students—Christopher Philippidis, Christopher Dewdney, Peter Holland, Fabio Frescura and others—Bohm and Hiley renewed work on the causal interpretation.[8] From the early 1970s on, Bohm developed his philosophy of wholeness, which was also featured in a key paper he wrote with Hiley in 1975.[9] In this paper Bohm and Hiley returned to the causal interpretation, highlighting the non-locality or "entanglement" issue referred to above, which had been brought out by John Bell: "the most fundamentally different new feature [of quantum mechanics] of all; i.e., the intimate inter-connection of different systems that are not in spatial contact."[10] This point was made while experiments were still continuing on the issue of Bell's theorem and entanglement, a long struggle documented by Freire[11] and culminating in the experiments of Alain Aspect in 1981–82.[12] Thus, Bohm reinterpreted the 1952 papers in terms of his distinctive ontology of the *Implicate Order*,[13] and worked with Hiley on their ontological interpretation of quantum mechanics in order to eventually produce the ground-breaking textbook *The Undivided Universe*.[14]

The development of Bohm's ideas after the period covered by the letters collected in this book is, of course, a vast subject which I can only touch on here. Bohm seems to have engaged with many different people in areas that seem so disparate that it makes it hard to take in his multi-faceted thinking about the world. For example, with

---

[3]Pylkkänen (1999).

[4]See Peat (1996), Chaps. 11–13.

[5]See, for example, Peat (1996), pp. 190–2.

[6]See http://www.bbk.ac.uk/lib/bohm/bibliography-publications-by-david-bohm, which verifies this point. From 1961, after David Bohm met Hiley, there is only one paper in the 1960s on the causal approach, and this is published jointly with de Broglie and de Broglie's former research student and Communist Party activist Jean-Pierre Vigier, who retained their commitment to the 1952 approach.

[7]https://www.youtube.com/watch?v=q_jHmoxuxsY.

[8]Hiley expanded on this to Olival Freire in an interview (Freire Jr. 2015), p. 61 and made the same points to me in an interview, January 25th, 2015.

[9]Bohm and Hiley (1975).

[10]I am indebted to Olival Freire for pointing this out.

[11]Freire Jr. (2015), Chap. 7.

[12]See also Whitaker (2011).

[13]Bohm (1980).

[14]Bohm and Hiley (1993).

David Peat he wrote *Science, Order and Creativity*,[15] which is in line with the view of the infinite possibilities/resources of humankind that he refers to in his letters. It is interesting to read the review of it by Detlef Dürr, one of the "Bohmian Mechanics" group of physicists we shall touch on later[16]:

> Only a few writers would be able to cover such a broad landscape of ideas and themes without condemning themselves to shallowness. Bohm, who was one of the greatest thinkers and physicists of the last century, shows in this discourse with Peat a tremendous depth of understanding which makes the book a helpful resource for all those who have the urge to inquire into human understanding of our physical world, our behavior, and the development of society.

For those interested in the scientific problem of "consciousness", an area which seems to be of growing interest to many physicists, the attempt of Paavo Pylkkänen to understand the implication of Bohm's later views for the philosophy of mind is well worth studying.[17] For "hard" physicists, repelled by Bohm's involvement in "spiritualistic" or "metaphysical" areas, I would recommend a look at the philosophical, psychological and political outpourings of the founding fathers of quantum mechanics, as discussed by historian of science Mara Beller in response to the smugness of many scientists in the so-called "Science Wars" in the 1990s,[18] when Alan Sokal and Jean Bricmont, after getting a "spoof" article published in a social science journal, lambasted the postmodern trends in the humanities.[19]

References

Beller, M. (1998). The sokal hoax: At whom are we laughing? *Physics Today*, *51*(9), 29. Also at http://www.mathematik.uni-muenchen.de/~bohmmech/BohmHome/sokalhoax.html.

Bohm, D. (1980). *Wholeness and the implicate order*. London: Routledge and Kegan Paul.

Bohm, D., & Hiley, B. (1975). On the intuitive understanding of nonlocality as implied by quantum theory. *Foundations of Physics*, *5*(1), 93–109.

Bohm, D., & Hiley, B. (1993). *The undivided universe: An ontological intepretation of quantum theory*. London: Routledge and Kegan Paul.

Bohm, D., & Peat, D. (2010). *Science order and creativity*. London: Routledge.

Düerr, D. (2012). The wrong, the good and the better, book review of bohm and peat, science, order and creativity. http://www.mathematik.uni-muenchen.de/~duerr/BohmPeat.pdf.

Freire, O., Jr. (2015). *The quantum dissidents: Rebuilding the foundations of quantum mechanics (1950–1990)*. Heidelberg: Springer.

Peat, D. (1996). *Infinite Potential*. New York: Basic Books.

Pylkkänen, P. (Ed.). (1999). *Bohm-Biederman correspondence*. London: Routledge.

Pylkkänen, P. (2007). *Mind matter and the implicate order*. Heidelberg: Springer.

Radner, M., & Winokur, S. (Eds.). (1970). *Minnesota studies in the philosophy of science, IV*. Minneapolis: University of Minnesota Press.

---

[15] Bohm and Peat (2010).

[16] Düerr (2012).

[17] Pylkkänen (2007).

[18] Beller (1998).

[19] Sokal and Bricmont (1998).

Sokal, A., & Bricmont, J. (1998). *Intellectual impostures*. London: Profile Books.
Whitaker, A. (2011). *The new quantum age: From Bell's theorem to quantum computation and teleportation*. Oxford: Oxford University Press.

Chapter 4
Growing Recognition of Bohm's Causal Interpretation

In terms of Bohm's contribution to physics—at least in the narrow sense of his original 1952 papers and of the extension of that work to include spin and special relativity—the last 60 years have seen a slow but significant change. Let us first remind ourselves that although standard quantum mechanics was developed in Copenhagen in the late 1920s, many of its pioneers moved to the United States to flee Nazi Europe in the 1930s, and theoretical quantum physics was at a world high point in America when Bohm was recruited by Oppenheimer. But despite the very visible application in the atomic bomb and the horrific bombings of Hiroshima and Nagasaki in 1945, the applications of quantum physics were only beginning in the 1950s. The first transistor was made in 1947, and transistor radios first went on sale in the US in 1954. The first transistor computer was built at Manchester in the UK in 1953, and integrated circuits were developed at Texas Instruments and Fairchild Semiconductors in the US in 1958, but it took until the 1980s before the PC we all know went into mass production. The first laser did not appear until 1960 and the CD in 1982. But nowadays, according to a recent interesting book by Brian Clegg,[1] about 35 per cent of GDP in the "advanced" countries comes from technology using quantum physics.

David Bohm's causal interpretation now provides an increasingly recognised area in the vast research output of theoretical physics that underpins this "Quantum Age". It could be said that the pragmatic criticisms of Bohm's work in the 1950s by other physicists, namely that he had failed to produce "results", are beginning to be answered. Citations of Bohm's two 1952 papers never reached more than 20 per year as late as 1975, yet by 2000 they never fell below a hundred.[2]

Of great importance here is a relatively new type of research—the application of Bohm's approach to different problems in physics, using the particle trajectories that can be computed by the Bohmian methodology.[3] In their introduction to this

[1]Clegg (2014).

[2]Oriols and Mompart (2012), p. 8.

[3]Oriols and Mompart (2012), Sanz and Miret-Artés (2012) & Sanz and Miret-Artés (2014).

C. Talbot (ed.), *David Bohm: Causality and Chance, Letters to Three Women*,
DOI 10.1007/978-3-319-55492-1_4

area, Oriols and Mompart stress that it "is not at all devoted to the foundations of quantum mechanics, but only to discuss about the practical application of the ideas of de Broglie and Bohm to understand the quantum world." They give "examples of such practical applications written by leading experts in different fields, with an extensive updated bibliography", addressing "students in physics, chemistry, electrical engineering, applied mathematics, nanotechnology, as well as both theoretical and experimental researchers who seek new computational and interpretative tools for their everyday research activity." The authors cite Steven Weinberg as giving the typical objection of physicists to Bohm:

> In any case, the basic reason for not paying attention to the Bohm approach is not some sort of ideological rigidity, but much simpler—it is just that we are all too busy with our own work to spend time on something that doesnt seem likely to help us make progress with our real problems.

But to this they can reply by pointing out that "in contrast to the Copenhagen formulation, the Bohmian formulation allows for an easy visualization of quantum phenomena in terms of trajectories that has important demystifying or clarifying consequences" and that "[i]n some systems, Bohmian equations might provide better computational tools than the ones obtained from the orthodox machinery."[4]

Another interesting new area of research, which is not, strictly speaking, quantum physics, and perhaps not as widely known as Bohmian trajectories, is in the study of experiments involving oil droplets bouncing on a vibrating tray of oil, or computer simulations of such experiments. Oil droplets behave like particles interacting with the wave on the surface of the oil they create (in effect this is a "pilot wave"). The phenomenon resembles very closely the Bohmian approach to quantum mechanics.[5]

Such new developments in computing, physics, chemistry, engineering, etc. relating to the Bohmian view of quantum mechanics, taking this in a broad sense, are relatively recent. Since they mainly relate to computer techniques, it is perhaps not surprising that a useful introduction to Bohm's approach has been given in a Cambridge (UK) course by Mike Towler.[6] Towler participates in the Cambridge Monte Carlo Quantum Computing group and hosts conferences at the Towler Institute, his sixteenth century monastery in Tuscany, Italy.

There is now also a possibility that the Bohm-de Broglie version of quantum mechanics will actually receive support from astronomical observations. In a series of papers over the last two decades, Anthony Valentini has attempted to show that the de Broglie theory[7]—there is a small technical difference between the original de Broglie theory and Bohm's later version, and Valentini has co-authored a book on the history of this[8]—gives slightly different results to standard quantum mechanics when applied to the early evolution of our universe. But information from that early

[4]Oriols and Mompart (2012), Introduction.

[5]Wolchover (2014).

[6]Towler (2009).

[7] Everyhope-Roser (2013).

[8]Bacciagaluppi and Valentini (2009).

stage is available to us in the microwave radiation that was emitted at that time, the so-called background radiation. This radiation contains small anomalies or fluctuations, corresponding to "lumpiness" formed under gravitational attraction that would eventually give rise to the galaxies that we see today. Valentini is hoping that the latest measurements of the background radiation by the Planck satellite will be accurate enough to test the de Broglie theory against the standard interpretation.

The rather optimistic picture I am painting of current attitudes of physicists towards the Bohm (or de Broglie, according to Valentini) version of quantum theory should probably be tempered by stressing that virtually all of the physics literature involved is overwhelmingly technical, using a lot of mathematics and computation, possibly looking at related experiments, but in general, this research would not dream of referring either to the philosophical ideas that motivated Bohm in the 1950s or to his later philosophy of the "Implicate Order". In today's academic climate, and even in the 1950s (as we see Bohm complaining in his letters), such "metaphysical" considerations are excluded from physics by the research funding process, by peer-reviewed journals, and generally, by the desire to preserve one's career.

There is, however, an area of physics known as "Foundations" which developed in the 1970s, together with small specialised areas in the philosophy and history of science, where it is possible to discuss alternatives to the standard "Copenhagen" interpretation quantum mechanics, or to discuss and develop the many other interpretations which have been put forward since Bohm's 1952 papers. It has now been well established in the historical study of Mara Beller, amongst others,[9] that there was really no consistent viewpoint to quantum mechanics developed by Bohr, Heisenberg, Schrodinger and others in the 1920s, but a rather botched together compromise. But although this philosophy, history and foundational physics is regarded as academically respectable, unlike Bohm's philosophical work of the 1950s or his later "ontological" developments, it seems to have had little impact on academic physics, although a number of introductory quantum mechanics texts do now refer favorably to Bohm's "causal" interpretation.[10]

One contribution to the Foundations area should perhaps be particularly noted. The recognition that Bohm's version of quantum mechanics was as valid as the standard interpretation and would appear to give the same results in every application was at the center of James T. Cushing's book in 1994,[11] which remains a good technical introduction to Bohmian quantum physics. Cushing, both a physicist and a philosopher of science in the analytic tradition, was highlighting what philosophers of science call the "underdetermination" of scientific theories by experiments—meaning that both the standard and Bohm's version of quantum mechanics describe the same experimental results, and so far, no experimental test has been devised to verify one of them and refute the other.

It is fitting here to mention the "Bohmian Mechanics" group, which developed in the 1990s and is by far the largest and best known of the researchers in the field of

[9]Beller (1999).

[10]For example Baggott (1992); Ghirardi (2005).

[11]Cushing (1994).

physics relating to Bohm. On their website,[12] they list 13 full Professors from the United States, France, Germany and Italy who, together with their research students, make up this group (Its most well-known members are Sheldon Goldstein, Detlef Dürr and Nino Zanghi, and it also includes Jean Bricmont, of "Science Wars" fame). It is impossible to summarize here all the research carried out by this group. They list dozens of papers and several books published under the Bohmian Mechanics imprimatur on a wide range of subjects in theoretical physics, and clearly pride themselves on mathematical rigor and a "no-nonsense" approach to philosophical issues, striving to convince the majority of physicists of the validity of Bohm's causal interpretation. It is still an uphill struggle.

A readable introduction is given by the philosopher of the group, Tim Maudlin.[13] Perhaps the most notable feature in Maudlin's account is the rejection of the conceptual importance of the "quantum potential". Bohm and Hiley stressed that the quantum potential explains "a number of strikingly new features which do not cohere with what is generally accepted as the essential structure of classical physics".[14] In contrast, Maudlin states that "the deeper defense against criticisms of the quantum potential is that it is superfluous".[15] Thus, the defense against physicists' accusations concerning unnecessary metaphysical baggage is to drop many of the conceptual ideas from Bohm's work, including the quantum potential, while preserving what is seen as the core of Bohmian physics. To return to Detlef Dürr's review of *Science, Order and Creativity*[16]:

> Bohmian mechanics is a robust theory which leaves no place for quantum romanticism or quantum mystery, and in the present cultural period I feel the need to state that as clearly and absolutely as possible. Such a statement does of course go against the spirit of Bohm's philosophy which, for all it may be humbly presented, is at the same time very ambitious in that it tries to grasp eternal truth, in the midst of which Bohmian mechanics is nothing but a tiny event. . . . .when it comes to physics urgent matters have to be dealt with, and Bohm's theory, Bohmian mechanics, is the best possible way to make, if not a better world, better physics. To some, like this reviewer, that is enough for a life's work.

In contrast to this "anti-metaphysics" tendency, Basil Hiley, now an octagenarian, continues to provide us with a prodigious output that staunchly defends what he sees as Bohm's contribution to physics.[17] Hiley is applying the mathematics of algebras, especially Clifford Algebras, to the *Implicate Order* conception which he and Bohm developed out of the original causal approach to quantum mechanics in the 1970s. Clifford, also known as Geometric Algebras, are an increasingly popular way of doing mathematical physics. First put forward by mathematicians Sir William Hamilton, Hermann Grassmann and William Kingdon Clifford in the 19th century, it was pushed aside by physicists such as Willard Gibbs at the beginning of the 20th

[12] http://www.bohmian-mechanics.net/.

[13] Maudlin (Maudlin (2011)).

[14] Bohm and Hiley (1993), p. 31.

[15] Maudlin (2011), p. 110.

[16] Düerr (2012).

[17] Hiley (2001).

century, who employed the now familiar vector approach to physics and engineering.[18] Popularized by American physicist David Hestenes in the 1960s, it was taken up by Bohm and Hiley as a way of doing fundamental physics which would move away from traditional conceptions of space and time. It was also studied in the 1950s by Mario Schönberg (or Shenberg), the Brazilian physicist referred to above, whom we shall meet again in Bohm's letters (in one of his papers, Hiley acknowledges the influence of Schönberg on him and Bohm on this topic[19]).

In a collection of essays dedicated to Paavo Pylkknen,[20] Hiley explains the evolution of his own and Bohm's thinking, and gives a muscular opposition to the Bohmian Mechanics approach, with its rejection of the quantum potential ("not the approach that Bohm originally proposed, nor is it the theory that our group at Birkbeck worked on with Bohm for three decades"). Hiley clearly hopes that, by his creative development of mathematics—which is certainly impressive, if difficult to follow, even for someone who has had some mathematical training—he can better establish Bohm's approach as central to theoretical physics. He insists that the "holist" conception of the *Implicate Order* as well as Whitehead's "Process Philosophy" can help provide a way forward in understanding the quantum domain.

References

Bacciagaluppi, G., & Valentini, A. (2009). *Quantum theory at the cross-roads: Reconsidering the 1927 Solvay conference*. Cambridge: Cambridge University Press.

Baggott, J. (1992). *The meaning of quantum theory*. Oxford: Oxford University Press.

Beller, M. (1999). *Quantum dialogue*. Chicago: University of Chicago Press.

Bohm, D., & Hiley, B. (1993). *The undivided universe: An ontological intepretation of quantum theory*. London: Routledge and Kegan Paul.

Clegg, B. (2014). *The quantum age: How the physics of the very small has transformed our lives*. London: Icon Books.

Cushing, J. T. (1994). *Quantum mechanics: Historical contingency and the Copenhagen hegemony*. Chicago: University of Chicago Press.

Düerr, D. (2012). *The wrong, the good and the better* (book review of bohm and peat, science, order and creativity). http://www.mathematik.uni-muenchen.de/~duerr/BohmPeat.pdf.

Everyhope-Roser, J. (2013). *Glimpse magazine*. Clemson university: Quantum heretic. http://glimpse.clemson.edu/quantum-heretic/.

Fletcher, J. (2014). *Clifford algebra*. http://wiki.c2.com/?CliffordAlgebra.

Frescura, F. A. M., & Hiley, B. J. (1984). Algebra, quantum theory and pre-space, volume especial, os 70 anos de mario schonberg. *Revista Brasileira de Fisica*, 6, 49–86. See also http://www.bbk.ac.uk/tpru/BasilHiley/P12FrescandHiley3.pdf.

Ghirardi, G. (2005). *Sneaking a look at god's cards, unravelling the mysteries of quantum mechanics* (Revised ed.). Princeton: Princeton University Press.

Hiley, B. (2001). Recent publications. http://www.bbk.ac.uk/tpru/RecentPublications.html.

[18] Fletcher (2014).

[19] Frescura and Hiley (1984).

[20] Hiley (2010).

Hiley, B. J. (2010). Some remarks on the evolution of bohms proposals for an alternative to standard quantum mechanics. http://pdfs.semanticscholar.org/8a28/9b354f22f6b7af05b4d6cb3f4faf1b00d22d.pdf.

Maudlin, T. (2011). *Quantum non-locality and relativity* (3rd ed.). Chichester: Blackwell.

Oriols, X., & Mompart, J. (2012). *Applied Bohmian mechanics*. Singapore: Pan Stanford (The first chapter is available at http://arxiv.org/abs/1206.1084).

Sanz, A. S., & Miret-Artés, S. (2012). *A trajectory description of quantum processes I*. Berlin: Springer.

Sanz, A. S., & Miret-Artés, S. (2014). *A trajectory description of quantum processes II*. Berlin: Springer.

Towler, M. (2009). *Pilot wave theory*. http://www.tcm.phy.cam.ac.uk/~mdt26/pilot_waves.html.

Wolchover, N. (2014). *Quanta magazine: Fluid tests hint at concrete quantum reality*. https://www.quantamagazine.org/20140624-fluid-tests-hint-at-concrete-quantum-reality/.

Chapter 5
Introduction to the Letters

Turning to the archival material included in the second part of this book, there are, firstly, copies of 23 brief letters from David Bohm to his former girlfriend, Hanna Loewy, 12 of which were written in 1950 before Bohm went to Brazil. One (also brief) letter from Hanna to Bohm survives, not included here, dated June 7, 1950, telling Bohm that although she loves him she does not want to marry him. A letter from Bohm specifically replying to this June 7 letter explains that he has completed his book, Quantum Theory (Bohm 1989). Loewy was the daughter of Lilly, who had married Erich Kahler, a Jewish intellectual and writer on social philosophy who had fled Nazi Europe. Bohm had lived as a lodger in Kahler's house while he was in Princeton, which was then renowned as a centre of intellectual life, attracting visitors that included Albert Einstein, Thomas Mann and Jacob Bronowski.[1] The folder numbers in the Birkbeck archives are C37-40 and C97. Here, they are letters 1–12 (1950) in Chap. 14, and letters 13–23 (1951–53) in Chap. 15. Also, two letters from David Bohm to Lilly Kahler from December 1953 are included as letters 24 and 25 in Chap. 15 (folder number C96 in the Birkbeck archives). It seems that the C37-40 material was obtained by Bohm himself, probably when he needed evidence to back his application for an American passport, whereas C96 and C97 were obtained by David Peat.

Next, there are copies of 30 letters to Bohm's friend and fellow physicist Melba Phillips (two of which are incomplete). Phillips, like Bohm, had been a student of Oppenheimer, and also refused to testify under the McCarthy witch-hunt. In her case, it was the McCarran internal security commission, whereas with Bohm it was the

[1] Peat (1996), pp. 84–5.

C. Talbot (ed.), *David Bohm: Causality and Chance, Letters to Three Women*,
DOI 10.1007/978-3-319-55492-1_5

House Un-American Activities Committee, HUAC. She was sacked from Brooklyn College in 1952 and, although a talented physicist who co-authored a textbook that is still in use, remained unemployed for five years. Melba died in 2004, so I spoke to Melba's niece, Ellen Vinson (Wolfe 2015), but unfortunately she has no knowledge of any material relating to David Bohm, including the originals of these letters. The folder numbers in the Birkbeck archives are C46-49, and they are included here as letters 26–55 in Chaps. 16–19.

Finally, there are photocopies of 68 letters (10 of which are incomplete, plus five loose pages from missing letters) from Bohm to Miriam Yevick, a mathematician who was married to physicist George Yevick but had a personal relationship with Bohm. Yevick had fled from Nazi Europe with her family at the age of 15, which she has detailed in her book (Yevick 2012). She obtained a PhD in Mathematics at MIT in 1947, the first woman to obtain this qualification.

These photocopies were obtained by David Peat from originals held by Yevick. Peat explains how his wife persuaded Miriam to stuff letters and documents into plastic bags and accompany her to a photocopying shop. After a while, Miriam said "That's enough", and took the bags back (Peat 2005). Consequently, only a part of this valuable material is available, sometimes with poor quality copies.[2] Repeated emails from me, attempting to obtain access to Miriam Yevick's letters, have met with no reply. Only one letter from Miriam Yevick to Bohm is in the archives, and it is not included here. It is from late 1951, mainly of a personal character, but also explaining a mathematical point (on a "strong law" in probability theory). Additionally, there is one letter to George Yevick, probably from early 1952, on Bohm's ideas about the "ether" or "substratum". The folder numbers in the Birkbeck archives for the Yevick letters are C115-126, and they are included here as Letters 56–124 in Chaps. 20–33.

It is hard to make sense of the development of Bohm's ideas just from reading through all the letters. Scientific and philosophical issues such as quantum theory, statistical mechanics, causality and determinism are juxtaposed with personal concerns about the possibility of sexual relationships in Brazil and problems of repeated diarrhoea and sickness. Mixed up with philosophical discussions of the qualitative infinity of levels, one finds endless political analysis from a pro-Soviet viewpoint, not to mention the agonies of suffering from severe depression. In order to give some assistance to the reader (hopefully!), I have outlined some themes, giving appropriate references to the letters in each case. The themes are Philosophy, Mathematics, Development of the Causal Approach to Quantum Mechanics, Probability and Statistical Mechanics, Feminism, Politics, Soviet Physics and Philosophy and Psychological Issues.

[2]See Figs. 9.1–9.3 in Chap. 9 for an example.

References

Bohm, D. (1989). *Quantum theory*. New York: Dover (Reprint of New Jersey original).
Peat, D. (1996). *Infinite potential*. New York: Basic Books.
Peat, D. (2005). *Pathways of chance*. Hoboken, NJ: Pari Publishing.
Wolfe, A. J. (2015). *Belt magazine, physics from the farm*. http://beltmag.com/physics-from-the-farm/.
Yevick, M. (2012). *A Testament for Ariela*. Accord, NY: Blue Thread.

Chapter 6
Bohm's Philosophy as Revealed by the Letters

The letters demonstrate that Bohm was exceptional for a physicist in that he took his philosophy very seriously.[1] More than that, he is even prepared to develop his own original philosophical ideas. One can attempt to divorce his scientific achievements from his philosophical approach, as do the Bohmian Mechanics group, but only if one ignores the actual development of his ideas as revealed in these letters.

Bohm based his philosophy on Marx and Engels, but their ideas had been interpreted by their followers, especially by Russian Marxists, such as Lenin. Thus Lenin's ideas were clearly an influence on Bohm, as we shall see. Additionally, there was a further important development of serious philosophical study and debate in the USSR in the 1920s, after the 1917 revolution. After 1930, Stalin took a personal interest in philosophy, suppressed all debate and banished all material from the 1920s so that he could present himself as the direct heir to Lenin in philosophical leadership (see Tucker 1990, Chap. 7). As a result, a debased form of philosophy was imposed under Stalin and exported to the Communist Parties of the world. This background is important in understanding Bohm, and we will return to it in Chap. 12 on Soviet Physics and Philosophy.[2] In discussing Marxism with his physicist friends in the 1940s, although his politics remained at a crude Stalinist level, as we shall see below, Bohm's philosophy linked up with the ideas of Marx and Engels while providing a framework to his attempt to deal with quantum mechanics. He would probably have

[1]See his letter admonishing Melba for ignoring philosophical questions (18, **45**, pp. 167–169). He argues that "[a]lmost no great physicist worked entirely without a philosophy", whereas today all physicists "want are "results" that "pay off" immediately in higher jobs, recognition, job security, navy contracts, etc.". Note also his comment to Miriam: "But I am afraid that my first love is the philosophical problem and not the detailed dry scientific problems. Yet it is the latter which supply the means by which the philosophical point of view comes close to reality and demonstrates its fruitfulness." (27, **104**, p. 344).

[2]In a 1953 letter, Bohm clearly shows some awareness of the "mechanical" approach of the Soviets, see (26, **97**, p. 320).

C. Talbot (ed.), *David Bohm: Causality and Chance, Letters to Three Women*,
DOI 10.1007/978-3-319-55492-1_6

read some of Engels' pamphlets, and we know he read the notes written by Engels on philosophy and science in *Dialectics of Nature*[3] in Portuguese,[4] so presumably he was familiar enough with the English version. In his interview with Wilkins[5] he noted that with his friend Weinberg in the 1940s "we were discussing on a rather superficial level, but even that level was enough to give overtones that were enough to arouse my energy."

We can assume that the "overtones" related, at least to some extent, to the distinctive ontology, or theory about the "stuff of the world" developed by Marx and Engels. This is a difficult subject, and its exposition is based on scattered parts of Marx and Engels' fifty-volume collection of writings, most of which are about their chief concerns with economics and politics. Also, much of their work, especially from the earlier formative period, was not available in English at the time of these letters. Here I would recommend the books of Allen Wood[6] and Scott Meikle[7] for far more complete expositions than the sketch I can give here. The first author is not a Marxist, but offers the sympathetic account of an expert on classical German philosophy; the second author is a Marxist, and unfortunately writes in the disputative style of left-wing groups, but has the distinct advantage of being an expert on Aristotle. Both authors would agree that Marx supports an ontology that is materialist. Wood takes the basic tenet of Marx's materialism to be *Naturalism*, which "says that the sole reality is the natural world, and this world is made up solely of matter." Naturalists, "deny that the world was created by anything outside it, and that natural motion requires God (or any other supernatural agency) as its cause". Naturalism thus defined implies realism, namely the "thesis that material things are not dependent for their existence or nature on any mind or minds."[8] Bohm is not usually very precise, but we may assume he was a materialist of this type in the period covered by the letters.

Marx's ontology, and Bohm seems to have agreed with this, in some sense, for most of his life, is also "organicist" or "essentialist". The world is made up of things or entities that have "qualities", that are not unchanging but have self-movement and internal tendencies, and that display novelty, coming into being and passing away. Both Wood and Meikle regard the entities of this philosophy as "goal directed" or "teleogical". They insist that the latter term need not mean, as critics suggest, directed by a "higher" power, or that the future can somehow cause the past. Aristotle used the term "final cause", distinguishing the tendencies in organic entities from "efficient causes", i.e. the usual causes in the natural sciences. Both authors see no fundamental conflict between efficient and final cause explanations, they may in

[3]Marx and Engels (1988).

[4](20, **57**, p. 201).

[5]Wilkins (1986), Vol III.

[6]Wood (2004).

[7]Meikle (1985).

[8]See Wood (2004), Chaps. 7, 8, 11 and 12, for a fuller explanation, as well as a refutation of the post-war academic trend wishing to deny Marx's materialism.

fact be deemed to be complementary,[9] but see below for the problem of "mechanistic" ontologies. Both authors wish to treat Aristotle seriously, despite the tendency, since the 17th century Enlightenment, to denigrate him by identifying his philosophy with Scholastic interpretations of his work developed in the Middle Ages.

Such a "teleological" approach as referred to above has been widespread in evolutionary biology, though not without controversy.[10] Marx and Engels themselves used such an organicist approach in their analysis of society, which later became known as "historical materialism". The same approach was also deployed by Marx for economics in his *Capital*. Working within the German philosophical tradition, especially that of Hegel, Marx and Engels considered that social and economic entities developed in a "dialectical" manner. In simple terms,[11] this means such entities contained opposing forces, also known as "contradictions", which could lead to their transformation into something new. Unlike some of their followers, they only used this approach with subject matter in which they were very knowledgeable—Marx spent 25 years studying economic theories and data. As they point out several times in their polemics, nothing is easier than using dialectics as sophistry, something we will find Bohm guilty of in Chap. 11 on politics.

There is, generally, a problem of emphasis in interpreting Marx and Engels on the question of "mechanism"[12] or a "mechanistic philosophy": the approach which says the world can be reduced to basic entities or "simples" which are fixed, or changing only quantitatively, normally using causality in the "efficient" sense of causal determinism only, although Bohm is not restricted to this. Especially in the *Dialectics of Nature*, Engels introduces the idea of levels, or what he calls "forms of motion". Wood examines the various references and reconstructs a list beginning with mechanical motion and going from the branches of physics, chemistry and biology through to the levels of life and consciousness, the latter including a tentative materialist theory of mind. Engels is especially opposed to attempts at mechanical reductionism, which "blots out the specific character" and "qualitative difference" of non-mechanistic forms of motion.[13] Wood regards this aspect of Marx and Engels's

---

[9]See Wood (2004), pp. 107–8 and footnote 8, and Meikle (1985), p. 171.

[10]A similar distinction to Aristotle's was made by the great biologist Ernst Mayr, who used the term "ultimate cause" in relation to Darwinian natural selection and that of "proximate cause" for immediate, physiological and mechanical causes. For a recent critical discussion of Mayr's approach with references see Laland et al. (2013).

[11]See Wood (2004), Chaps. 13 and 14 on the complex issue of dialectics in Hegel and Marx and the differences between them.

[12]This is Bohm's terminology. Wood uses the term "mechanistic reductionism" and Meikle uses the term "atomism".

[13]Marx and Engels (1988), pp. 527–532.

philosophy as a "characteristic wrinkle",[14] based on Hegel and Schelling, and does not seem to take it as a significant trait of their work. Meikle, however, goes in for a strong denunciation of mechanical reductionism, or what he terms "atomism", though he bases himself largely on Aristotle.[15] The issue played a crucial role in the debate over "reductionism" between Deborinites and Mechanists in the 1920s, as we will see in Chap. 12 on philosophy and science in the USSR. Despite having no direct knowledge of the 1920s dispute, Bohm clearly saw "mechanism" as a major problem and apparently has some familiarity with the *Dialectics of Nature* arguments.

Besides mechanism, the other major philosophical issues which Bohm raises are "positivism" and the related question of "idealism." He took up ideas from Lenin's work *Materialism and Empiriocriticism*,[16] or most probably a Soviet publication based on Lenin.[17] Lenin's earlier philosophical work is often contrasted unfavourably to his later study of Hegel in the *Philosophical Notebooks*, and has had many criticisms from its first appearance,[18] but the important points for our purposes are: (1) it was directed against the positivist philosophy which was beginning to gain influence at the beginning of the 20th century, due especially to Ernst Mach, and which, to Lenin's horror, was supported in his own Bolshevik Party[19]; (2) it advocated that, philosophically speaking, all the new discoveries in science at the time – electromagnetic radiation, electrons, radioactivity, etc. – were all different forms of matter and the "levels" of matter were infinite. For the dialectical materialist, "matter" should not be limited to the corpuscular matter of 19th century science.

Relating to (1), by "positivism"[20] we mean the philosophical approach that merely correlates sensory perceptions, or experimental results, opposing the claim that such correlations reflect actual relationships among real things, which exist independently of observation.[21] Lenin thought that Mach's positivism led to "idealism", in the subjective sense, i.e. not in the "absolute" sense of Hegel or Schelling. Lenin's idealist is a "methodological solipsist", like the 18th century philosopher Bishop Berkeley, who thinks we can begin only with our experience and that we can never attain knowledge from a world beyond experience.[22] We assume Bohm is using the

[14]Wood (2004), pp. 169–170.

[15]Meikle (1985), especially Chap. 7.

[16]Lenin (1962).

[17]Wilkins (1986), Vol VI.

[18]See Bakhurst (1991), Chap. 4, for a more sympathetic treatment.

[19]For a good introduction to Mach and positivism see Holton (1993), Chap. 1. It is well known that Mach influenced Einstein in his earlier period, see Holton (1993), Chap. 2.

[20]In the sense of Lenin and Bohm. It is now often called anti-realism. David Joravsky uses the term differently to mean limiting "knowledge to the methods and results of the empirical sciences". Joravsky (2009), p. xi.

[21]See the definition Bohm gives in his 1952 paper Bohm (1952b) and the definition in Bohm (1957), Chap. III, Sect. 8.

[22]See Bakhurst (1991), Chap. 4, for more detail. Bakhurst draws the "modest conclusion that there are reasonable, though perhaps not conclusive, grounds to associate Empiriocriticism [i.e. positivism in Bohm's sense] with the doctrines of Lenin's idealist" (ibid. p. 102).

term "idealist" in the same sense as Lenin, and that he sees it as the opposite of "materialism". On point (2), we note that the idea of "levels" does not explicitly appear in Lenin. Although Lenin always bases himself on Engels, giving extensive quotes, *Dialectics of Nature* was not available at that time. Lenin writes only that "[t]he electron is as inexhaustible as the atom, nature is infinite, but it infinitely exists."[23] As we shall see, Bohm expands this idea into the "qualitative infinity of levels."

To understand how Bohm developed his distinctive philosophical approach, one must start from the issues arising from standard "Copenhagen" quantum mechanics and his attempt to expound it in the book "Quantum Theory".[24] Let us review some key features of standard or "Copenhagen" quantum mechanics, developed by Niels Bohr, Werner Heisenberg, Wolfgang Pauli and others in the second half of the 1920s.[25] "Copenhagen" quantum mechanics assumes that quantum systems do not have definite properties until measurements are carried out, and that the theory can only predict the probabilities of the results obtained by these measurements. Previous physical theories using probability had assumed an underlying reality whose details, such as the positions and velocities of very large numbers of atoms in the theory of "ideal" gases developed by Maxwell and Boltzmann, were knowable in principle, if not in practice. However, in standard quantum mechanics, there is assumed to be no such underlying reality at the quantum level until measurements have been carried out.

Certain pairs of properties of quantum systems are called "complementary", such as a particle (which is associated with position) and a wave (which is associated with corresponding momentum or velocity). Unlike the properties of "classical", 19th century physics, according to Heisenberg's uncertainty relation, the more accurately position is measured, the less accurately the momentum can be measured and vice versa, and similarly with other complementary physical variables.[26]

The probabilities in quantum mechanics are given by the Schrodinger wave function that uniquely describes the state of the system.[27] Since there is no other information about the system, the theory is *essentially* random, or to paraphrase Einstein, "God *does* play dice." After measurement, the wave function somehow "collapses" to correspond to the value measured. There is ambiguity in forming the division (or "cut") between the non-classical quantum system and the classical measuring apparatus. Since the observer can be regarded as part of the measuring apparatus, it

---

[23] Lenin (1962), p. 262.

[24] Bohm (1989).

[25] For a more detailed historical but non-mathematical treatment see Bohm (1957), Chap. III. For a good, more mathematical treatment see Baggott (1992). It should be said, however, that the outline given here must be qualified by the recognition that there was a lack of clarity and agreement among the founding fathers (Beller 1999).

[26] Other examples of such properties are energy and duration, as well as the "spin" of a quantum particle along different axes.

[27] Usually denoted by the Greek letter "psi", ψ, which is assumed to be a solution of Schrodinger's wave equation.

is possible for the measurement process to be interpreted as involving a conscious observer (and hence to support an idealist philosophy).

In practice, the emphasis in the standard approach was, and still is, on mathematical calculations and applications to experiments, in which outstanding successes could be reported, from the late 1920s onward.

In *Quantum Theory* Bohm attempted to develop a rational approach to this strange quantum world. Using the Schrodinger wave function, or usually a kind of concentrated piece of it, a "wave packet", he takes the student through a range of experiments and develops enough mathematics to analyse what is happening. As he writes in a letter to Miriam, "a qualitative "plausibility" argument is more valuable at an early stage in the text than a precise argument full of a forest of symbols."[28] He was basing himself on notes from Oppenheimer's lectures and was influenced by the approach, as he understood it at that time, of Niels Bohr.[29] The book was widely acclaimed for its conceptual treatment rather than the usual formal, mathematical exposition, and even the greatest critic of quantum mechanics, Einstein, thought that "It is the best that could be done with the usual interpretation."[30]

Bohm thus wanted to reveal what was happening in the "collapse of the wave function". He opposed positivistically inclined physicists who "say that after all, only the results of measurements need to be treated, and what happens to the apparatus when nobody looks at it is a "meaningless" question."[31] Opposing this "basically idealist" position, Bohm developed the theory of the measurement process.[32] It is obvious, then, that combating positivism and idealism were central to Bohm's approach.

One would have thought that this was a great achievement and that Bohm could now settle down to a career in teaching and researching theoretical physics. It did not work out to his satisfaction, however. Bohm had argued that there did indeed exist a quantum mechanical reality, whether measurements were being carried out on it or not. He suggested that causal determinism (or "complete" determinism as he called it) should be replaced by "statistical laws". To understand the "complementary" properties of this reality, he had used the idea of "potentialities"[33] But these did not allow him to confidently claim that here was a genuinely "dialectical" feature,

---

[28](21, **65**, pp. 225–229). $\Delta x \Delta k \geq 1$ is mathematical shorthand for the Heisenberg relation. Further discussion on *Quantum Theory* is undertaken at (21, **66**, pp. 234–236) and (22, **67**, pp. 238–240). We will leave Bohm's very critical views on mathematics and the role of mathematicians in holding back theoretical physics to the next chapter.

[29]Later he considered it was Heisenberg's approach rather than Bohr's. Bohm and Hiley (1993), p. 18.

[30](21, **65**, p. 226).

[31](22, **67**, p. 239).

[32]Bohm (1989), Chap. 6, Sect. 2 and Chap. 22. In this latter chapter, he analyses the apparatus as well as the quantum system under investigation and concludes that "this problem can be solved without carrying the analysis as far as the stage in which the apparatus interacts with a human observer" (Bohm 1989, p. 606).

[33]See Bohm (1989), Chap. 8 and the reference in his letter to Hanna in early 1950 (14, **1**, pp. 99–100).

something that a Marxist could point to, as a verification of his or her philosophy in nature. As he explained it to Miriam, Bohr's approach had seemed incredibly vague, but it had offered something more, "(i)t seemed progressive because it broke the old mechanist materialist determinism, which left no room for growth and development of something new."[34]. But after all his years of toil, Bohm was unable to reveal genuine dialectics: "dialectically opposing concepts are made just vague enough so that the contradictions between them are avoided."[35] Instead of the synthesis of opposites that a Marxist view requires, "contradictions leading to something new at another level" were thereby lost.

With this disappointing outcome to the completion of his book in the summer of 1950,[36] Bohm spends the next year, or at least part of it, working on the causal interpretation, or the hidden variable interpretation, as he called it then.[37] He submits his papers based on this work to the Physical Review in July 1951.[38] Bohm had clearly decided by then that he could take the complementarity approach no further, and that only a causal determinist approach could place quantum mechanics on materialist foundations. What influenced him to make this turn? Olival Freire notes that Bohm later refers to at least two influences in his sudden shift to the causal interpretation: discussions with Einstein and the "reading of a paper by a Soviet physicist criticizing the complementarity view for its idealistic and subjectivist inclinations."[39]

The influence of Einstein is very clear. For example, in a letter in 1953, Bohm wrote: "In fact, you may remember that after writing a whole book on the usual interpretation of the quantum theory, I abandoned it when presented with arguments which convinced me."[40] As Freire points out, there is no evidence in the letters of a paper written by a Soviet physicist. We review the situation in Soviet physics and philosophy as well as Bohm's responses to it below in Chap. 12. However, since the attack on the idealist interpretation in standard quantum mechanics came from the highest levels in the USSR, which Loren Graham describes as "the most intense ideological campaign in the history of Soviet scholarship,",[41] to the extent that even the term "complementarity" was banned from 1948 to 1960,[42] it would be naive to believe that, given his politics, Bohm was not influenced by this in some way,

---

[34](21, **66**, p. 235).

[35](21, **66**, p. 235).

[36]See the letter to Hanna (14, **10**, p. 110). Note that Bohm is hoping to work with Niels Bohr on a philosophy book!

[37]In (14, **11**, p. 112), he is "getting interested in the problem of the electron again", which may well relate to this.

[38]Bohm 1952a, b.

[39]Freire (2015), p. 27.

[40]Letter to Einstein, Feb 27, 1953 in Folder C12 in the Birkbeck archives.

[41]Graham (1971), p. 74.

[42]ibid, p. 80.

probably through Communist Party publications, whether he read material from a Soviet physicist or not.[43]

What Bohm views as the essential characteristics of his causal interpretation is explained in a letter to Melba, responding to Phil Morrison, a leading physicist and Communist Party member, who thought Bohm's philosophy was correct, but the standard theory was "simpler".[44] The probabilistic aspect of his theory, Bohm explains, is "a result of chaotic collisions with atoms, molecules, etc., undergoing random thermal motion", and so the "probabilities have the same origin as those of classical statistical mechanics", i.e. the result of an extremely large number of causally determined processes and not some unexplained, essentially random, process.[45] In Bohm's theory, quantum systems have both a wave function and a particle or particles. Therefore two atoms, for example, can have the same wave function but may not be identical, and there exist also particles which, though unobserved, can be at different positions. To emphasize his point, Bohm notes that in the apparently "simpler" standard theory, a uranium atom exploding tomorrow and another one exploding in two billion years have exactly the same wave function and so are indistinguishable. In contrast, in his causal interpretation, "the two uranium atoms are not "physically identical" because each of them has a particle in it in a position that will determine when it will disintegrate." So whilst admitting that this is not "simpler", in that it makes for an additional hypothesis, "namely that there exists an as yet unobserved particle", Bohm can clarify "things that were previously arbitrary".[46]

Bohm's new causal interpretation certainly seemed to deal with the positivism and idealism of the standard theory. But hadn't he also re-opened the door to that very "old mechanist materialist determinism" which he had once thought Niels Bohr had

[43]See (26, **94**, p. 311): here Bohm states that a "vigorous criticism of the foundations of quantum mechanics is going on in the "East"". Loren Graham refers to a 1951 paper by the Russian physicist D.I. Blokhintsev, which was critical of the standard interpretation. It advocated his own, distinctive statistical ensemble approach, which is not at all similar to Bohm's. However, according to Graham, Blokhintsev dismisses the usual objections to "hidden variable" theories and acknowledges that such a "theory of quantum mechanics might at some future date permit a numerical description of the individual microparticle, although at the present time he considered such a description to be impossible." Graham (1966). It may be that an earlier version of Blokhintsev's work had reached the US and an English translation was obtained by Bohm.

[44](17, **37**, pp. 151–153).

[45]He clarifies this in a paper following up the original 1952 papers (Bohm (1953)), referred to in mathematical notation by $P \to |\psi|^2$ and discussed in a number of letters (23, **74**, p. 257), (23, **78**, pp. 264–265), (24, **83**, p. 277), (25, **88**, p. 293), (26, **93**, p. 308), (26, **94**, p. 312), (17, **37**, p. 151) and (17, **38**, pp. 154–155). Another version, with input from Vigier, was published in Bohm and Vigier (1954).

[46]Note that in explaining his theory to Morrison, Bohm only uses the wave function and not the "quantum potential" concept derived from it, which features in the 1952 papers. He only discusses the "quantum potential" aspect of his theory in the letters once, in (25, **90**, p. 298), explaining how hypotheses are put forward. In Bohm (1957), Chap. 4, Sect. 4 he considers a general "quantum force" rather than a quantum potential, the nature of which, it is presumably hoped, will be clarified in the relativistic generalizations of the theory. See also Hiley and Peat (1987), p. 37, where Bohm explains that the instantaneous "entanglement" of distant particles was seen to contradict relativity and was regarded as a "serious difficulty to be resolved with the aid of further new orders.".

successfully dealt with, in other words, was this not a reiteration of the problem of a mechanist ontology referred to above? Had his introduction of causal laws brought in the "nightmare of a mechanically determined universe that follows an inevitable course"[47]? It is to counter this threat that Bohm developed his own distinctive philosophy of an infinite number of levels.[48] With such qualitatively distinct levels "we can have complete causality at every level, in the sense that we can use this causality to change the world in a predictable way, with the error in the predictions dependent only on our level of knowledge;" but with the addition that "we can in no sense conceive of the world as completely determined".[49] In the early 1952 period we find a number of letters to Miriam setting out Bohm's infinite levels philosophy in relation to physics[50]; note also the letter to Hanna on this issue.[51] The reader may also find Hans Freistadt's 1956 paper of use here.[52] Freistadt was a Marxist physicist who lost his job in the McCarthy witch-hunt. He was supportive of Bohm's views, and was apparently in contact with Bohm via Melba Phillips.[53] Note that Freistadt has a reference to the 1951 paper, in Russian, by the Soviet physicist Blokhintsev, already referred to above.

In these letters, Bohm wants to explain the concept of matter and some aspects of dialectics, from his infinity of levels standpoint. He proposes that "in some aspects at least, matter is indestructible and uncreatable" and "matter as a whole in its infinity of properties and potentialities is eternal". This would seem to make change and transiency impossible. But with the levels approach:

> The things at each level, are made up of smaller elements at a more fundamental level, and it is the motion of these more fundamental elements (not usually directly visible to us, except with the aid of elaborate scientific research) which causes the appearance and disappearance of the things existing at a higher level.[54]

By considering his work on plasma physics, Bohm points to examples where the behaviour of an "individual" at one level is "collectively conditioned" by a higher level. Thus:

> The universe cannot be analyzed into a series of components, each of which are the constituents of the next higher level, and each of which determine the higher levels in a purely analytic way. For the higher levels will also always help determine the character of things that may exist at the lower levels. Thus, every level is in a sense, just as real as every other, since the "whole picture" cannot be deduced by starting at the "lowest level" and working upward.[55]

---

[47](22, **73**, p. 254).

[48]In a November 1951 letter to Miriam, (20, **58**, p. 205), it is clear that Bohm expects Miriam to know about his infinity of levels approach, so presumably he first developed the idea at Princeton.

[49](22, **73**, p. 255).

[50](21, **65**, pp. 227–229), (22, **68**, pp. 245–246) and (22, **73**, pp. 254–255).

[51](15, **20**, pp. 123–124).

[52]Freistadt (1956).

[53]See the distribution list in (19, **52**, p. 180).

[54](21, **65**, p. 227).

[55](22, **68**, p. 246).

New things can come into existence:

> For if we have a finite number of causal levels, then the future is already contained logically in the present, but not if we have an infinite number. The appearance of qualitatively new things with time is possible with an infinite number, because the effects of the limitless number of lower levels can always surge up into a higher level (and vice versa) producing qualitative [missing words] describable as a rearrangement of things already in existence.[56]

It should be noted that in this earlier period Bohm uses the concept of causality entirely in a narrow sense, which, in philosophical terms, may be called "efficient causality".[57] He also seems to be using an ontology of "things" or "elements" in a mechanistic sense, in line with the physics of particles, employing the infinite levels approach in order to get beyond this restriction. In the chapter below on probability and chance, we will see a similar restriction at this earlier stage, treating causality as fundamental and chance, or "chaos", as derived from it. There is a distinctive change in Bohm's philosophical approach after 1953, although this is only clearly brought out in two letters to Melba[58] and in the later letters on probability and chance. Almost certainly, this change is due to the visit of philosopher-physicist Mario Bunge, who came over from Argentina to São Paulo as Bohm's postdoctoral student from April to October in 1953.[59] It is possible that some of the change in Bohm's views, especially on probability and chance, is also due to Professor Mario Schönberg, as detailed in Chap. 9.

Bunge's views in that period were presumably in a state of flux, between the Marxism of his 1951 paper[60] and the "analytical", though still materialist approach of his later *Causality and Modern Science* book.[61] The outcome of the Bohm-Bunge interaction was a definite clarification of Bohm's views on mechanism as well as on causality and determinism. It should be noted that Chap. I of *Causality and Chance* has similar material to Bunge's *Causality and Modern Science*. The ideas of one-to-many and many-to-one causality, for example, appear in both,[62] but do not appear at

---

[56](22, **73**, p. 255).

[57]The "causal laws" in (22, **73**, pp. 254–255).

[58](18, **43**, pp. 163–165) and (18, **46**, pp. 170–173).

[59]Dates given by Mario Bunge in an email to the author, 21.09.2015. There is little to indicate Bunge's visit in the letters: Bohm refers to Bunge and his wife as "nice people" in (28, **110**, pp. 360–361) and refers to Bunge's view of George Yevick, (29, **111**, p. 370). Bohm also visited Bunge in Argentina on his way to Israel, early in 1955 (18, **49**, p. 175).

[60]Bunge (1951).

[61]Bunge (2009). In his email to me, Bunge says he had "just been weaned from Marxist philosophy." Bunge also writes that he "asked him [Bohm] why he was wasting his time reading that garbage" (i.e. Hegel). Bohm replied "Because it inspires me." Bunge ruefully adds "Fortunately his physics did not suffer". It is difficult to reconcile this story, based on hindsight, with the many references to Hegel in Bunge (2009). There are more fulminations against Hegel in Bunge's autobiography (Bunge 2016), as well as attacks on Bohm's later "New Age" views, and the comment about his own philosophical development: "I wasted too much time trying to decipher Hegels riddles, and it took me several years to realize that he had started the Counter-Enlightenment and had invented the trick of passing off absurdity as depth" (ibid p. 102).

[62]Bohm (1957), Chap. 1, Sect. 7, Bunge (2009), Sect. 5.1.

all in the letters. Bunge's *Causality* book, though clearly moving away from his earlier Marxism, should be recommended for its materialist critique of the still dominant philosophical view on causality, which dates back to the 18th century empiricist philosopher Hume, proposing that causality is merely the "constant conjunction" of observations, and denying the existence of such a thing as objective necessity.[63]

Bohm sets out his view of mechanism in the second of the two letters to Melba[64]:

> (1) Everything is made of certain basic elements which themselves never change in essence (i.e. qualitatively). (2) All that these elements can do is to undergo some quantitative change according to some fixed laws of change. For example, if they are bodies, they can move in space. If they are fields, they can change their numerical values, etc., etc. But the basic elements themselves never undergo qualitative change.

Mechanism is thus criticized from the standpoint of an *ontology*, relating to the fundamental nature of reality, in line with Marx and Engels' view of it. In earlier letters,[65] Bohm refers to "deterministic mechanism", especially when he discusses quantum mechanics and the "infinity of levels", and this means that he understood the "basic elements" to be governed by causal determinism. He now wants to "sharpen the distinction between causality and mechanism". The earlier approach would seem to relate only to corpuscular matter, but here the ontology is broadened to include electromagnetic radiation or "fields".[66] In the first of the two letters to Melba,[67] Bohm has already extended the definition of mechanism to include the "mechanistic indeterminists" of standard quantum theory "who insist that in the quantum of action, we have reached an ultimate, indivisible, and unanalyzable entity, which will never be found to have a structure understandable in terms of a deeper level."[68]

Note that Bohm makes the important point that mechanical laws are not to be simply rejected, but that "[i]n certain limited domains (the limits of which of course, cannot be known a priori) mechanical laws will surely be good enough approximations."[69]

The two letters to Melba are thus an all too brief outline of a position much more fully elaborated in the first three chapters of *Causality and Chance*. We may presume that in 1954, with work on the causal interpretation vigorously continuing, as set out in Chap. 8 below, Bohm was hoping that his approach would come to be seen as going beyond the mechanism which had so far characterized physics, and as leading to a possible way out of what some saw, with a nod at Lenin, as the "Current Crisis

---

[63]For a readable summary of the range of modern philosophical views, mainly Humean, see Mumford and Anjum (2013).

[64](18, **46**, p. 170).

[65]e.g. (21, **66**, pp. 235–236).

[66]See more detail in Bohm (1957), Chap. 2. This differs from the view of some advocates of "organism" and "holism", who would include fields under that particular heading, e.g. Dusek (1999).

[67](18, **43**, p. 164).

[68]As explained further in Bohm (1957), 3.

[69](18, **46**, p. 171). Similarly in Bohm (1957), Chap. 5, Sect. 4, he writes that his infinite levels philosophy enables us "to retain all the positive achievements that were made possible by the development of mechanism."

in Microscopic Physics".[70] That did not happen, and Bohm admits in 1957 that his approach had also often been criticised as too mechanical. But it still, he argues, "may be a good starting-point from which qualitatively new developments are likely to arise."[71]

Bohm also gives a brief indication of how to go beyond mechanism in a philosophical sense, and this is the subject of his last chapter in *Causality and Chance*. It is partly, of course, by means of the "qualitative infinity of levels". But he also wants to extend the type of change that matter can undergo, moving beyond the more limited "efficient" causality of mechanism:

> But more general types of change are possible. Thus, we may have qualitative change, as in evolution or in embryology. The appropriate causal laws then govern the qualitative changes, and tell which things will change into what and under which conditions.[72]

This broader usage is developed in *Causality and Chance* and is an essential part of Bohm's attempt to move beyond mechanism in physics. Of course, it is always permissible to extend a definition. But in order to avoid confusion, it is important to note that "causality", used in this broader sense by Bohm, refers not to "efficient" cause but to the "teleogical" change in the "organicist" ontology of classical Marxism, as referred to above.[73]

There is clearly a problem facing Bohm that refers to the key area of natural science, biology, to which, as we pointed out at the beginning of this chapter, such an organicist ontology is highly relevant. Although criticizing Lysenko's methods, as late as March 1955 Bohm thinks he was basically correct against geneticists.[74] Bohm's reference to evolution and embryology in these letters to Melba is, therefore, only a passing one. Perhaps because of his isolation in Brazil, he was unable to grasp Lysenkoism's fraudulent character, or to make any reassessment of Darwinian evolution from a Marxist standpoint at this stage, although, by the early 1950s, many Communist Party members and supporters were becoming aware of the complete disaster in Soviet genetics.[75] By the time he writes *Causality and Chance*, Bohm appears to have given Darwinian evolution more thought.[76]

---

[70]The title of Bohm (1957), Chap. 4, Sect. 7. Lenin (1962), Chap. 5, Sect. 1, is entitled the "Crisis in Modern Physics", following the lead of Henri Poincaré.

[71]Bohm (1957), Chap. 4, Sect. 9.

[72](18, **46**, p. 171).

[73]In *Causality and Chance*, Bohm refers to causal laws related to the "mode of being" of things, in Chaps. 1 and 6, and the "Process of Becoming" in Chaps. 5 and 8.

[74]In (19, **50**, pp. 178–179) he writes of "the extremes to which Lysenko went in criticizing backward trends in biology". Previously, in March 1953 (26, **97**, p. 320), he wrote of "Lysenko's excessively dogmatic presentation of his basically correct point of view".

[75]See Brown (2012) for details of and on the response of the famous Communist physicist J.D. Bernal. Although Bernal wrote an appalling eulogy to Stalin after his death in 1953, he did recognize the importance of the discovery of DNA in the same year.

[76]See the references to the theory of evolution and "natural selection" in Chaps. 1 and 6, and the "well-known evolution of the species" in Chaps. 5 and 8, where Bohm also considers evolutionary processes in geology, astronomy and cosmology.

This chapter has concentrated on the relatively small number of letters relating to the philosophy of physics and to the development of the ideas in *Causality and Chance*. An examination of Bohm's numerous letters concerning probability and the relation between causal and statistical laws is left to Chap. 9. However, we should also note that the philosophical concepts that Bohm uses in his physics, and that can be explained fairly precisely as we attempted above, are employed by Bohm in a much looser sense and quite extensively throughout these letters, when referring to social and political issues, issues relating to science and society, and so on.

A few examples will illustrate Bohm's type of socio-political analysis. Firstly, an example on positivism, where Bohm is concerned, not with the application of philosophy to quantum mechanics, but with what he sees as a problem with physicists:

> The flexibility of positivism is amazing, for among [experimental] physicists in [the] U.S., there is a belief that physics flows solely from empirically observed data, or "operations", which is also combined with a belief that theorists take these numbers, and with the aid of a few geniuses like Dirac, produce equations that fit these numbers.[77]

For Bohm, positivism is not just confined to scientists, but is a philosophical problem of society in general, with a related view of "relativism":

> Thus, a characteristic attitude of people toward life is a cynical one, "relativistic", in the sense that morals and responsibilities are said to be determined only by the prevailing society. If one happens to be in a Nazi society, then one naturally adopts Nazi morals, etc. In America, one adopts the prevailing "American Way of Life".[78] This is the counter-part of positivism, for it says that there is no objective material basis for morals, but that all is determined by a commonly agreed upon convention, which introduces "order" into the system of behaviour.[79]

Secondly, Bohm sees mechanism or mechanical materialism as dominating current social thought:

> It was in dealing with nature that man was forced to produce his first objective and clearly thought out concepts, and in this way developed a form of thinking that we hope can now be applied to human beings and to society. But the experience gained in this pursuit up to now has largely been seen in terms of the distorted idea that "things are what they are, and nothing more". Mechanical materialism is a form of this idea.[80]

Such scattered and quite numerous philosophical remarks relating to society are usually made in a pejorative sense. The "infinity of levels" idea, however, is used by Bohm in a more optimistic sense:

> But in this regard, human nature is no different from Nature in general; for according to the ∞ of levels, all properties can be altered with sufficient changes in conditions. Thus, the ∞ of levels is an integral part of a better view of Nature in general, and of human nature in particular.[81]

---

[77](17, **35**, p. 148).

[78]Bohm has a profound hatred of the "American Way of Life", to which he repeatedly refers.

[79](20, **59**, pp. 206–207).

[80](21, **66**, p. 233) See also (32, **123**, pp. 430–431).

[81](27, **101**, pp. 332–333).

One could argue that these social or political aspects of Bohm's philosophical thought, as shown in these few examples – there are many more – are not separate from his thoughts on physics, and also can be seen to provide a motivation for his scientific work. I have taken the view that there are, in fact, differences between social criticism and philosophy of physics, and that Bohm's profound knowledge of physics does give more validity to his scientific philosophy. It can also be placed firmly in the Marxist tradition that was largely suppressed by Stalin. I have, therefore, only briefly considered his far less well-defined social philosophy in Chap. 11, in the context of looking at his politics as a whole.

References

Baggott, J. (1992). *The meaning of quantum theory*. Oxford: Oxford University Press.

Bakhurst, D. (1991). *Consciousness and revolution in soviet philosophy, From the Bolsheviks to Evald Ilyenkov*. Cambridge: Cambridge University Press.

Beller, M. (1999). *Quantum dialogue*. Chicago: University of Chicago Press.

Bohm, D. (1952a). A suggested interpretation of the quantum theory in terms of hidden variables I. *Physical Review*, *85*(2), 166–179.

Bohm, D. (1952b). A suggested interpretation of the quantum theory in terms of hidden variables II. *Physical Review*, *85*(2), 180–193.

Bohm, D. (1953). Proof that probability density approaches $|\psi|^2$ in causal interpretation of the quantum theory. *Physical Review*, *89*(2), 458–466.

Bohm, D. (1957). *Causality and Chance in Modern Physics* (1st ed.). London: Routledge and Kegan Paul. (Second edition with new preface by Bohm).

Bohm, D. (1989). *Quantum theory*. Dover. (Reprint of New Jersey original 1951).

Bohm, D., & Hiley, B. (1993). *The undivided universe: An ontological intepretation of quantum theory*. London: Routledge and Kegan Paul.

Bohm, D., & Vigier, J.-P. (1954). Model of the causal interpretation of quantum theory in terms of a fluid with irregular fluctuations. *Physical Review*, *96*(1), 208–216.

Brown, J. R. (Ed.). (2012). *Philosophy of science: The key thinkers*. London and New York: Continuum.

Bunge, M. (1951). What is chance? *Science and Society*, *15*(3), 209–231.

Bunge, M. (2009). *Causality and Modern Science* (4th ed.). Transaction Publishers: New Brunswick, New Jersey.

Bunge, M. (2016). *Between two worlds, memoirs of a philosopher-scientist*. Switzerland: Springer International Publishing.

Dusek, V. (1999). *The holistic inspirations of physics, the underground history of electromagnetic theory*. New Brunswick, New Jersey and London: Rutgers University Press.

Freire, O., Jr. (2015). *The quantum dissidents: rebuilding the foundations of quantum mechanics (1950–1990)*. Heidelberg: Springer.

Freistadt, H. (1956). Dialectical materialism: A friendly interpretation. *Philosophy of Science*, *23*(2), 97–110.

Graham, L. R. (1966). Quantum mechanics and dialectical materialism. *Slavic Review*, *25*(3), 381–410.

Graham, L. R. (1971). *Science and philosophy in the soviet union*. London: Allen Lane.

Hiley, B., & Peat, D. (Eds.). (1987). *Quantum implications*. London and New York: Routledge and Kegan Paul.

Holton, G. (1993). *Science and anti-science*. Harvard University Press.

Joravsky, D. (2009). *Soviet marxism and natural science, 1917–1932*. London and New York: Routledge, Taylor and Francis.

Laland, K., Odling-Smee, J., Hoppitt, W., & Uller, T. (2013). More on how and why: Cause and effect in biology revisited. *Biology Philosophy*, *28*(5), 719–745.

Lenin, V. I. (1962). *Collected works* (Vol. 14). Moscow: Foreign Languages Publishing House.

Marx, K., & Engels, F. (1988). *Collected works* (Vol. 25). Moscow: Progress Publishers.

Meikle, S. (1985). *Essentialism in the thought of Karl Marx*. London: Duckworth.

Mumford, S., & Anjum, R. L. (2013). *Causation: a very short introduction*. Oxford University Press.

Wilkins, M. (1986). *Interviews with David Bohm*. College Park, MD USA: Niels Bohr Library & Archives, American Institute of Physics. (In 12 parts: https://www.aip.org/history-programs/niels-bohr-library/oral-histories/32977-1 to https://www.aip.org/history-programs/niels-bohr-library/oral-histories/32977-12).

Wood, A. W. (2004). *Karl Marx* (2nd ed.). London and New York: Routledge, Taylor and Francis Group.

Tucker, R. C. (1990). *Stalin in power: The revolution from above, 1928–1941*. New York and London: W.W. Norton and Company.

References

Chapter 7
Bohm on Mathematics

In considering Bohm's views on mathematics, we are still mainly in the realm of philosophy, but this time epistemology. The question of how knowledge develops has a particular significance for Marxists, since they seek to root knowledge in a material world. It is not possible to expand on Marxist epistemology in any depth here, but a good introduction is given by Allen Wood.[1] We have seen that Bohm is especially opposed to positivism, which states that theories merely correlate sensory perceptions, or experimental results. How, then, do concepts and theories develop? Bohm expounds his ideas on this in *Causality and Chance*, linking the issue to his theory of infinite levels, in Chap. 5, especially in Chaps. 9 and 11.[2] In the letters, there is a brief mention of how to oppose positivism in the letter to Melba, which we mentioned in the last chapter, where Bohm explains his causal interpretation in response to Phil Morrison.[3] Thus, according to Bohm, every element in a hypothesis need not correspond to something observable. This was once the case with the atomic hypothesis, and as is now the case with the causal interpretation. Not admitting such hypotheses "is to cripple the imagination, when the time comes to investigate new fields, and to confine ourselves to the existing domain of concepts". In order to prevent the multiplication of hypotheses, the latter should also be required to "unite facts or domains of facts which had hitherto been arbitrary."

However, apart from this brief aside on hypotheses, Bohm's considerations on epistemology in the letters are mainly concerned with mathematics and the relation

---

[1] Wood (2004), Chap. 12.

[2] See also the early 1952 letter to Hanna (15, **20**, p. 124).

[3] (17, **37**, p. 153).

C. Talbot (ed.), *David Bohm: Causality and Chance, Letters to Three Women*,
DOI 10.1007/978-3-319-55492-1_7

of mathematics to physical theory. Following Engels,[4] Bohm considers all concepts and scientific laws to be "abstract". In the Marxist tradition, to say that something is abstract means that it is taken out of context, in other words, that it has been abstracted.[5] Concepts and laws are abstract because they can only give an approximation to the truth, as there are always processes and levels in reality that are not taken into account. Bohm wants to employ abstractions in order to give a theory that is sufficiently close to reality and likely to represent what is essential in the problem under consideration.

The problem with mathematicians, in Bohm's opinion, is that they tend to utilise abstractions that are too far removed from physical reality. This is because their propensity will always be to choose that mathematics which offers a tractable solution. Thus, in a letter to Miriam on mathematics he writes: "Of course, there are certain advantages in abstracting the problem in the way you mathematicians do, but in this case [statistical mechanics] I have hopes that the treatment of something approaching a real physical problem may suggest new ideas that will even be useful in the more abstract problems thus far treated in mathematics."[6]

Bohm expects the inter-connections in the world to be rationally understandable, but clearly also recognizes that this can lead to difficult problems. Skill is always involved in choosing between moving closer to reality, in order to bring out the more complex interconnections in the world, and taking a more abstract approach that is simpler to deal with.[7] Too much abstraction, however, can obscure the real problem. Mathematicians tend to have lost this skill and to be lazy, opting for those abstractions that can be given a simple solution, rather than "breaking their heads on real problems that might lead to new concepts and new modes of inter-connection of these concepts".[8]

The point about choosing the abstractions in order to give solutions is further expanded upon in Bohm's critique of von Neumann's work in relation to quantum mechanics. It is quite dangerous to use a number of abstract postulates in physics, and then to deduce "a great many things in an impressively ironclad way, including for example, that no causal interpretation of quantum theory can possibly lead to all of the results given by the usual probability interpretation".[9] It is not possible to check that the postulates are correct, and, even worse, nobody is clear (even von Neumann himself) "just what is being assumed." Bohm is clearly angry at the way mathematical virtuosity, towards which he admits feeling some jealousy, has held back conceptual clarification of the subject matter, perhaps by as much as 20 years. Hard, clear mathematical thinking is needed, but the postulates must be "clearly defined and well established". In relation to physics, it is possible that "with the aid

[4] Part I of Anti Dühring, Marx and Engels (1988).

[5] See Bohm (1957), p. 2.

[6] (21, **64**, p. 223).

[7] In Chap. 9, we will see Bohm attempting to follow this approach in developing a model for "chaos".

[8] (21, **64**, p. 223).

[9] (22, **67**, pp. 237–239).

of carefully reasoned analysis plus imagination, new solutions can be found, often having unsuspected properties, and thus a genuinely creative element is introduced." At the same time, finding precise mathematical solutions is a great service, as "it facilitates comparison with experiment."

Bohm explains that he is not opposed to mathematical thinking,[10] and he doesn't think it just has to be aimed at physics, but has a creative role in itself. However, with no objective basis to it, "it degenerates into the spinning of would-be-closed systems of thought, aimed mainly at raising the position of the individual in the pecking order."

Bohm is willing to admit that he finds mathematical notation difficult. In studying statistical mechanics and working with Miriam through Khinchin's, book he complains that "I have struggled for 20 years to learn what notation I now know, and I dont like to have to double my fund of notation just to learn about one subject."[11] His way of thinking is "not step by step, but rather through the inter-connection of various aspects of the whole."[12] He may then find a result by "intuition", and see other results that follow. But this qualitative approach, although "almost always correct", would not convince others, and was certainly not accepted in publications. Bohm felt he needed people who could detect logical "weaknesses in my arguments without attacking the basis", even if it sometimes infuriated the students correcting his manuscripts.

Clearly, Bohm does not think, initially, that he has adequately thought through the question of "intuition" versus logical deduction (or "instinct" versus "rigor", as he later puts it).[13] He devotes two pages to it,[14] followed by a long discussion, which then extends to other mathematical questions.[15]

Bohm agrees with the "intuitionist" that new ideas cannot be said to just come from "geniuses." But neither should intuition be seen as something mysterious, external to the thinking process. It includes observation and practice, but also, considerably more. In his case (and one could say, in general) it required "a lot of work to do it, thinking day by day, and "living" with a long series of scientific problems". It is also guided by a certain philosophy—in his case, the materialism that led him to look for causal laws—but we have noted that Bohm argues that all development of new concepts requires some kind of philosophy.[16] However, Bohm does not dismiss the role of abstract logical rigour in thinking, and with new hypotheses, "logic often plays a genuinely creative role in leading to unexpected conclusions", as it did in his work on the causal interpretation. It would seem, therefore, that the correct way

---

[10](23, **74**, pp. 285–259).

[11](21, **64**, pp. 223–224).

[12](21, **65**, p. 229).

[13]See the comment in (26, **92**, pp. 306–307) where Bohm recalls he had thought out these ideas while on vacation.

[14](25, **89**, pp. 295–296).

[15](25, **90**, pp. 297–302).

[16](18, **45**, pp. 167–169).

of looking at the thought process is a dialectical one, in which ""intuition" + logic were bound into an inseparable unit."

There are some four pages[17] discussing Bohm's ideas of dialectics in relation to mathematics and logic, and also referring to Hegel on "being and not being, which are resolved in the concept of becoming."[18] But we will not consider any further here the ideas Bohm sketches out, as that would take us into the specialized area of the foundations of mathematics.

Bohm concludes this comment on mathematics[19] with a broadside against "formalism", by which he aims to convey that the dominant trend in theoretical physics was to concentrate on lengthy mathematical calculations and equations (which are supposed to "fit" the numerical results from experiments). Bohm thinks that this move away from the conceptual approach he is trying to develop in his work has turned theoretical physics into "a grim dull business", spreading the idea that the world is an irrational place. Science has been turned into an "instrument of obscurantism", undermining confidence in the possibility of solving social problems, and spreading cynicism. In philosophical terms, the formalist approach is idealist, concerned only with abstract standards of elegance, rather than with ability to understand the real world. In sociological terms, the formalist approach informs a critique that tends to assess scientists (and lesser mortals!) with respect to how they meet the formal mathematical standards of a small elite group.

References

Bohm, D. (1957). *Causality and Chance in Modern Physics* (first ed.). London: Routledge and Kegan Paul (Second edition with new preface by Bohm).

Marx, K., & Engels, F. (1988). *Collected works* (Vol. 25). Moscow: Progress Publishers.

Wood, A. W. (2004). *Karl Marx* (2 ed.). London and New York: Routledge, Taylor and Francis Group.

[17](25, **90**, pp. 299–301).

[18]This, along with the discussion on "a stone is a stone" in (21, **65**, p. 228), would seem to support the idea that Bohm had been studying Hegel's philosophy from quite early in his Brazil period.

[19](25, **90**, pp. 301–302).

Chapter 8
Development of the Causal Interpretation

A number of letters in 1951 and 1952 refer to the responses that Bohm received to his causal interpretation. Some have been noted by Olival Freire,[1] and the reactions, mainly negative, from the major theoretical physicists of the day, have been discussed. The letters give a blow by blow account of Bohm's campaign to get a response to his ideas, and often result in anger or depression. To review all the issues involved would take us too far afield, so we merely indicate the various relevant parts of the letters, before proceeding to outline how Bohm develops his scientific work.

Feynman's eventually positive response after discussion[2] is clearly important to Bohm, and he is particularly pleased, after initial opposition, at the response from de Broglie and others, such as Vigier, Régnier and Schatzman, in France.[3] As mentioned in the opening paragraphs, Pauli reviewed the 1952 papers, and after the revisions had been accepted, agreed they were logically consistent but he opposed the philosophical approach.[4] Bohm sent material for publication in the *Nature* magazine to the Australian physicist H. Massey, who was in England.[5] The reason it never appeared, according to Freire's research, was because of the campaign against Bohm by the physicist Leon Rosenfeld, who strongly disagreed with Bohm's version of Marxism. There is little mention of Rosenfeld in the letters, even though Rosenfeld visited

[1] Freire Jr. (2015), Chap. 2.

[2] (20, **57**, p. 203), (21, **64**, p. 224) and (17, **38**, p. 155).

[3] In (20, **58**, p. 205) and (20, **62**, p. 216) he has not heard from them, then an initial response from Vigier, Régnier and Schatzman in (20, **63**, p. 219) and (16, **32**, p. 143). Vigier is then referred to in (21, **65**, p. 230). In (17, **35**, p. 147), we learn that de Broglie has become friendly, and then that his criticisms were answered in (21, **64**, p. 224), though a critical seminar was held with Pauli (17, **37**, p. 151) and on the single page (33, p. 439). In (24, **84**, p. 284) it is noted that the causal interpretation is discussed in the left-wing magazine La Pensée.

[4] (20, **57**, p. 203), (20, **62**, p. 216), (20, **63**, p. 219), (21, **64**, p. 224), (15, **21**, p. 125) and (16, **32**, p. 143).

[5] (20, **58**, p. 205), (20, **62**, p. 216), (16, **32**, p. 143) and (17, **34**, p. 145).

C. Talbot (ed.), *David Bohm: Causality and Chance, Letters to Three Women*,
DOI 10.1007/978-3-319-55492-1_8

São Paulo in 1953.[6] Bohm was apparently aware of Rosenfeld's opposition and of his support for complementarity, but not of the extent to which he was going to oppose him.[7] Bohm is concerned that the "big-shots" will ignore or at least play down the significance of his work, and notes that Oppenheimer expects that there will be less causality at the nuclear level than at the atomic level.[8] The responses of von Neumann, Niels Bohr and his son, Aage Bohr (the latter reacted favourably to Bohm), as well as of Erwin Schrödinger are reported to Melba and to Miriam[9] Bohm was pleased at the response from experimental physicists,[10] he was critical of someone called Keller from Courant's group,[11] and particularly upset by those he thought should have supported him, like Philip Morrison,[12] discussed in Chap. 6, and someone called Jehuda, who apparently didn't respond.[13] By about April 1952, he reports 15 favourable comments and 30 requests for reprints.[14]

In July 15–29, 1952, Bohm takes part in a conference entitled "New research techniques in physics", held in Rio de Janeiro and São Paulo.[15] In attendance were 8 American physicists, including Eugene Wigner, Herb Anderson and Donald Kerst, 8 Latin Americans, and a few Europeans. Bohm concentrates on the response to his presentation by Isidor Isaac Rabi, who, as a Nobel Prize winner, Bohm would have definitely classified as a "big-shot": "As yet, your theory is just based on hopes, so why bother us with it until it produces results. The hidden variables are at present analogous to'angels' which people introduced in the Middle Ages to explain things". Bohm complains at the double standards: when a mainstream physicist puts forward a new theory of elementary particles, hoping for a result in the future, even if the theory is not self-consistent, people say "fine". But the response to Bohm was: "[w]e are interested only in results". Not surprisingly, Bohm feels "somewhat discouraged". In the second letter to Miriam, Bohm comments that it "brought home to me with a shock how little interested most physicists are in the questions that interest me." Note that he includes the work on probability and on the infinity of levels philosophy here, and concludes that books and articles have little effect. The problem is that "customary ideas are so thoroughly interwoven into most peoples thinking".

Bohm's feeling of isolation was compounded in the first months of 1953. The problem he identifies is philosophical, there is a "lack of response to what seem to me important indications of what is wrong with our ideas, which explains why there have been such few results in the past 20 years." The domination of the mainstream approach, namely "Only results count the ideas behind them are just "window-

[6]Freire Jr. (2015), Sect. 2.4.

[7](15, **21**, p. 125), (17, **35**, p. 147), (22, **69**, p. 247) and (17, **38**, p. 154).

[8](21, **64**, p. 224).

[9](17, **35**, p. 147) and (22, **69**, p. 247).

[10](26, **95**, p. 315) and the review by Dancoff, (21, **65**, p. 226).

[11](17, **38**, p. 154).

[12](23, **78**, p. 265), (26, **95**, p. 315), (17, **37**, pp. 151–152) and (17, **38**, p. 155).

[13](22, **69**, p. 247) and (23, **77**, p. 264).

[14](17, **37**, p. 152).

[15](23, **80**, p. 271), (23, **81**, p. 273) and (17, **39**, p. 156).

dressing""", had not in fact produced much in the past 20 years. He begrudgingly admits Feynman and Schwinger had produced some "resultlets" (a somewhat peevish dismissal of what turned out to be the basis of much of modern theoretical physics) after "20 years of labor by the mountain of thousands of theoretical physicists."[16] However, now, according to reports from Miriam, the very people who were supposed to be supporting him, Eugene Gross, Bohm's friend and former research student, and George Yevick, had apparently banged on the table, saying of Bohm, "Let him give us results". As a result, Bohm lashes out, feeling that he is being asked to produce a new revolution in physics in a year or two, an equivalent to "Newton, Einstein, Schrödinger + Dirac, all rolled into one", otherwise there is no need to think about the deeper questions. He thought that at least George and Gene appreciated this, but now he finds this is not the case, and even has doubts about Miriam.[17]

Though expressed in terms of Bohm's often rather emotional comments, the letters give some idea of the declining interest in the causal interpretation one year after from the publication of the 1952 papers. Clearly, this experience put Bohm under considerable pressure to get "results". Let us now turn to the development of Bohm's scientific work, much of it done with various collaborators.[18]

We have already noted Bohm's point in approximately April 1952, namely that in his causal interpretation "probabilities have the same origin as those of classical statistical mechanics"[19] and that he can therefore explain the usual probabilistic wave function by statistical mechanical techniques.[20] There is a downside to this, however: the "ensemble" and "probability distribution" techniques employed, as discussed in the next chapter, can be applied only to special cases, as in the 1953 paper. Bohm, however, had another approach in mind. As he explained in a January 1952 letter to Melba[21]:

> I have become convinced that the time has come to reconsider the concept of an "ether" that fills all space. For in the causal interpretation of the quantum-mechanics, the negative energy electrons, protons, and neutrons are all supposed to be really there, forming a medium of incredible density. For this reason, the word "ether" is bad, as the medium is anything but "ethereal". A better word would be "substratum".

Where does this leave Einstein's theory of special relativity, which appeared to have made the ether a nineteenth century relic? As with all scientific theories, "it is always conceivable that in some domain, the laws of relativity may fail". Consequently, Bohm is hypothesizing that, in as yet unexplored levels of reality, with

---

[16](26, **93**, pp. 308–309).

[17](26, **94**, pp. 311–312).

[18]Jayme Tiomno, Ralph Schiller, Jean-Pierre Vigier, Louis de Broglie, Mario Schönberg and others. Tiomno was a Brazilian physicist who worked with Bohm in São Paulo from his arrival, but them moved to Rio in March, 1952 (20, **59**, p. 206). Ralph Schiller, formerly Peter Bergman's student, was Bohm's postdoc for 2 years from September, 1952 (24, **83**, p. 277).

[19](17, **37**, p. 152).

[20]As in Bohm (1953).

[21](17, **34**, pp. 145–146).

techniques as yet unknown, perhaps down at distances of 10^{-13} cm, an ether will be found, and also, hopefully, evidence for the causal interpretation.

There is more on the history of the ether, and a recognition that in Einstein's general theory of relativity "something like" an ether is reinstated, in the early 1952 letter sent to George Yevick.[22] The point is also made that in the causal interpretation of quantum mechanics "one is led necessarily to the conclusion that "empty" space contains real particles moving in every conceivable direction and really fluctuating electromagnetic fields", and that the density of this matter can be expected to be very high.[23] Bohm then argues that our current "matter" can be viewed as "small localized deviations from uniformity in this substratum". At the same time, the "ether" or substratum can carry electromagnetic and gravitational waves, and the processes of quantum field theory—the "annihilation" and "creation" of matter as well as the transformation of energy into matter and vice versa—can be understood as fluctuations in the substratum.

It seems that Bohm now takes this study of the "ether" to be the way to develop the causal interpretation, rather than following the statistical mechanics approach of Bohm (1953). He works with Tiomno and then Ralph Schiller, so that in the later part of 1952 he writes that "I wouldnt be surprised if something resembling Einstein's dream of deducing qu. mechs from general relativity could eventually be achieved."[24] His approach is to first tackle spinning particles in the non-relativistic version of the causal interpretation (the so-called Pauli theory), and then, over the next period, he attempts to deal with the difficult case of relativistic spin (the Dirac theory). As we shall see in the next chapter, probability and statistical mechanics are not abandoned, but are considered according to their own specific problems.

In April/May 1953, Bohm writes to Melba[25]:

> I am making some progress on the quantum theory, having found a relation between the Pauli spin theory (causally interpreted) and Diracs new electrodynamics, which introduces a theory highly reminiscent of the old ether theories. The spin appears definitely to be related to vortex motion in the "ether". I am now trying to extend this theory to cover the Dirac equation. If this could be done, then it would be an important step, (as well as a victory in my life-long struggle to understand the Dirac equation). I'll keep you posted about any further progress.

Later that year, Bohm is visited by George Yevick and Jean-Pierre Vigier, and it seems that it was then that a new "hydrodynamical" or fluid model approach to the causal interpretation was developed, perhaps a result of previous work done in France.[26] In Bohm and Vigier (1954), the publication which subsequently came out, the word "ether" is not mentioned, but the fluid model, with random fluctuations, is supposed to apply to a "sub-quantum mechanical level", and it is claimed that

---

[22](33, **124**, pp. 37–42).

[23]Arguments that are still relevant today with regard to "dark matter and energy".

[24](17, **40**, pp. 157–158).

[25](18, **42**, p. 161).

[26]In the letter of July 1, 1954 to Miriam (31, **119**, p. 406), Bohm comments that "you may tell George that the general picture that we developed when Vigier was here is beginning to take shape.

this approach also leads to the usual probability interpretation of the wave function, apparently more general than that in Bohm (1953). We will not attempt to deal with this highly technical material in any depth, but note that in June 1954, Bohm writes to Melba,[27] explaining that the Pauli spin case has now been dealt with, and giving a summary of his intended approach to the Dirac case, with "[a]n "ether" composed of spinning body-like elements." The idea is that there is a connection between the spin of a particle and the turbulent motion of the fluid. There is a "coordinated motion" between the spin motion and the motion due to displacement in space, with "matter", as before, conceived of as an inhomogeneity in the ether.

Bohm gives a similar explanation of this work to Miriam in July 1, 1954.[28] The "coordination" between spin and mass motion "is so strong that particles do not collide with each other, but get out of each others way, adjust to each other, etc." In both this letter and the letter to Melba, Bohm explains the "coordination" as more like a ballet dance than the motion of a crowd of people. He is clearly hoping to have developed a satisfactory theory of elementary particles.

References are given in Bohm and Vigier (1954) to work by Madelung in the 1920s, and to recent work by the Japanese physicist Takabayashi, but also to Mario Schönberg (written also as Schenberg), who returned to São Paulo in April 1953. There is a whole series of denunciations of Schönberg by Bohm (see Chap. 13) in 1953, but by May 1954,[29] Bohm is apparently having useful discussions with him on "dialectics and on the notion of a turbulent ether explanation of the quantum theory". By the time of the above mentioned letter of July 1st 1954, Schönberg is "working on this problem too and he knows a great deal of the math involved".[30] At this point, Bohm decided to delay leaving Brazil, given that he thought that this work was going well.

Finally, we note that while in Israel, in October 12, 1955,[31] and then either later that year or in the beginning of 1956.[32] Bohm writes two brief reports to Melba on the development of the "turbulent ether" approach, which he had made on visits to France. Prophetically, Bohm writes at the end of the second letter: "Things seem to be working out, but the details are still puzzling. It may be a long time before this theory is really ready." However, in 1954, Bohm was still optimistic that the causal interpretation would lead to new developments.[33] As we point out in Chap. 6, by 1957, in *Causality and Chance* he has retreated to suggesting that it " may be a good starting-point".

In fact, the Dirac problem was not solved in this earlier period, and as we noted above in Chap. 3, Bohm had lost interest in it by the 1960s. Even with the revival from the 1970s onwards, it took another 60 years from 1952 for the relativistic Dirac

---

[27](18, **47**, pp. 173–174).

[28](31, **119**, pp. 405–406).

[29](31, **118**, p. 404).

[30](31, **119**, p. 405).

[31](19, **52**, pp. 180–181).

[32](19, **53**, pp. 181–182).

[33]Note the footnote in (32, **123**, p. 431), just before leaving Brazil, where he writes about Vigier's defence of causality being cheered by students.

case to be more satisfactorily dealt with by Hiley and co-workers,[34] and also, by the Bohmian mechanics group,[35] each using their distinctive approach. The hydrodynamical models were eventually abandoned, even by the French group. We will not attempt to deal with the technical details of Bohm's joint work with Vigier, de Broglie and others, published in papers up to the early 1960s, which would take us a long way from these letters.

References

Bohm, D. (1953). Proof that probability density approaches $|\psi|^2$ in causal interpretation of the quantum theory. *Physical Review*, *89*(2), 458–466.

Bohm, D., & Vigier, J.-P. (1954). Model of the causal interpretation of quantum theory in terms of a fluid with irregular fluctuations. *Physical Review*, *96*(1), 208–216.

Dürr, D., Goldstein, S., Norsen, T., Struyve, W., & Zanghì, N. (2013). Can Bohmian mechanics be made relativistic? In *Proceedings of the Royal Society A* (Vol. 470, No. 2162).

Freire, O., Jr. (2015). *The quantum dissidents: Rebuilding the foundations of quantum mechanics (1950–1990)*. Berlin, Heidelberg: Springer.

Hiley, B. (2012). *Weak measurements: Wigner-Moyal and Bohm in a new light*. https://www.semanticscholar.org/paper/Weak-Measurements-Wigner-Moyal-and-Bohm-in-a-New-Hiley/23fb5baa7c235b476921640bfea671f34ff86e77.

34 Hiley (2012).

35 Dürr et al. (2013).

Chapter 9
Notes on Probabilty and Statistical Mechanics

From the very first letters to Miriam, we find that Bohm is working on probability and statistical mechanics and asking Miriam to use her mathematical expertise in order to help him. In the period up to and including February 1953, there are 14 letters or parts of letters on this topic, most of them quite technical. Then, in 1954, there is a second group of 7 letters, several of them quite long (more than 10 pages).

Bohm and Miriam begin[1] by working through Khinchin's quite mathematical book (Khinchin 1949). Note that from the start, Bohm uses the term "chaos", meaning something similar to today's term "deterministic chaos", although, of course, there have been many new discoveries since 1951. Thus[2]: "the condition for development of chaotic instability i.e. a small change of initial conditions produces a big change in the final position," which we recognize as the well-known "butterfly effect". Bohm is also using a standard physics text by Tolman.[3]

We will just look at some of the conceptual issues in the letters and will not attempt to follow the detailed technical and mathematical aspects of Bohm and Miriam's joint work: this would require too much specialization, and in any case, many documents, as well as Miriam's letters, are unavailable. Bohm explains his concerns[4] about the abstract nature of the probability concept used in mathematics. This follows, as it were, from his attitude to mathematics discussed above. The hypothesis is that the probability of an event (in the "frequency" approach used by Bohm) consists in the fraction of times this event occurs in an "infinite" series of events. But if, in practice, we get a completely different fraction of events occurring (in a finite total number of events, however large), we can only conclude that the hypothesis is "probably" wrong. As Bohm comments, "(w)e are led around and around in a conceptual circle,

---

[1](20, **59**, pp. 208–209).
[2](21, **64**, pp. 221–222).
[3]Tolman (1979) in (20, **61**, pp. 212–215).
[4](20, **61**, pp. 212–215).

C. Talbot (ed.), *David Bohm: Causality and Chance, Letters to Three Women*,
DOI 10.1007/978-3-319-55492-1_9

and the real meaning of none of the concepts can be pinned down." This lack of conceptual clarity becomes a problem when we deal with "improbable" events in nature, i.e. systems that are far from equilibrium. Bohm concludes that "(t)he purely statistical point of view becomes inadequate here, and leads to a fantastic description in which, for example, the earth and everything on it (oceans, waterfalls, people, etc.) are a highly "improbable" fluctuation away from statistical equilibrium." Thus, his approach will be "to use causal laws, combined with the concept of the development of chaos." Bohm considers the standard Maxwell-Boltzmann velocity distribution in an ideal gas[5] and argues that as time increases, "the possible range of initial conditions that lead to a deviation from the Boltzmann distribution approaches a set of measure zero". A set of measure zero has a mathematical definition, but basically means vanishingly small. In other words, the Boltzmann case is what "typically" happens, something we will discuss later.

Throughout 1952, Bohm is developing various mathematical models of chaos and communicating them to Miriam, asking for her response. He presumably thinks that, even though these models are abstracted from the complexity of real systems, they will help him to grasp the essentials of his causal approach. He starts[6] by looking at a "rough-walled" container, and at the molecules bouncing off an irregular wall. This continues with further calculations.[7] Bohm's hope is that he can demonstrate "all the results of statistical mechanics in a dynamic way, without making any assumptions, other than that the laws of mechanics hold". There is a brief comment[8] where Bohm re-iterates his solution to dealing with those "unusual" initial conditions (of measure zero) which do not lead to "chaos". They need exactly prescribed numerical values, and all experience shows that "no physical system can have a certain set of parameters with exactly prescribed numerical values." Work on the rough-walled container continues,[9] and, as Bohm notes in passing, it was the subject of a discussion with von Weizsäcker. There is an interesting diagram here,[10] which clearly demonstrates (on the left) one of those special cases of initial conditions whereby "chaos" does not result, but which Bohm sees as a mathematical artifact, not arising in experience. The particle here bounces to and fro between the two rough walls, rather than having a "chaotic" trajectory that will fill the whole container. There is a development of the rough wall problem[11] as Bohm starts looking at the angles of deflection from the wall.

Clearly, the rough wall approach proved to be too difficult, and a change of tack follows[12] whereby Bohm looks at an array of scattering centres (sometimes known as a Lorentz gas). He considers the deflection or change in angle of a particle's trajectory

[5](21, **64**, pp. 221–222).

[6](21, **64**, pp. 221–224).

[7](22, **67**, pp. 240–242).

[8](22, **68**, p. 246).

[9](23, **75**, pp. 259–261).

[10](23, **75**, p. 260).

[11]In the incomplete letter (23, **76**, pp. 262–263).

[12](23, **78**, pp. 265–270).

($\Delta\theta$) before and after collision, and he offers an account of what happens using a series of diagrams. Then,[13] Bohm hits on the simple model that he will repeatedly refer to for the whole of the period in Brazil. He works only with angles (mathematical notation θ) before and after a collision, and assumes the angle after the collision can be obtained from the angle before the collision by multiplying its value by a large constant number (mathematical notation K, assumed to be larger than 1000). This means the $(n + 1)$st angle is obtained from the nth angle in a sequence of collisions by multiplication by K (in mathematical notation $\theta_{n+1} = K\theta_n$). Increasing the value of n is equivalent to increasing the "time". Of course, this is a simplification of what happens with the angle in real collisions, but Bohm considers the model contains the "essential element of the real physical problem; i.e. the extreme dependence of final angle on initial position".[14] To see why this is so, we have to remember that an angle is always assumed to be between 0° and 360°, so that if the angle obtained by multiplying by K is too large, we must subtract multiples of 360° until it is in the usual range of 0° and 360°. For example, suppose $K = 2000$ and we start at 30°. The next angle in the sequence is then $30 \times 2000 = 60000°$. A little division tells us we must subtract $60000 \div 360$, i.e. 166 (to the nearest whole number below the answer) times 360°, which gives us a result of 240°. It is not difficult to see how a slight change in the initial angle, to 30.1° for example, using the same kind of calculation, gives a result of 80°, so a slight change of 0.1° in the initial angle gives a very large change in the next and subsequent angles.

A number of letters follow, discussing this model ($\theta_{n+1} = K\theta_n$). Bohm stipulates[15] that K should be a whole number (integer) and embarks on two possible proofs that the result of a large sequence of angles generated by this model should be uniformly distributed throughout the range of 0°–360°.[16] This is for "most" starting values of the angle: there will be a small set (of "measure zero") of starting values for which this does not happen. As in the "rough-wall" model, he assumes that those special angles that don't give rise to "chaos" will not arise in physical experience. This is the aim of Bohm's efforts here: the angles in the sequence in the model are considered to be "causally" related to each other in increasing time, yet the outcome can be regarded as a "chaotic" distribution, even if it is rather a simple case of a uniform distribution (so that "chaos comes from causality").

When he further discusses this new model,[17] Bohm notes that "equidistribution", i.e. a uniform distribution of angles, was proved by H. Weyl in 1916,[18] but he considers that he has found "a much simpler proof, based on a method whose meaning can easily be seen physically." Unlike Miriam's earlier approach, Bohm did not start with

[13](24, **83**, pp. 277–281).

[14](24, **83**, p. 279).

[15](24, **84**, pp. 281–284).

[16]See Figs. 9.1, 9.2 and 9.3 for the original copies of this letter.

[17](24, **86**, pp. 284–285).

[18]Actually it was not Weyl, but the mathematician E. Borel, who in 1909 first looked at the so-called "normal" numbers, beween 0 and 1, which would give rise to a uniform distribution. See Niven (2005), Chap. VIII.

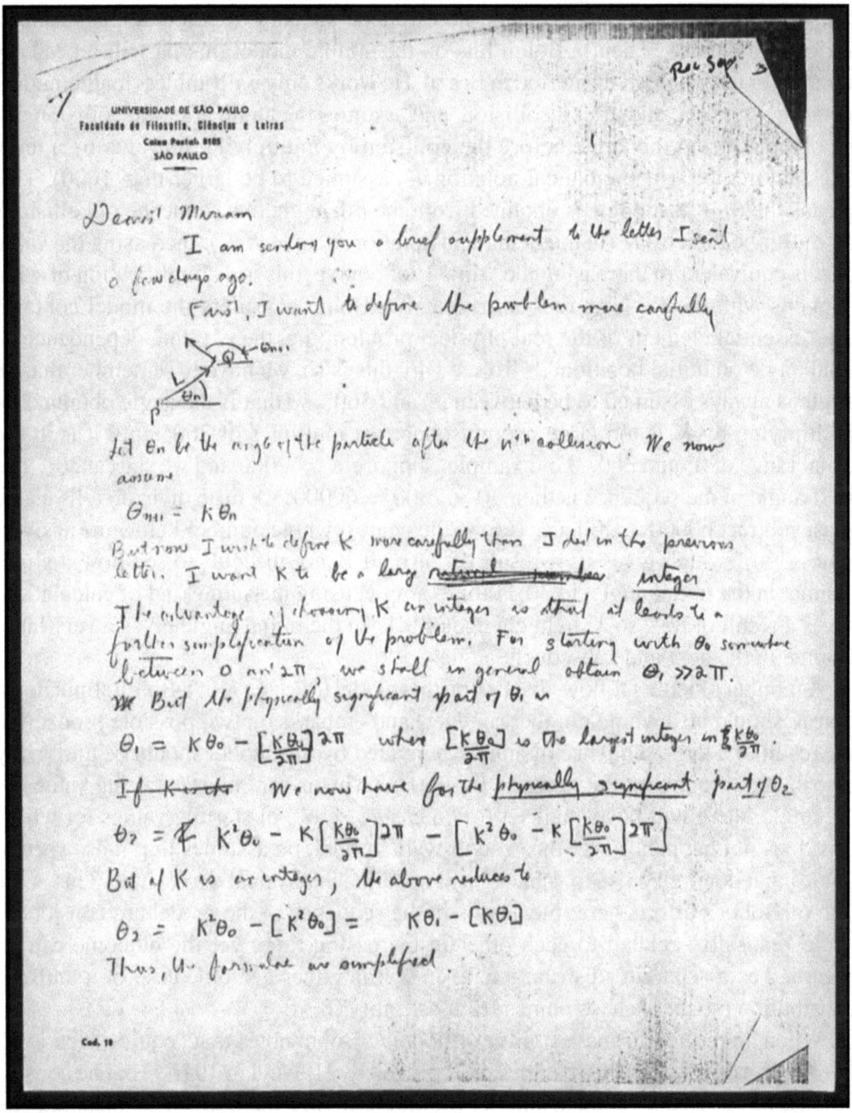

UNIVERSIDADE DE SÃO PAULO
Faculdade de Filosofia, Ciências e Letras
Caixa Postal: 8105
SÃO PAULO

Dearest Miriam

I am sending you a brief supplement to the letter I sent a few days ago.

First, I want to define the problem more carefully

Let θ_n be the angle of the particle after the nth collision. We now assume

$$\theta_{n+1} = K\theta_n$$

But now I wish to define K more carefully than I did in the previous letter. I want K to be a large ~~natural number~~ integer

The advantage of choosing K an integer is that it leads to a further simplification of the problem. For starting with θ_0 somewhere between 0 and 2π, we shall in general obtain $\theta_1 \gg 2\pi$.

But the physically significant part of θ_1 is

$$\theta_1 = K\theta_0 - \left[\frac{K\theta_0}{2\pi}\right]2\pi$$ where $\left[\frac{K\theta_0}{2\pi}\right]$ is the largest integer in $\frac{K\theta_0}{2\pi}$.

~~If K is~~ We now have for the physically significant part of θ_2

$$\theta_2 = K^2\theta_0 - K\left[\frac{K\theta_0}{2\pi}\right]2\pi - \left[K^2\theta_0 - K\left[\frac{K\theta_0}{2\pi}\right]2\pi\right]$$

But if K is an integer, the above reduces to

$$\theta_2 = K^2\theta_0 - [K^2\theta_0] = K\theta_1 - [K\theta_1]$$

Thus, the formulae are simplified.

Cod. 18

Fig. 9.1 Original of letter **84**, page 1.

a probability distribution of angles (which he sees as assuming chaos to begin with) but, he claims, he has shown that a "causal sequence of numbers leads to chaos". Afterwards,[19] Bohm is using the same model, but with the angle measured by real

[19](24, **87**, pp. 286–289).

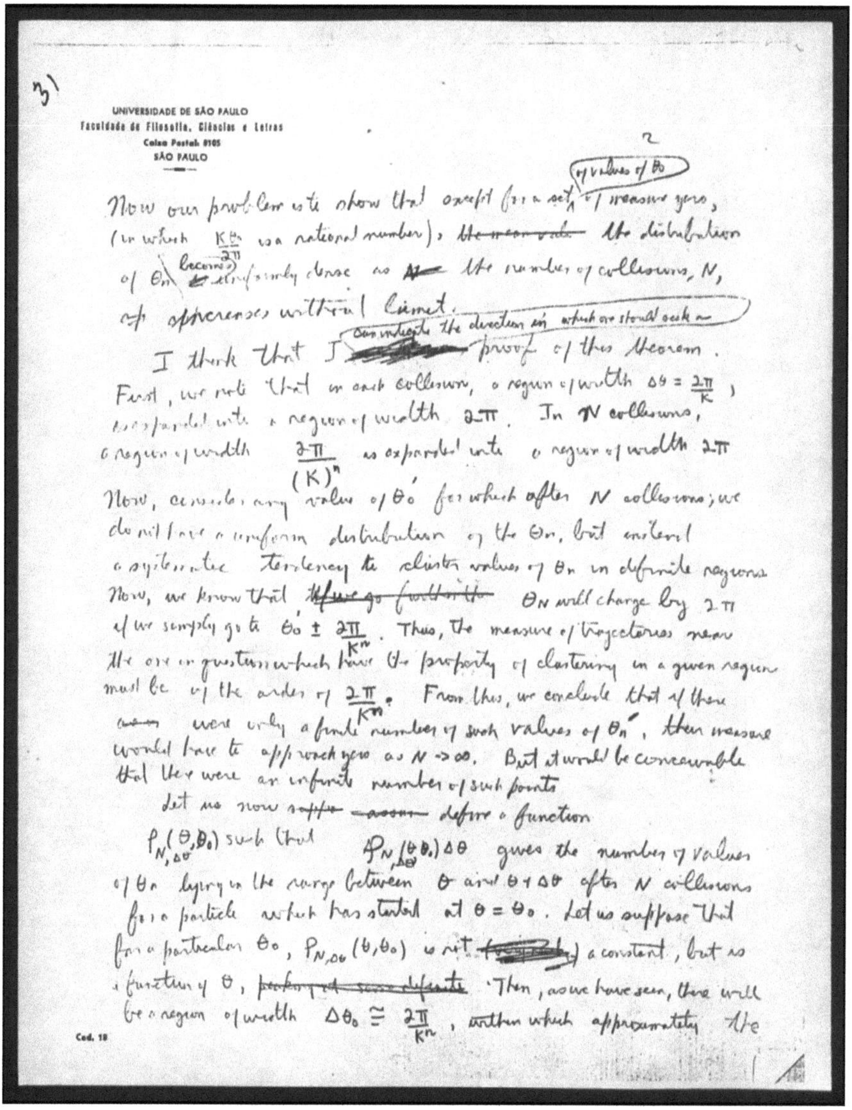

UNIVERSIDADE DE SÃO PAULO
Faculdade de Filosofia, Ciências e Letras
Caixa Postal 8105
SÃO PAULO

2

Now our problem is to show that except for a set of values of θ_0 of measure zero, (in which $\frac{K\theta_0}{2\pi}$ is a rational number), the distribution of θ_n becomes uniformly dense as the number of collisions, N, approaches without limit.

I think that I can indicate the direction in which one should seek a proof of this theorem.

First, we note that in each collision, a region of width $\Delta\theta = \frac{2\pi}{K}$, is expanded into a region of width 2π. In N collisions, a region of width $\frac{2\pi}{(K)^n}$ is expanded into a region of width 2π.

Now, consider any value of θ_0' for which after N collisions; we do not have a uniform distribution of the θ_n, but instead a systematic tendency to cluster values of θ_n in definite regions. Now, we know that θ_N will change by 2π if we simply go to $\theta_0' \pm \frac{2\pi}{K^n}$. Thus, the measure of trajectories near the one in question which have the property of clustering in a given region must be of the order of $\frac{2\pi}{K^n}$. From this, we conclude that if there were only a finite number of such values of θ_n', their measure would have to approach zero as $N \to \infty$. But it would be conceivable that there were an infinite number of such points.

Let us now define a function $P_{N,\Delta\theta}(\theta,\theta_0)$ such that $P_{N,\Delta\theta}(\theta,\theta_0)\Delta\theta$ gives the number of values of θ_n lying in the range between θ and $\theta+\Delta\theta$ after N collisions for a particle which has started at $\theta=\theta_0$. Let us suppose that for a particular θ_0, $P_{N,\Delta\theta}(\theta,\theta_0)$ is not a constant, but is a function of θ. Then, as we have seen, there will be a region of width $\Delta\theta_0 \cong \frac{2\pi}{K^n}$, within which approximately the

Cod. 18

Fig. 9.2 Original of letter **84**, page 2.

numbers between 0 and 1 (easily obtained by dividing the angle in degrees by 360), and gives another proof. Using numbers between 0 and 1 is the approach used by mathematicians.[20]

---

[20]See for example Stewart (1990), p. 112, where this model is used with $K = 10$, so that the change from one angle to the next in the sequence is given by shifting the decimal point to the right, and then dropping the whole number before the decimal point. Stewart's book is a useful introduction to Chaos Theory.

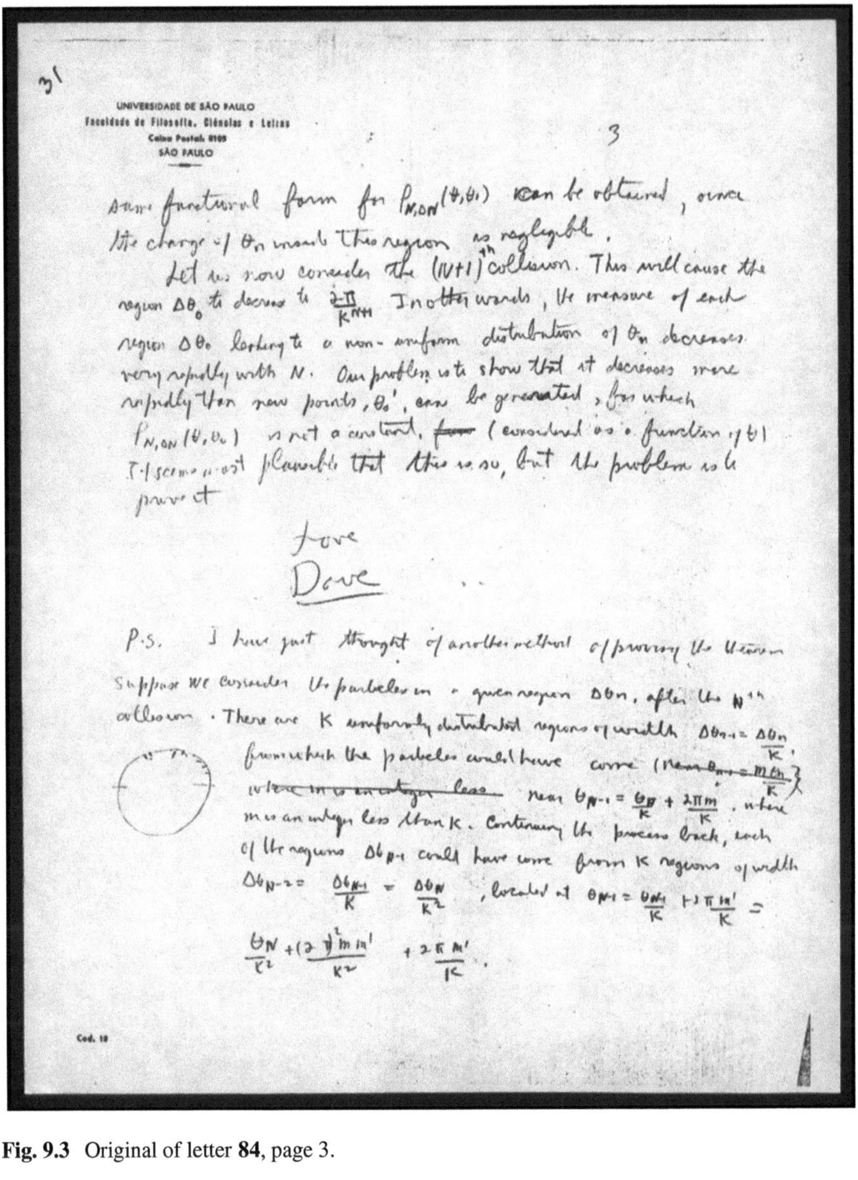

31

UNIVERSIDADE DE SÃO PAULO
Faculdade de Filosofia, Ciências e Letras
Caixa Postal 8105
SÃO PAULO

3

same functional form for $P_{N,\theta N}(\theta,\theta_0)$ can be obtained, once the change of θ_n inside this region is negligible.

Let us now consider the $(N+1)^{th}$ collision. This will cause the region $\Delta\theta_0$ to decrease to $\frac{2\pi}{K^{N+1}}$. In other words, the measure of each region $\Delta\theta_0$ leading to a non-uniform distribution of θ_n decreases very rapidly with N. Our problem is to show that it decreases more rapidly than new points, θ_0', can be generated, for which $P_{N,\theta N}(\theta,\theta_0)$ is not a constant, ~~for~~ (considered as a function of θ). It seems most plausible that this is so, but the problem is to prove it.

Love
Dave

P.S. I have just thought of another method of proving the theorem. Suppose we consider the particles in a given region $\Delta\theta_n$, after the Nth collision. There are K uniformly distributed regions of width $\Delta\theta_{N-1} = \frac{\Delta\theta_n}{K}$, from which the particles could have come (~~near $\theta_{N-1} = \frac{m\theta_n}{K}$ where m is an integer less~~) near $\theta_{N-1} = \frac{\theta_N}{K} + \frac{2\pi m}{K}$, where m is an integer less than K. Continuing the process back, each of the regions $\Delta\theta_{N-1}$ could have come from K regions of width $\Delta\theta_{N-2} = \frac{\Delta\theta_{N-1}}{K} = \frac{\Delta\theta_N}{K^2}$, located at $\theta_{N-1} = \frac{\theta_{N-1}}{K} + \frac{2\pi m'}{K} = \frac{\theta_N}{K^2} + \frac{(2\pi)^2 m m'}{K^2} + \frac{2\pi m'}{K}$.

Cod. 10

Fig. 9.3 Original of letter 84, page 3.

Bohm's model, a type of "deterministic chaos", generates what are now called "pseudo random numbers", usually distinguished from actual "random numbers", which are considered to be generated by physical quantum processes.[21] Producing such randomness with a computer has led to a massive growth in such mathematical

[21] Note that this assumes the "essential randomness" of standard quantum mechanics. Bohm's causal interpretation would assume only a difference of degree between the two cases.

techniques as well as many elaborate statistical tests to test them for randomness.[22] Random numbers are widely used in cryptography by the finance industry and state security services.

Bohm's "proof" that his model gives rise to a uniform distribution[23] relies on approximating the circle by a series of small segments (known as "coarse graining") and looking at all the possible paths that converge on a chosen segment over a long time period. Bohm thought his method gave him an "essential view" of the problem, based on him being "more familiar with the physical background", thus also giving him an advantage over Miriam, who tried to apply "ready-made theorems".[24] Unfortunately it is unlikely that his method, as it stands, could satisfy the rigorous scrutiny of mathematicians. They would certainly use "Lesbesgue Measure Theory" and the "Strong Law of Large Numbers".[25] Miriam almost certainly stressed the importance of these areas, but she probably found them very difficult to explain in letters.[26]

It should be noted that Miriam clearly played an important role in this joint work, despite Bohm downplaying it here. Unfortunately, as I have stressed, her letters do not survive, but we can assume she was trying to develop the "chaos" approach with her knowledge of mathematics. It seems she received short shrift from the mathematical experts in the field, Kai Lai Chung and Joseph L. Doob. Bohm describes the letter she received from Chung as "disgraceful",[27] and makes a swipe at Chung and Doob as "short-sighted".[28] Later, he admits "they are very intelligent and capable", but "they would get a lot further if they were a little less conceited, so that they could, without condescension, work on problems a bit closer to reality."[29]

Apart from this somewhat boastful reference to his successful treatment of the chaos model in the above letter[30] dated May 12, 1953, there is little mention of probability and statistical mechanics following the long burst of activity in 1952. There is, in February 28, 1953,[31] an emphasis on randomness being derived from causal laws, stating that "all chaos comes from causality" and "empirically observed properties of "randomness" should also be deducible as necessary consequences of the causal laws applying in each particular case". See also the letter to Melba, around the same time,[32] on "The Causal Interpretation of the Theory of Probability. There

---

[22]For a nice introduction see Bennett (1999). For Bohm's model as a pseudo random number generator for large K, see Broer and Takens (2011), p. 99 (Transition from chaotic to stochastic).

[23]See also Bohm and Schützer (1955), Sect. 7.

[24](27, **104**, pp. 344–345).

[25]See Dürr and Teufel (2009), Sect. 4.3.

[26]For the Strong Law see (20, **58**, p. 205), (21, **64**, pp. 223–224) and Bohm thanking Miriam for her help on this topic in (23, **77**, p. 264). For Bohm's dismissal of Lesbesgue integration see (23, **75**, pp. 259–260).

[27](23, **77**, p. 263).

[28](25, **89**, p. 295).

[29](27, **104**, pp. 344–345).

[30](27, **104**, pp. 344–345).

[31](26, **96**, pp. 317–318).

[32](18, **42**, pp. 161–163).

is then a considerable change in Bohm's approach to causality and its relation to chance in the 7 letters of 1954.[33]

The change must be put down, firstly, to the influence of Mario Bunge. Secondly, there were discussions with Mario Schönberg, after Bohm seems to have overcome his hostility to him and carried out collaborative work from at least May 1954, as noted in the previous chapter. Note that in September, 1953,[34] Bohm was indignant because Schönberg believed that explaining chance in terms of causality is "reactionary" and "undialectical", and that there is somehow "pure chance" which can propagate from level to level, but is not explained, except in terms of itself.

We know that Bohm read Mario Bunge's article "What is Chance"[35] before inviting him to São Paulo[36] in 1953. Following Engels, from whom he gives several quotations, Bunge opposes the view of mechanical determinists, or "metaphysical" determinists in Engels' terminology, namely that chance is not objective but merely reflects a temporary lack of knowledge which will disappear as knowledge develops. Although Bohm, with his infinity of levels approach, is by no means a simple mechanical determinist, this was clearly a criticism of his view that "chaos" (presumably here the same as chance) is a by-product of causality. Bunge goes further and criticises the "indeterministic metaphysics", which was then the dominant view among current quantum physicists (he cites John von Neumann as an outstanding example). As we attempted to show above, standard quantum mechanics does treat the probabilities that arise as *essential*, and thus it can be regarded as a form of mechanical indeterminism. We saw that Bohm had accepted this extended view of mechanism put forward by Bunge, especially from his 1953–4 letters to Melba.[37]

Bunge follows Engels in proposing that chance is just as objective as causality, and that rather than placing full weight on either causality or chance, as the determinists and indeterminists do, respectively, they should be seen as dialectical opposites. Accidental or chance phenomena, things or events are found to have causes, but they, in turn, can be the cause of something else.

As we also noted above, in Chap. 6, when he wanted to extend his view of causality to non-mechanical and qualitative changes, Bohm had difficulties in considering evolutionary biology, because of Soviet Lysenkoism. In 1951, Bunge clearly felt obliged to pay homage to the charlatan Lysenko. He attacks geneticists for maintaining that "descent is strictly and exclusively by predetermined chromosomes"[38] as opposed to "Lysenko's claims to have modified this fatal destiny in a constructive way".

---

[33](30, **114**, pp. 378–386), (30, **115**, pp. 391–393), (31, **117**, pp. 397–403), (31, **119**, pp. 406–407), (31, **120**, pp. 408–414), (32, **121**, pp. 417–423) and (32, **122**, pp. 423–427).

[34](28, **106**, p. 351).

[35]Bunge (1951).

[36]See Bunge (2016), p. 90.

[37](18, **43**, pp. 163–165) and (18, **46**, pp. 170–173).

[38]This is not true, a tradition of Marxist work on biology was developed in the USSR and suppressed under Stalin. See, for example, the article written by the American geneticist and Nobel prize winner, H.J. Muller, Lenin on Genetics, reproduced in Graham (1971), pp. 451–472. Muller supported the Soviet Union to the extent of moving to Russia in 1933. Stalin's opposition to him and the growing support for Lysenko forced him to leave by 1937.

This is particularly serious, as the main example of the importance of chance given by Engels in *Dialectics of Nature* relates to Darwin's work, very highly regarded by both Marx and Engels. Bunge omits this key quotation.[39]

Bohm took on board this dialectical approach to causality and "objective" chance but, looking at the dates of the above letters, which start in March, 1954, it took some time to absorb its implication for physics. We can also assume that Schönberg also supported such a dialectical approach to causality and chance, recommending Bohm to study Lenin's philosophical notebooks.[40]

Thus, eventually we see Bohm insisting, for example,[41] that "the irregular or random side of things is a basic thread which must be woven together with the regular [i.e. causal—CT] side to come to a more nearly complete concept of reality". This was integrated with the theory of infinite levels: "not only do all causal laws follow eventually from the regular trend in random motion arising at other levels, but also we can never reach the infinite depth and breadth and height needed to exhaustively trace all random fluctuations to causal origins."[42] Chance or accidental events "are events that are contingent on causal factors existing outside the data specified by the physical conditions of the problem."[43] Contingent means here "that which could be otherwise",[44] the opposite of necessity. Bohm is not concerned with single chance events but with the specific problems in physics involving a series of chance events, usually ordered in time, though there can also be ordering in space or with respect to other physical quantities.[45] In these 1954 letters, he now uses the term "randomness" for such irregularly fluctuating events, rather than the earlier term "chaos", which no longer appears. He clarifies the concept of randomness further[46]:

> . . . mere contingency is not enough. There must exist irregular fluctuations in these contingencies (i.e. irregular relative to the data specified in the problem). For example, if we specify the macroscopic properties of a system, the microscopic behaviour of a given atom is not determined, but remains contingent on factors existing at lower levels (e.g., positions and velocities of other atoms, quantum fluctuations originating at a deeper level, etc.). These contingent factors show an irregular or "random" fluctuation. One of the most important properties of truly random fluctuations is that one can count on their mutual cancellation in the long run, when many objects or events are involved, to permit the development of regular mean trends, which are treated by the theory of probability.

It is this randomness that gives rise to statistical laws, and Bohm now recognizes that statistical laws are just as important as, and cannot simply be reduced to,

---

[39]Marx and Engels (1988), p. 498. The example of "Darwin, in his epoch-making work" is on p. 501.

[40]In Wilkins (1986), VI, Bohm says "I thought Schönberg had a deeper view of these things than most of the left-wing people. In a way, he helped to show me that I had been approaching the thing in a narrow way, by just looking at causality, without bringing in the opposite side of chance."

[41](31, **117**, pp. 402–403).

[42]Ibid.

[43](31, **119**, p. 407).

[44]Bohm (1957), p. 2.

[45]See (30, **114**, p. 379).

[46](31, **119**, p. 407).

individual causal laws. Many aspects of our world depend on such randomness, for example, a book lying on a table is held down by gravity, but also depends on the random motion of the molecules in the book cancelling out and not simultaneously all moving upwards: "The statistical law which states that this will practically never happen is just as much at the basis of our predictions as the individual law of gravitation."[47] We could say that the concept of statistical law that Bohm brought into his book *Quantum Theory* to attempt to explain standard quantum mechanics has returned, but now dialectically related to the causal laws he later stressed in his own interpretation of quantum mechanics.

To understand further (1) the dialectical relation between individual, causal laws and statistical laws, and (2) to investigate the criteria for randomness that he develops, Bohm is not satisfied with generalities but requires specific examples. He has only three where he can work things out in detail. The first, and most important, is the model he has developed above, in 1952, of multiplying an angle θ repeatedly by a large integer K, or in mathematical notation $\theta_{n+1} = K\theta_n$. In the joint paper with research student Walter Schützer,[48] published in 1955 but presumably completed before Bohm left Brazil at the end of 1954, there is also the consideration of the statistical mechanics of an ideal gas giving rise to the Maxwell-Boltzmann velocity distribution.[49] Finally, Bohm considers systems with a number of harmonic oscillators, showing how random phases develop. This aspect is discussed both in the paper[50] and in the letters.[51]

Bohm discusses the relationship between statistical and individual laws using the example of the simple model $\theta_{n+1} = K\theta_n$.[52] The model does help to show what Bohm means. For the first few steps, $n = 0, 1, 2, \ldots$, which may be considered as increasing time, a small range of angles which Bohm calls the "precision", presumably thinking in terms of measurement, denoted mathematically by $\Delta\theta$, will remain within the circle 0° and 360°.[53] Then comes the point, explained above, where the range of angles becomes greater than 360°, and we would have to subtract 360°, or later multiples of 360°, to keep within the usual range. In doing this, however, we lose track of "precision" and the ability to predict where an angle ends up after a "large" number of time steps.[54] Bohm points out that this is precisely where the individual "causal" law, which relates each angle to the next one in time, by multiplying by K, breaks down. After a much longer time period, statistical regularities will occur and the proportion of times out of the total time when the angle from the model θ_n enters a given range of angles will be proportional to the fraction of that range out of

---

[47](32, **121**, p. 418).

[48]Bohm and Schützer (1955), Sect. 6.

[49]As in (21, **64**, pp. 221–222) above.

[50]Bohm and Schützer (1955), Sect. 10.

[51](31, **120**, pp. 414–415) and (32, **122**, pp. 425–427).

[52](31, **120**, pp. 408–413).

[53]Bohm uses radian measure, so 360° is 2π radians.

[54]To be more precise, one can always predict forward in time by using $\theta_{n+1} = K\theta_n$, but one cannot predict backwards or retrodict.

the full 360°. At that point, we have a "statistical law", at least for most trajectories. To make the statistical law valid, the extra hypothesis must be added, namely that none of those special initial angles of "measure zero" are included that don't give rise to randomness.[55] This means there is no necessity to "reduce" the statistical laws completely to causal laws, the former have their own validity and it is enough to show that they are compatible. It is clear that statistical laws, by means of averaging,[56] can give rise to individual and causal laws as larger and larger time periods are taken.[57] Thus, a dialectical relationship is established between causal laws and statistical laws, inasmuch as both are abstractions from a reality with infinite levels and both are needed for a scientific understanding.

Bohm thought that the transition from causal laws, where calculus is often employed, to statistical laws, using probability theory, and vice versa, calls for a new more general approach and its associated mathematics. Causal laws and statistical laws would be the extreme "poles" of such an approach, but there would also be cases of combinations of both types of laws.[58] There is no indication that he got any further with the applications of this idea.

In these later letters[59] and also, in the paper with Schützer,[60] Bohm considers the history of probability theory, especially concentrating on the two main twentieth century schools of von Mises and of Kolmogorov.[61] Bohm is clearly attracted to von Mises' frequentist approach[62] as opposed to the axiomatic approach of Kolomogorov. The latter is a worked out mathematical theory, now almost universally used by natural and social scientists and engineers, which is quite independent of the discipline in which it is applied. If it works successfully, which, with sufficient skill, is often the case, all well and good. But why it works in a particular case and what is its range of validity when it does work, this is often unknown. Despite originating in the USSR, this sparked off all of Bohm's bitter experiences with relying on mathematical axioms in physics. However, the former approach of von Mises, developed by Abraham Wald, had its own problems, particularly with respect to knowing when a sequence of numbers, such as data collected over a time period, could be regarded as "random" or not.

---

[55] Or "chaos", as in the above discussion relating to (24, **84**, pp. 281–284).

[56] Bohm refers to a key theorem in statistics, the "law of large numbers".

[57] As, for example, the statistical mechanics of an ideal gas give rise to thermodynamical laws.

[58] See (30, **114**, pp. 380–382) for examples.

[59] (30, **114**, pp. 378–381).

[60] Bohm and Schützer (1955), Sects. 3–5.

[61] For a more thorough history, see Plato (1994). Von Plato (p. 231) notes the extraordinary support for Kolmogorov's approach and the dismissal of von Mises by none other than J.L. Doob, who Miriam had dealings with, as noted above. Doob wrote that Kolmogorov "transformed the character of the calculus of probabilities, moving it into mathematics from its previous state as a collection of calculations inspired by a vague non-mathematical context."

[62] In which, as noted above, the ratio of the number of times a particular event occurs to the total number of events is used, taking larger and larger totals, in order to approximate a probability.

Bohm therefore applies considerable effort in the letters,[63] in order to develop a "correlation" method to test for randomness. If a long sequence of numbers are not correlated in a certain well-defined mathematical sense, can they be said to be random? Bohm also considers higher-order correlations, which can be calculated as well. Unfortunately, and Miriam points this out,[64] even if all these correlations are zero, this is not a sufficient condition for randomness. There may also exist, for example, varying types of correlation not detectable by this method. It is certainly useful in science and engineering when a correlation *is* found, as it will often show the presence of a signal amongst a lot of "noise". To test for randomness, however, which, as we noted above, has become an important business in cryptography, a range of more sophisticated statistical tests are now used.

Finally, in line with his critical view of the "abstract" nature of probability, which we noted at the start of this chapter,[65] Bohm opposes its use in statistical mechanics inasmuch as it would be used to give a probability distribution for the initial values of the position and velocity of the atoms under investigation. In his model $\theta_{n+1} = K\theta_n$, he made a point of ignoring those initial conditions of "measure zero" that would not contribute to the long term uniformity of the distribution of angles. He rejects the approach of Poincaré, who, at the beginning of the twentieth century, had begun the investigation of such "chaotic" models,[66] applying them to the motion of asteroids. Poincaré took into account a probability distribution of initial values in his approach to the problem. Bohm rejects this. (Not only with the $\theta_{n+1} = K\theta_n$ model, but also with the statistical mechanics of an ideal gas.) He insists, with an ideal gas, that only a very small proportion of initial conditions would not lead to the desired Maxwell-Boltzmann statistics. Thus, one could begin not with a probability distribution, but with causal laws for the atoms, aware that the vast proportion of "typical" initial values would lead to the required result.[67] This conception of *typicality* was put forward in a seminal paper by Lebowitz,[68] who notably thanks his teachers Melba Phillips and Peter Bergmann (Ralph Schiller's supervisor). Typicality is now a key part of the approach of the Bohmian mechanics group,[69] though note that it is not without detractors among mathematical purists.[70]

---

[63](30, **114**, pp. 382–386), (30, **115**, pp. 391–392), (31, **117**, pp. 398–402), (31, **119**, p. 407), (32, **122**, pp. 424–427), as well as in Bohm and Schützer (1955), Sect. 9.

[64](31, **119**, p. 407).

[65]In (20, **61**, pp. 213–215).

[66]Bohm and Schützer (1955), pp. 1030–1031.

[67]Bohm and Schützer (1955), pp. 1018–1020.

[68]Lebowitz (1993).

[69]Dürr and Teufel (2009), Sect. 4.1.

[70]Frigg (2011).

References

Bennett, D. J. (1999). *Randomness*. Cambridge: Harvard University Press.
Bohm, D. (1957). *Causality and chance in modern physics* (First ed., p. 1984). London: Routledge and Kegan Paul (Second edition with new preface by Bohm).
Bohm, D., & Schützer, W. (1955). The general statistical problem in physics and the theory of probability. *Il Nuovo Cimento, Supplemento AI, 2*, 1004–1047.
Broer, H., & Takens, F. (2011). *Dynamical systems and Chaos*. Berlin: Springer.
Bunge, M. (1951). What is chance? *Science and Society, 15*(3), 209–231.
Bunge, M. (2016). *Between two worlds, memoirs of a philosopher-scientist*. Switzerland: Springer International Publishing.
Dürr, D., & Teufel, S. (2009). *Bohmian mechanics, the physics and mathematics of quantum theory*. Berlin: Springer.
Frigg, R. (2011). Why typicality does not explain the approach to equilibrium. In M. Suárez (Ed.), *Probabilities, causes and propensities in physics* (pp. 77–93). Berlin: Springer (Chap. 4).
Graham, L. R. (1971). *Science and philosophy in the Soviet Union*. London: Allen Lane.
Khinchin, A. I. (1949). *Statistical mechanics*. New York: Dover Publications.
Lebowitz, J. L. (1993). Boltzmann's entropy and time's arrow. *Physics Today, 46*(9), 32–38.
Marx, K., & Engels, F. (1988). *Collected works* (Vol. 25). Moscow: Progress Publishers.
Niven, I. (2005). *Irrational numbers*. Washington, D.C: The Mathematical Association of America.
Stewart, I. (1990). *Does god play dice? The mathematics of Chaos*. London: Penguin Books.
Tolman, R. C. (1979). *The principles of statistical mechanics*. New York: Dover Publications.
von Plato, J. (1994). *Creating modern probability, its mathematics, physics and philosophy in historical perspective*. Cambridge: Cambridge University Press.
Wilkins, M. (1986). *Interviews with David Bohm*. Niels Bohr Library & Archives, American Institute of Physics, College Park, MD USA (In 12 parts: https://www.aip.org/history-programs/niels-bohr-library/oral-histories/32977-1 to https://www.aip.org/history-programs/niels-bohr-library/oral-histories/32977-12).

References

Chapter 10
Bohm and Feminism

There are a number of letters that are devoted, in part, to Bohm's views on the "woman question" or feminism, in today's parlance.[1] All these letters are to Miriam, and all are written in 1953, except for a short section in the long letter **111** of 1954. Though largely "abstract", as Miriam points out, they are presumably concerned, at least to some extent, with assessing their relationship before her possible visit to Brazil, and then with responding to her decision not to come and to start a family in the US instead. Bohm could possibly have developed some interest in the question of feminism when he was at Berkeley, though there is no evidence of this in the letters. As Sean Mullet points out, both Bohm and Lomanitz dated a graduate psychology student named Bettye Goldstein in 1942–3[2] Under the name Betty Friedan, she later wrote the ground-breaking *The Feminine Mystique*. At Berkeley, Goldstein was interested in left-wing politics, a commitment that she shared with Bohm, according to her biography.[3] She was a gifted and assertive student, working on Freudian-type theories, but did not seem to be especially interested in feminism. As her biographer puts it: "whether what she later called the feminine mystique was actually affecting her in 1942–43 is open to question."[4] The problems which Friedan later identified were certainly present for many women. They faced hostility from male colleagues and struggled to combine family with a career.

Bohm was not unaware of how these questions affected Miriam. We noted in the previous chapter that Bohm had clearly been angry at the way Miriam had been treated by her mathematical advisors. He is also sympathetic to her difficulties in combining a career with having children.[5]

---

[1](27, **99**, p. 329), (27, **102**, pp. 336–340), (27, **103**, p. 342), (28, **105**, pp. 347–350), (28, **107**, p. 352), (28, **108**, p. 354), (28, **109**, p. 355) and (29, **111**, p. 370–371).

[2]Mullet (2008), pp. 45–46.

[3]Horowitz (1998), Chap. 5.

[4]Horowitz (1998), p. 99.

[5](23, **77**, p. 263).

C. Talbot (ed.), *David Bohm: Causality and Chance, Letters to Three Women*,
DOI 10.1007/978-3-319-55492-1_10

Despite showing his support for Miriam's situation, however, Bohm seems to have little sympathy for feminist views in the replies he makes to her. He criticizes[6] Simone de Beauvoir's *The Second Sex*[7] for making a separate issue of women being controlled by men, wanting instead to make domination of both sexes by "bourgeois" society the key question. It is unclear how Bohm got his theory[8] that we all have a desire to find something or someone above ourselves to which we can be devoted. It is, surely, a crude and simplistic theory. Apparently, he thinks that it applies "symbolically" in sexual intercourse. His idea that we should rather submit ourselves to the "future of humanity as a whole, in its struggle with nature, and with disruptive trends within humanity itself" seems to tie in with his Stalinist politics, which we review in the next chapter, as does his view of sex helping to "deviate humanity from its natural goal". In any case, submitting to a "common end" is clearly an avoidance of the issue of the sexual oppression, which does exists in society and which is reflected in Miriam's emotions around what she identifies as her own "insidious tendency to submission".

In the next paragraph, Bohm seems to be contradicting himself in admitting that ""femininity"is a response conditioned by society." He takes the view, perhaps common in that period, when the "nurture versus nature" debate was prominent, that pain in childbirth and menstruation are social in origin and can be reduced by conditioning. This may have been connected with the "psychoprophylactic method" of childbirth, promoted in the Soviet Union at that time and taken up by the pro-Soviet French obstetrician Fernand Lamaze. It is now referred to as the "Lamaze method" in America, apparently without any knowledge of its Soviet origins.[9] I could find no discussion of menstrual pain in de Beauvoir, though she certainly sees feelings of "uncleanness" as a social construct.[10]

Bohm does seem to have quite progressive views on women combining a career with bringing up children, particularly if we consider that it was quite common during that period to insist on women giving up work. At the same time, his conception of the father as playing an equal role with the mother in raising children was unusual for that period. Demands for nursery care and collective restaurants were a normal part of socialist campaigning propaganda. As for Bohm's idea of the need to "enhance, intensify, and symbolise in the sex act" the desire for domination or submission, and so on, I will not elaborate any further except to note that it could, perhaps, relate to his own psychological issues.

In his comments,[11] Bohm opposes the current stereotypes of women and men and thinks he agrees with Miriam who "cannot really accept the current role of women in society." He would like a woman with "some independent personality of their own" and not just passively reflecting the will and desire of a man, a point on which he

---

[6]In (27, **99**, p. 329).

[7]de Beauvoir (1988).

[8]In (27, **102**, pp. 336–340).

[9]Michaels (2007).

[10]de Beauvoir (1988), pp. 340–341.

[11](28, **105**, pp. 347–349).

agrees with de Beauvoir. He also considers that Miriam's struggle to raise the baby may cost "some of your achievements in math, but doesn't have to mean the loss of all possibility of work." Bohm admits his failings in relationships,[12] he "shall have to try to care more for the next one", and thanks Miriam for sending him the Kinsey report,[13] which he hopes will help him better understand women. The reports were, of course, landmarks in making known, by applying scientific statistical studies, the many aspects of human sexuality that had been covered over by bigotry and conservative obscurantism.

Bohm offers his advice[14] on bringing up a child, assumed to be a male. Miriam should try to arouse "a spirit of courage and love for the possibilities in human beings as individuals, and in humanity as a whole", her son should be ready to face risks and not "become rotten and corrupt inside, as so many of our modern liberals are." Finally,[15] Bohm once again rejects Miriam's advice on finding a woman in Brazil. He doesn't want to have sex with a woman "who does not already attract me in other ways" and disapproves of George treating women as sex objects, although he admits that with George, this was only a "superficial veneer".

References

de Beauvoir, S. (1988). *The second sex*. UK: Picador.

Horowitz, D. (1998). *Betty Friedan and the making of The Feminine Mystique : The American left, the Cold war, and modern feminism culture, politics, and the Cold war*. Massachusetts: University of Massachusetts Press.

Michaels, P. (2007). *Childbirth pain relief and the Soviet origins of the Lamaze method*. The National Council for Eurasian and East European Research.

Mullet, S. K. (2008). *Little man: Four junior physicists and the red scare experience*. PhD thesis, Harvard University.

[12](28, **108**, p. 354).

[13]Presumably, the 1951 female report. The male one came out in 1948.

[14](28, **109**, p. 355).

[15](29, **111**, pp. 370–371).

Chapter 11
Bohm and Politics

Up to a quarter of the letters written to Miriam between 1951 and 1954 are on the subject of politics. Here, I am taking politics in both a narrow sense of current affairs and issues relating to the Stalinist Communist Parties, especially in the USSR, but also in a more general sense that refers to how Bohm thought that science and philosophy could impact on people's ideas and play an important role in a socialist transformation of society. As usual, Bohm tends to write in an unstructured manner, so the classification can only be an approximate one.

Although there is no distinct division of this material such as there is in the material on Probabilty and Statistical Mechanics in Chap. 9, there is still something of a distinction to be made between the period before and the one after the visit of George Yevick and Jean-Pierre Vigier in July/August 1953. Before that time, there is not so much material referring to politics in the narrower sense[1] and a tendency to discuss socially-related science and philosophy.[2] After Yevick and Vigiers' visit, however, the letters mainly concentrate on political commentary in the narrower sense,[3] with a fair amount of criticism, some of it rather unpleasant and patronizing, of Miriam's politics of the "anti-communist" left. One could speculate that this change in the later period follows Miriam's decision to end her personal relationship with Bohm, but the change also coincides with a decline in support for the USSR in left politics, especially after Stalin's death and as more revelations about the regime

[1]The exceptions are discussions in (20, **58**, pp. 204–205), (22, **70**, pp. 248–250), (125, **89**, pp. 203–205), (26, **97**, p. 320), (26, **98**, pp. 321–323), (27, **102**, p. 336) and (27, **103**, pp. 340–342); there are some discussions on Brazil in (23, **78**, pp. 264–266), (23, **82**, pp. 274–275) and (27, **99**, pp. 325–326).

[2](20, **57**, pp. 202–203), (20, **59**, pp. 206–208), (20, **60**, p. 211), (20, **63**, p. 219), (21, **65**, pp. 229–230), (21, **66**, pp. 230–236), (22, **68**, pp. 244–245), (23, **79**, pp. 270–271), (26, **92**, pp. 307–308), (26, **94**, p. 311), (26, **96**, pp. 316–317), (27, **99**, pp. 327–329) and (27, **101**, pp. 331–333).

[3]These are (28, **106**, p. 351), (28, **107**, pp. 352–353), (28, **108**, p. 354), (28, **109**, pp. 355–359), (28, **110**, pp. 360–362), (29, **111**, pp. 363–369, 371 and 373), (29, **112**, pp. 374–375), (30, **115**, pp. 388–391), (31, **118**, p. 404) and on Brazil (30, **114**, p. 378).

C. Talbot (ed.), *David Bohm: Causality and Chance, Letters to Three Women*,
DOI 10.1007/978-3-319-55492-1_11

were coming out. All this certainly seemed to make Bohm increasingly desperate to defend his Stalinist views.[4] The final letter to Miriam in the archives contains both a Stalinist political assessment of the world and a discussion of the role that a philosophy of science could have in changing attitudes. Bohm refers to this letter[5] as "more or less a statement of my principles and aims, as they stand at present."

Reading through these letters leaves us with no doubt regarding Bohm's political convictions. Admitting "grave defects" and "a servile careerist point of view in many people" in the USSR, as well as many similar provisos, Bohm supports the Stalinist version of "Socialism". Since there is a "constant threat of military invasion" and all the people cannot "be trusted to resist the temptation of a promise of an immediate improvement in conditions",[6] the Moscow dictatorship appears as essential in carrying out socialist ideals. The response to Isaac Deutscher shows that Bohm supports Stalin's leadership, and "the whole theory of Communism is that the rise in the [economic] level resulting from the "dictatorship of the proletariat" will eventually make this dictatorship unnecessary."[7] In his philosophical work, Bohm bases himself on the legacy of the Marxist classics rather than following the debased philosophy put forward by Stalin in the 1930s. As we have pointed out, he makes a disparaging reference to the "mechanical" approach taken in the USSR.[8] Nevertheless, in politics, Bohm seems woefully ignorant of the conceptual framework of Marxism, and shows no real sign of making a serious Marxist analysis of current politics and economics. He accepts without comment the alleged "socialist" nature of the USSR which, in an underdeveloped country, was inconceivable in the Marxist tradition. The Bolshevik leaders Lenin and Trotsky never conceived of the possibility of socialism in Russia without revolutionary transformations in more advanced countries like Germany. Bohm's experiences in and around the Communist Party, however, seem to have resulted in his uncritical acceptance of Stalinist ideology and propaganda.

Parenthetically, the point should be made that I am accepting the historical evidence for a very significant rupture between the politics of the Bolshevik party led by Lenin and Trotsky, that took power in Russia in 1917, and Stalin's "revolution from above", which began in the USSR in the late 1920s. This follows, for example, the historical work of Stephen F. Cohen and Robert C. Tucker (Tucker 1990; Cohen 1975). There are, of course, historians of a "Cold War" persuasion, who consider that Stalinism was the outcome of Marx and Lenin or, perhaps, "Marxism-Leninism in action".[9] There were other historians, some of a more left wing persuasion, like Isaac Deutscher, writing before the evidence used by Tucker and Cohen became

---

[4]As well as increasing in number, the letters on politics in the narrower sense tend to get longer, with letter **111** the longest of all.

[5](32, **123**, pp. 427–434).

[6]Quotes from (28, **107**, pp. 352–354).

[7](28, **110**, p. 361).

[8](26, **97**, p. 320).

[9]See Tucker (1999), where this view is put forward by Leszek Kolakowski and countered by other historians.

available, who saw more continuity between the 1920s and 30s, with the possibility of an eventual removal of Stalinist dictatorship.[10]

One can understand why, in the 1940s, thousands of left wing intellectuals throughout the world were attracted, as Bohm was, to the USSR. It had played a tremendous role in defeating the Nazis in World War Two, making far greater sacrifices than the European and American Allies with 27 million dead. Capitalism had resulted in fascism, slump and war, and hardly seemed to be offering stability in the post-war period. Bohm fears a renewal of war, resulting from the Korean conflict, and he thinks McCarthyism could lead to fascism in the United States. But by the 1950s, more and more questions were being asked about Stalin's regime and many were quitting the Communist Party.[11] Miriam sent Bohm critical books to read on the subject, and he seems to have leapt to the defence of the USSR in an increasingly extreme manner. It has to be said that, as well as using the arguments of the Stalinists, he is sinking to a low level in using their kind of propaganda techniques. What else can one make of the comment:

> Now that I realize that you do not feel it necessary to maintain logical consistency, I shall be more at ease when you repeat ideas that might have come from Koestler or Orwell. Perhaps you wish to take advantage of your being a woman, (especially feminine, since you are now in the process of producing a child) to excuse your inconsistency?[12]

This is quite astonishing when one compares it with the supportive comments he sent Miriam only a few months earlier. It should be pointed out that Arthur Koestler and George Orwell are part of the demonology of Stalinism. It would be easy to conclude that this was due to their initial support, and then to their later criticism of the USSR.[13] *Darkness at Noon*, published by Koestler in 1940, was a major attack on the USSR; *Animal Farm* was published by Orwell in 1945.[14] These books on their own would have been enough in order to get their considerable literary talents rubbished by Communist Party members. The reality goes much deeper, though. In both cases, these writers provided reliable inside information on how the Stalinist dictatorship functioned. In Koestler's case, this came from his own experiences, but also particularly through his friend, Alexander Weissberg[15] Koestler wrote the preface to Weissberg's book, *The Accused*, which gives a devastating account of the Stalinist show trials and gulags. Weissberg was an Austrian physicist who went to

---

[10]See Tucker (1999). Deutscher's position is explained in the article on Trotskyism by Robert H. McNeal, pp. 48–51.

[11]John Gates, one of the 11 American communist leaders jailed in 1949, explains in his book Gates (1958) how, when still in jail, he began asking questions about the Soviet regime after Stalin's death in 1953. According to Gates party membership fell from 75,000 in 1946 to 10,000 in 1957.

[12](29, **111**, pp. 363–364).

[13]To be more precise, Koestler was a Communist Party member in 1931–38. Orwell, as expressed in his *Road to Wigan Pier*, was always more critical.

[14]It should, perhaps, be noted that *Animal Farm* was regarded by some as ambiguous in relation to the USSR, but Stalinists always regarded it as hostile and Orwell himself regarded it as a satire on the USSR. Bowker (2003), p. 358.

[15]For more details on Koestler and Weissberg see Scammell (2009).

the USSR in 1931, thinking he would contribute to the building of socialism. He was arrested in 1937 with the typical false charges made in the Great Purges, spent three years in Soviet prisons, and was then handed over to the Nazis as part of a deal made in the Soviet-Nazi pact. The book describes in detail the torture and methods used by the GPU in order to extract false confessions. Weissberg shows how victims were forced to incriminate relatives and friends, creating mistrust that divided all possible opposition among those who were not sent to the gulags. In his book, he makes a rough statistical analysis of the extent of the Great Purges, estimating that some 9 million individuals were arrested by the GPU. It was Miriam who sent Bohm a copy of Weissberg's book[16] and it is astonishing that Bohm merely responds by saying that a great deal of the book must be true. Perhaps he knew that Einstein had campaigned on Weissberg's behalf, writing letters calling for his release, copies of which are included in the book. Orwell was damned for writing about his experiences in the Spanish civil war in his book *Homage to Catalonia*, which reveals how the Communists suppressed and murdered Trotskyists and Anarchists, preventing the socialist opposition to the republicans from developing.[17]

Bohm's lashing out at Edmund Wilson's *To The Finland Station* and Deutscher's *Russia, What Next?*, important books by writers of some standing, can hardly be regarded as a serious assessment of these works.[18] His attitude reflected the fact that both Wilson and Deutscher were ideological targets for the world's Communist Parties. Wilson was a member of the American Committee for the Defense of Leon Trotsky, which also set up the Dewey Commission, convened under the famous philosopher John Dewey in order to examine the Stalinists' case against Leon Trotsky, then in exile in Mexico. They exonerated him of all charges of working with foreign powers against the USSR. Deutscher had actually been a leader of the Trotskyist Left Opposition in the Communist Party of Poland. He was expelled and went to London in 1939, eventually becoming a historian. Bohm's first mention of Deutscher is not critical, so it may be that he asked for information from party members before making his intemperate attack. It should be pointed out that Deutscher's alleged "hypocrisy" in invoking "oriental traits" would have to be ascribed to two other writers who referred to the special features of "orientalism" or "asiatism" in Russia, namely Karl Marx and Frederick Engels. Again, the reason for Bohm's attack on "asiatism" lies deeper in Stalinist demonology. Stalin had banned all mention of asiatism in the 1930s. Even mentioning the asiatic mode of production, which appears in Marx's writings, was forbidden.

Bohm's general perspective on "socialism" can be pieced together from the letters. It consisted in applying Stalinist methods, perhaps not as severe as those in the USSR, in all countries separately. The reason is that "complete democracy would not work", as "the people as a whole, educated in the old system, do not really understand the need for reaching the goal of a radical transformation of society". It would be necessary "after a socialist govt takes power to make the people work harder than ever for 10

[16](28, **108**, p. 354).

[17]Bowker (2003), Chap. 11, The Spanish Betrayal.

[18](28, **110**, pp. 360–362).

or 20 years."[19] In this sense, Bohm urged Miriam to note that "middle class leftists" are just protesting "against the way in which the present social system interferes with their individual lives", that they would like "to do as they please in return for 4 hours a day of work", and that they are "nice people who only want to be left alone to their little plots of land, careers, or what have you, to raise their little children, etc."[20] Since they represent this "vague desire of the middle class for a better life",[21] Socialist parties will decay as "they are just not able to take the stern and difficult measures needed to solve the problems that they face." Therefore, "the Communists will be the ones to take over, because they are the only ones who have the steadiness of purpose and clarity of goal needed to keep them moving in the same direction over 10 or 20 years, in spite of the temptation to sell out."[22]

We have already referred to how dialectics can be used to support an argument with a degree of sophistry. We can now see an example of this in one of the letters he addressed to Miriam. In order to convince her of the deeply unattractive vision he is offering, Bohm employs precisely that false dialectic:

> When you look more deeply at the processes that are taking place, you discover that there are usually opposing trends or currents. In the long run, these currents will carry you in either one direction or another. . . . And I believe that the currents of today are such that in the long run, you will either be for socialism or against socialism, because these are the only two possible attitudes that can be maintained in the face of todays problems. And if you oppose socialism in Russia you will find yourself inevitably dragged to the side of capitalism, whether you like it or not.[23]

It is, of course, the "currents of the "middle class" liberal and "anti-communist left" people that Miriam is mixing with and that are carrying her towards capitalism. Bohm refers to influence of the "Weisskopf crowd", grouped around physicist Victor Weisskopf.[24] He makes a vicious personal attack on the man he sees as a prototype of those who are leading Miriam astray[25] Weisskopf is "a sort of namby-pamby, not to be taken seriously", with "no courage whatever", adopting "opinions which he thinks are relatively safe while still permitting him to pose as a liberal." It is difficult to believe that Bohm did not know more about Weisskopf and his personal history. He was, in fact, a friend of both Arthur Koestler and Alexander Weissberg, as a student in Berlin in the early 1930s, before fleeing Nazi Germany to the US later. Weisskopf even visited Weissberg at the institute he had set up in Kharkov, Ukraine. As he later recalled, in the winter of 1932–3, when millions of peasants were dying in the famine, he had argued with Weissberg and Koestler, both party members (Weisskopf was not a communist), that the famine was the result of Stalin's policies and not an "act of

---

[19](29, **111**, p. 365).

[20](26, **98**, p. 322).

[21](30, **115**, p. 391).

[22](32, **123**, p. 429).

[23](29, **111**, p. 369).

[24](28, **109**, pp. 357–358).

[25]Though later Bohm changes his mind on this (28, **110**, pp. 361–362).

God", as they maintained.[26] Presumably, Weisskopf was well placed to explain to Miriam the disasters of Stalinism, and may even have recommended Weissberg's *The Accused* for Bohm's reading.[27]

Bohm also follows the Moscow line in relation to the European politics of the period. Stalin's signing of a non-aggression pact with the Nazis in August 1939 and then refusing to believe in the existence of the Operation Barbarossa invasion by the Nazis of the Soviet Union until it was well under way, displayed gross and culpable stupidity.[28] After the war, Stalin made a 180 degree turn, regarding (West) Germany as a potential Nazi threat, and Communist parties waged a campaign against German rearmament. As the Cold War escalated, with the US involved in the Korean war from 1950, America wanted to deploy German troops in a NATO European alliance directed against the USSR. German rearmament was essential in order to create a force big enough to match that of the USSR. Bohm's reference to the "disgraceful" action of the French socialists forming "alliances with the Nazis"[29] refers to a recent vote in the French parliament, supported by only part of the socialists, for a European Defence Community (EDC). France had put this forward as an alternative to NATO. No German national army would be created, German troops would join the EDC only, and the popular idea of European unity would be promoted. However, French politics became increasingly fraught,[30] with four governments between 1952 and 1954, not only because of European issues, but also because of the war France was fighting in Indo-China and eventually lost at Dien Bien Phu in 1954. The French communists strongly opposed European unity and the EDC in the anti-Nazi campaign referred to by Bohm.[31] They defended French national autonomy, along with the Gaullists, and advocated an alliance with the USSR rather than the US, outside the "Germany-US orbit", as Bohm puts it. Following the French defeat in Indo-China and the division of Vietnam at the Geneva conference, the US entered the war, backing the South Vietnamese regime in 1955. The French parliament dropped the EDC in 1954, but this was hardly the victory for the Communists that Bohm hoped for.[32] There was to be no alliance with the USSR. The US gained what they wanted. France remained within NATO (until de Gaulle withdrew in 1966) and Germany was incorporated into NATO in 1955.

---

[26] Scammell (2009), Chap. 9.

[27] Another personal attack is made by Bohm on physicist and cosmologist George Gamow in a letter to Miriam, back in 1951 (20, **60**, pp. 210–211). No doubt, a valid scientific point is being made about the unjustified extrapolation of the mathematics of the general theory of relativity in order to deduce the existence of a "beginning of time", which the Pope can make so much of. But unfortunately, this none too subtle propaganda against religion also informed a Stalinist attack against George Gamow, calling him a "traitor to science". Gamow defected from the Soviet Union in the extreme oppression of the early 1930s, which put him on the Stalinist blacklist.

[28] See Beevor (2012), Chap. 12, Barbarossa.

[29] (22, **70**, p. 248).

[30] See Willis (1965) for more details.

[31] (28, **106**, p. 351) and (28, **109**, pp. 355–356).

[32] (30, **115**, p. 391).

Bohm also takes the Stalinist line on Britain. The British Communist Party, small but influential, tried to pressurize the left of the Labour party into a pro-Soviet, anti-NATO stance. Thus, Bohm denounces the Labourites for supporting German rearmament.[33] What prompted the denunciation of the leader of the Labour left-wing, Aneurin Bevan, for "pious neutrality"? Bohm had mentioned the Slansky case[34] to Miriam without understanding its significance. This was a show trial in 1952, in which Joseph Slansky, leader of the Czech Communist party, along with other leaders, who were mainly Jewish, were framed for a "Trotskyite-Titoite-Zionist" conspiracy. Most of them, including Slansky, were executed. Stalin was issuing a warning to the Communist leaders of the world not to defy Moscow's line, as Tito had done in Yugoslavia.[35] Bevan had given some support to Tito and even visited Yugoslavia in 1951. As a result, he was condemned by the Stalinists for "neutralism" and for not supporting Moscow.[36]

In case it should be thought that we are painting too black a picture of Bohm's Stalinism in this period, it could be pointed out, in his defence, that he never went public with the political views he expressed in the letters to Miriam. He seems to have had no involvement in politics and states that his "own temperament is against either the exercise or the acceptance of dictatorial authority."[37] More than this, one should consider his genuine attempt to expose the nature of Stalinism to Melba in his very important letters of 1956,[38] following what must have been a traumatic breakdown of his own increasingly futile attempts to defend the Communist Party ideology. At the end of 1954, he is still telling Miriam, in his "statement of principles and aims", that: "They are trying to build socialism under conditions, not only of powerful bourgeois opposition, but also of lack of understanding on the part of the peasantry and of a large fraction of the working class."[39] However, there is a spectacular reversal of all his previous political views in the 1956 letters. This, of course, follows the Khruschev revelations, but is also the consequence of other material becoming available, such as "Not by Bread Alone" and the report on the treatment of Jews in Russia by British Communists. The suppression of the Hungarian uprising also took place around this time. While in Israel, Bohm could also have access to a first-hand testimony of what life was like in a Communist regime. As Bohm notes:

> I have it on the word of a man (who is still a convinced Communist), who spent 10 years in one of these camps that the use of slave labor was a regular factor entering into the five year plans. The cost of this period in twisted lives, insanity, destruction of independently thinking people in the Communist movement, etc. is beyond accounting.[40]

---

[33](29, **111**, pp. 367–368) and (30, **115**, pp. 389–390).

[34](25, **89**, p. 296).

[35]It was also a purge of the Jews from the Stalinist bureaucracies, itself a symptom of Stalin's growing anti-Semitism.

[36]See Jenkins (2012) for more detail.

[37](29, **111**, p. 364).

[38](19, **54**, pp. 183–193) and (19, **55**, pp. 193–197).

[39](32, **123**, p. 429).

[40](19, **55**, p. 194).

We should also not forget that Bohm gained much emotional stability from his relationship with Saral. This probably enabled him to assess the wealth of material coming out of the USSR objectively. Melba, who had made considerable sacrifices for the Communist cause, seems to have wanted to "soldier on", playing down the significance of the revelations. Bohm clearly feels the responsibility to make clear to Melba that this cannot be done, given that the extent of the crimes of Stalinism is far too great.[41] Bohm goes systematically through a catalogue of indictments: the criticism of leaders was almost impossible, tantamount to signing a death warrant under Stalinism; people had been led to believe in superhuman leaders but in reality, they were subhuman in cruelty and stupidity; members had joined the Communist Party feeling the necessity for a fuller, more satisfying life, and had expected Communist leaders to have more dedication and selflessness—in fact the latter had proved to be far worse, they framed people, beat them up, tortured them, murdered them or sent them to slave labor camps, and they even did this to their own "comrades". Reading through all this, we remember that Bohm had only recently argued that whilst there were mistakes and excesses, the overall goal had, nevertheless, to be carried through. Bohm had once gullibly accepted the "stupid" conclusion that Stalin had come to, namely "that with the successful establishment of socialism, the struggle of the enemies within the country would intensify", and that this translated into the "implicit assumption that the country was full of potential spies and saboteurs". In a complete change of stance, Bohm now recognized that Stalin had a "psychotically suspicious temperament", was increasingly "unbalanced" and had used all the alleged threats on the country in order to eliminate his opponents. The key statement, explaining, if not completely justifying Bohm's own mistakes, is the following:

> The use of conscious and systematic falsification was I think the biggest crime, and the most unforgivable, committed by the Stalinists. And they are still doing their best to prevent the truth from being known, even at this late date.[42]

We will skip Bohm's deliberations, now very much dated, on the possible future for socialism in that period, and turn to the other aspect in the letters to Miriam, which relates to politics and the impact that Bohm thought his science and philosophy could have on society. The material is nebulous but builds on Bohm's conviction that his scientific and philosophical ideas could help to inspire support for socialism. He sees the problems of capitalist society principally as "ideological ones". This is referenced either in philosophical terms, such as with positivism and mechanical materialism, which extend to social thinking, as we have already pointed to in the chapter on philosophy, or, perhaps, in the many recurrent comments on what he sees as alienation and the problem of consumerism in American society, for example: "The fluorescent lamps and television screens have somehow come to symbolize for me the transformation of the U.S. into another country, because in their cold glaring

[41] Unfortunately, we do not know what he told Miriam, after all his Stalinist diatribes in the letters of 1953–4. Hopefully, more letters will become available.

[42] (19, **55**, pp. 195–196).

light, the emptiness of life is somehow emphasized; i.e., one has superficial polish, cleanliness, and brilliance, but it is totally empty and meaningless."[43]

Bohm hopes that his philosophy of the "infinity of levels", perhaps backed by a scientific breakthrough, could have an impact on social thought, for example:

> A clear, sharp, optimistic, materialistic point of view, promising infinite possibilities of development, and the possibility of growth of a common human purpose having objective existence within the human race, as a developing thing, would have far more appeal, particularly if it could be backed by some striking success in dealing with nature.[44]

In later letters,[45] Bohm even goes so far as to suggest that his philosophy of levels could help provide a "spiritual" approach to life. It could encourage "religious" feelings that were no longer directed at the supernatural, for "everyone wants to feel connected with something that goes beyond his immediate personal life."

The hope that his philosophical ideas will be a factor aiding the development of socialism is present throughout the letters to Miriam. In the final "statement of principles and aims" letter to Miriam he writes:

> Meanwhile I think that I can make some contribution by trying to clarify our thinking a little bit. . . . I feel convinced that it will have an important long-run effect to show that while the world is governed by rationally understandable causal relationships, these relationships do not imply the impossibility of qualitative changes, but rather, they imply instead the necessity of such changes under appropriate conditions. . . . I believe that an incorrect philosophical attitude to the world is at the root of our difficulties.[46]

It is interesting to note that, even after 1956, Bohm still wishes to stress an "ideological" and philosophical approach in dealing with the problems facing socialists:

> Up till now, theory has been too mechanical in its stress on economic factors, and its denial of autonomous causal action to what is, in Marxist terms called the "superstructure" of ideas, traditions, culture, myths, superstitions, religion, science, et al. as well as in the emotional make-up of the people. But we have seen how such factors almost wrecked socialism in Russia . . .[47]

It should be pointed out that this kind of social criticism, perhaps expressed in a more sophisticated language than Bohm's, became quite widespread in the post second world war period. It came from such diverse sources as the Frankfurt School, the New Left, and so on, all of them often subsumed under the term of "Western Marxism". Many, as Bohm does here, saw this as a philosophical alternative to the debasement of Marxism that had emerged under Stalin's rule. However, Bohm does not pursue this "Western Marxist" approach to socialist politics after 1956 and I have not considered it necessary to expand on it any further. As I indicated in Chap. 6 in my opinion it does not have the same validity as his philosophy of physics developed in *Causality and Chance*, which he did often refer back to in later life.

---

[43] (23, **79**, p. 271).

[44] (20, **59**, p. 207).

[45] (27, **99**, pp. 327–329) and (27, **101**, pp. 331–333).

[46] (32, **123**, pp. 430 and 432).

[47] (19, **54**, p. 189).

References

Beevor, A. (2012). *The second world war*. London: Weidenfeld and Nicolson.
Bowker, G. (2003). *George Orwell*. Boston: Little, Brown.
Cohen, S. F. (1975). *Bukharin and the Bolshevik revolution*. New York: Vintage Books.
Gates, J. (1958). *The story of an American communist*. Nashville: Thomas Nelson and Sons.
Jenkins, M. (2012). *Bevanism: Labour's high tide*. Nottingham: Spokesman Books.
Scammell, M. (2009). *Koestler*. New York: Random House.
Tucker, R. C. (1990). *Stalin in power: the revolution from above, 1928–1941*. New York and London: W.W.Norton and Company.
Tucker, R. C. (Ed.). (1999). *Stalinism essays in historical interpretation*. New Brunswick (USA) and London (UK): Transaction Publishers.
Willis, F. R. (1965). *France, Germany and the New Europe 1945–63*. Redwood City: Stanford University Press.

Chapter 12
The USSR, Philosophy and Science

In several letters, Bohm considers what was happening in physics in the USSR. Early in 1952[1] he asks "Why in 25 years didn't someone in USSR find a materialistic interpretation of quantum theory?" The answer, he thinks, lies in the existence of "an inadequate theoretical background, a background going back to concepts elaborated largely in the 19th century", and such a background is due, perhaps, to the relative backwardness of the USSR. Then, a bit later in 1952,[2] he suggests the problem is the conservatism of scientists: "Perhaps the you know whos [i.e. the Soviets] will work on these ideas, but I am even somewhat doubtful of this, because their physicists, like ours, take the lead from a few big shots, who might easily decide that they dont like it." Bohm then surmises, in 1953,[3] that the Soviet philosophers may eventually help: "I notice that a vigorous criticism of the foundations of quantum mechanics is going on in the "East", and in 5–10 years this may bear fruit, as a large number of people discuss the ideas, counter-discuss them, propose solutions, criticize them, etc." Finally, in 1955,[4] he gets a report on what was happening in physics and philosophy in the USSR. This report made the following points: (1) Soviet physicists are carrying out the same kind of long calculations that are carried out in the West, with little interest in philosophical problems. (2) Physicists' "orientation is strongly determined by the older men, such as Fock and Landau, who, in addition to their training background, are influenced by the fear of a sort of "Lysenko affair" in physics." (3) Physicists have "not thought much about problems such as the re-interpretation of qu. mechs., but tend[s] to take the word of the "big-shots" that ideas on this such as mine are "mechanistic." (4) There are "philosophers in Moscow who criticize the usual interpretation, but they haven't had much influence on the physicists."

[1](21, **65**, p. 230).
[2](23, **78**, pp. 265–266).
[3](26, **94**, p. 311).
[4](19, **50**, pp. 178–179).

C. Talbot (ed.), *David Bohm: Causality and Chance, Letters to Three Women*,
DOI 10.1007/978-3-319-55492-1_12

There is no doubt some truth in Bohm's 1955 assessment, but such an assessment fails to understand the impact of more than 20 years of Stalinism. The fears of the older scientists like Fock and Landau were very real; their "conservatism" cannot be accurately compared to that of the Western "big-shots". At the same time, to get a deeper understanding of the problems that Bohm identified, it is necessary to explore the way in which philosophy had been very seriously, if not irreparably damaged in the USSR during the Stalinist period. No serious support for Bohm could be expected from that quarter. Later, in the 1960s and 70s, there was perhaps some interest in Bohm's alternative causal approach to quantum mechanics in the narrow scientific sense but not, as far as I am aware, in the philosophy of *Causality and Chance*.

Bohm learned in Israel about the appalling crimes of Stalin's regime, but he could not have known the truth about the repression and bureaucratic treatment of science and philosophy and could not have grasped the reasons for which he was so completely isolated. Some information has in fact only become available in English fairly recently. We now attempt to sketch an outline of the historical background in Soviet science and philosophy, giving references for more details.

From the 1920s onwards, the leaders of the Russian revolution, Lenin and Trotsky, had paid particular attention to Marxist philosophy, as Yehoshua Yakhot explains.[5] The Sverdlov Communist University was set up in Moscow in order to train party members, most of whom had little knowledge in this area which, compared to politics and economics, took up only a small part of the Marxist tradition. They appointed former Mensheviks Liubov I. Axelrod (known as Orthodox) and Abram M. Deborin to teach philosophy, in which their expertise was recognized despite the fact that they were Menseheviks and thus had opposed the Bolsheviks before the revolution. Lenin and Trotsky also recognized the importance of the philosophical texts written by the principal founder of Marxism in Russia, Georgii Plekhanov. Axelrod had, in fact been Plekhanovs assistant, despite the latter's later opposition to the revolution. The working out of theoretical issues was the task of the philosophers, and during the 1920s, as is clear from both Joravsky and Yakhot,[6] they were *not* expected to engage in current political questions. Neither did the political leaders, including Stalin himself, enter the philosophical debates. This applied to both the Left faction, led by Trotsky, which was defeated and expelled from the party in 1927, and the Right, led by Bukharin, with whom Stalin was aligned until he defeated them, in 1929. The tradition which Lenin and Trotsky had established acknowledged that, while the class pressures of capitalist society could be expressed in philosophy, the relationship between philosophy and society was a complex one. Philosophy had logical methods of its own, which should be respected and could not be simply subsumed or transferred to politics. This outlook provided a relatively fertile and creative environment for philosophical discussion in the 1920s.

During this period of relatively open debate, the very issues which interested Bohm, namely the relation between dialectics and science and the problem of retaining a causal approach without sinking into mechanical materialism, were discussed

[5] Yakhot (2012).

[6] Yakhot (2012), Joravsky (2009).

by Soviet philosophers and scientists alike. However, there was no way that Bohm could have known that. The whole history was rewritten after Stalin assumed absolute power. For example, Yakhot explains the dispute which took place in 1924 between a philosopher, one of Deborin's group, Jan Sten, and a leading Bolshevik, Ivan. I. Svortsov-Stepanov (usually shortened to Stepanov), who was particularly concerned with the relation between Marxism and the natural sciences. Stepanov had written an afterword entitled "Historical Materialism and Modern Science. Marxism and Leninism" to a book on historical materialism he had translated.[7] In it, he advocated an approach which became known as "mechanism". Stepanov wrote on the example of plant life (or any other living organism) in biology. It was, as Yakhot paraphrased:

> extraordinarily complex and refined, but nevertheless, as a mechanism that absorbs energy from the external world and changes it from one form into another.... Nowhere, in Stepanov's opinion, are there any special and mystical forms of energy [i.e. vitalistic life forces CT] above those which are generally observed in chemical and physical processes.

Yakhot also explains how Sten responded to this:

> Engels ... protests against dissolving organic and chemical processes into mechanical ones, and he considers it necessary to discover the specific and qualitatively special laws belonging to each of these processes. . . . dialectical materialism protests against the transfer of the specific laws of a given type of process to other forms and types of processes, and demands a qualitatively concrete investigation of each particular portion of reality.

This was the essential character of the debate which lasted throughout the 1920s, between the "mechanists", led by Stepanov until his death, in 1928, but also by Axelrod and quite a number of scientists that were sympathetic to Marxism, and the "Deborinites" or "dialecticians", represented here by Sten.[8] The mechanists claimed that they were merely offering an overall philosophical account of the great strides being made in science. In return, Sten could attack Stepanov for belittling dialectical materialism and reversing the debate to the level of 18th century mechanical materialism. Stepanov countered that "mechanism" did not mean the mechanical materialism referred to by Engels, and claimed that Sten's response had amounted to just rehearsing a few pages from Engels. The year after this debate saw the publication of Engels' *Dialectics of Nature*, and the debate continued thereafter, with both sides claiming that Engels could be seen to support their position.

What is significant here in relation to Bohm is that the "portions of reality" referred to by Sten–Engels had actually used the distinctive term of "forms of motion", which he identified with different areas of natural science, transforming among themselves according to the law of the conservation of energy—are the "levels" which were central to Bohm's philosophy. Were such levels and issues of "quality" secondary to explanations in terms of basic physical and chemical entities, which natural science had been increasingly pursuing since Engels' day? Was "dialectics", perhaps, of relevance to social sciences only, as the mechanists seemed to be proposing? Or

---

[7] Yakhot (2012), pp. 21–27.

[8] Although recommending Yakhot's book, Bakhurst gives a somewhat different exposition of the debate Bakhurst (1991), Chap. 2, as he is particularly concerned with the philosophy of mind.

was it crucial to interpret the latest discoveries of science in dialectical materialist terms, enlisting science in the support for Marxist philosophy, as the Deborinites proposed? It is important to note that it was only after the Sten-Stepanov debate that the Deborin group moved away from concentrating only on the history of philosophy and the elaboration of Hegel's dialectics, beginning their involvement with natural sciences.

Joravsky[9] points to an interesting speech by the political leader Leon Trotsky on this very issue of levels. In his address to a Soviet congress of chemists in 1925, Trotsky wrote:

> Chemistry has its special approach to matter; its own methods of research, its own laws. If without the knowledge that chemical reactions are reducible in the final analysis to mechanical properties of elementary particles of matter, there is not and cannot be a finished philosophy linking all phenomena into a single system, so, on the other hand, the mere knowledge that chemical phenomena are themselves rooted in mechanics and physics does not provide in itself the key to even one chemical reaction. Chemistry has its own keys.
>
> This applies to all sciences. Chemistry is a powerful pillar of physiology, with which it is directly connected through the channels of organic and physiological chemistry. But chemistry is no substitute for physiology. Each science rests on the laws of other sciences only in the so-called final instance. (https://www.marxists.org/archive/trotsky/1925/09/science.htm)

Joravsky points out that, in general, political leaders kept out of the mechanist-Deborin debate, but considers that, although Trotsky didn't comment on the dispute, his views were "strikingly similar to those of the mechanists"[10] It was reductionist, even though qualified by "in the last analysis", claims Joravsky, to advocate a level of "elementary particles" that would unite all matter into a "single system."[11] It seems unlikely that Trotsky would conceive of "elementary particles" to be an ultimate or absolute level of matter as Joravsky claims, but clearly, some Soviet philosophers did think Trotsky was a mechanist. However, prior to the 1930s, no one made that identification. It was regarded as unacceptable to attempt to make any link between Trotsky's politics and mechanist philosophy, even during the height of the struggle against the Left. When, in 1927, A.A. Maksimov, at that time a Deborinite, attempted to link Stepanov's mechanist philosophy with Trotsky, his offending passages were withdrawn from publication. Bukharin, identified with the Right and at that time allied with Stalin, called it "not mighty clever."[12]

It is important to note that the mechanists were more diffuse than the Deborinites but contained a core group around the physicist A.K. Timiriazev (the son of the famous biologist K.A. Timiriazev), who was influential in the Bolshevik party and who was completely opposed to Einstein's theory of relativity. He claimed that Einstein's original support for Mach's philosophy made it a completely idealist and "bourgeois" theory. Timiriazev wanted a return to Newtonian classical mechanics.

---

[9]Joravsky (2009), p. 98.

[10]Joravsky (2009, p. 58.

[11]Joravsky (2009, pp. 98–99.

[12]Joravsky (2009, p. 55.

According to Joravsky, he saw reduction to classical mechanics as the only basis for all physics, and notably saw the search for "material causes" of phenomena in terms of classical mechanics.[13] Timiriazev was joined by V.K.Mitkevich, who wanted to identify Faraday-Maxwell force lines as a mechanical explanation for electromagnetism. Both of them wanted to bring back the 19th century mechanical version of the ether.

By the late 1920s, the Deborin group were able to take over the major philosophical institutions and journals. They used the ploy of labelling the mechanists "revisionists", implying that these were no longer Marxists and thus were not entitled to have their contributions published. This was an extremely unfortunate precedent, considering what was to happen in the 1930s. It may be easy to criticise Timiriazev but, among the mechanists, there were clearly serious philosophers like Stepanov, Axelrod and so on, who Yahkot clearly feels did not get the discussion that was required[14] and who produced serious problems in Soviet philosophy, even in the 1970s and 80s. Yahkot feels that, as well as the unresolved issues of "levels" and "reduction", there was also confusion over the problem of necessity and contingency, both of which are clearly very relevant to Bohm's work.[15]

Some mechanists, like S.Iu. Semkovskii, the leading philosopher in the Ukraine and a first cousin of Trotsky, who wrote a book attempting to show that Einstein's theory *could* be seen as a confirmation of dialectical materialism, switched to the Deborinites in the late 1920s. Semkovskii's approach was influential even among physicists, most of whom were not interested in philosophical questions but were horrified by Timiriazev's attacks on relativity. In 1927, a philosopher in the Deborinite faction, who was also trained as a physicist, Boris M. Gessen (or Hessen), took up Semkovskii's approach and led the Deboronites to embrace relativity and later quantum mechanics, interpreting them within the dialectical materialist tradition. The hard-line mechanists around Timiriazev were ousted from leading philosophy circles but were still active in science, continuing their campaign against relativity and demanding a return to 19th century physics throughout the 1930s and beyond.

Unfortunately none of the philosophical investigations into relativity and quantum theory made by Gessen and others in this period have yet been translated into English, and would have certainly not been available to Bohm. The brief extracts given by Paul Josephson show an interesting attempt to take a dialectical approach to standard quantum theory,[16] based on an analysis of the relation between "dynamic and statistical laws". Gessen also tried to develop an approach to a version of the "ether" that was not the mechanical ether of the 19th century[17] prefiguring the approach taken by Bohm. He also worked on probability theory, examining the approach of von Mises, and also on statistical mechanics, again suggesting he was thinking in ways similar to Bohm's approach above. Gessen (Hessen) is well known in the history

[13]Joravsky (2009, p. 279.

[14]Yakhot (2012), Chap. 5.

[15]Yakhot (2012), pp. 115–116.

[16]Josephson (1991), pp. 266–269.

[17]Josephson (1991), p. 244.

of science circles for the "externalist" approach he presented in a paper he gave at the International Congress of the History of Science and Technology in London in 1931.[18]

After 1930, the Deborinite philosophers came under a huge attack. Yakhot shows in some detail how the process developed. Stalin gave a speech in December 1929, "On the Problems of Agrarian Policy in the USSR", intending to show he was a "great theoretician" in philosophy.[19] In his speech, Stalin had focused on his claim that the theoretical front was lagging behind the successes of "practical construction." The method of criticism and self-criticism of experts was now developed. For more than a year, "bourgeos specialists" had come under attack, and there were a number of highly publicized trials of "wreckers". Several thousand untrained workers and peasants were brought into the universities, and professors had to be elected to their posts. In most university areas, the process produced few results and was soon dropped, but in the Institute of Red Professors, Deborin and his colleagues came under particular criticism. Party leaders began to demand the immediate linkage of philosophy to politics, and aspiring recent graduates such as M.B. Mitin and P.F. Yudin, later to be called the "Bolshevizers", were encouraged to criticise the Deborinite professors. They attacked the professors for not having "partisanship" in philosophy. In 1930, Mitin's group attempted to pass a resolution attacking the Deborinite philosophers for their failure to link Trotsky's philosophy of science to his politics. Was it because leading Deborinite philosophers like Jan Sten and Nikolai Karev were known supporters of Trotsky? Their resolution on "the well-known separation of Communist philosophical thought from the most pressing political problems that have stood and stand before our Party" was voted down, so they sent it to the main Communist newspaper Pravda. Here, it was published "for discussion" with a note by the editors, who stated that they were in full agreement with its "basic views" and signalled that it had Stalin's personal support. During 1930, the full weight of the party was placed behind Mitin and Yudin as Deborin was denounced and forced to make a "confession" of his errors, thereby ending up a broken man. Yahkot gives several pages detailing how the Bolshevizers operated, commenting that Mitin, Yudin et al. "brought so much harm to Soviet philosophy that to this day it has not been able to recover from the destructive impact of those years". In December 1930, Stalin himself spoke to the bureau of the party cell at the Institute of Red Professors, told Mitin and Yudin and the others that they had done a good job, and described the Deborinites as "Menshevizing Idealists", a term of abuse that was then used for years by the Stalin regime against its opponents.

In the purges from 1936 onwards many of the leading philosophers were arrested and executed, or died in the camps. Yakhot gives brief biographical details in an appendix to his book. For example, Sten was arrested on 3 August 1936 and executed on 20 June 1937. Semkovskii was arrested on 2 March 1936 for allegedly belonging to a "Trotskyist terrorist organisation", and was shot on March 9 or 19, 1937. Gessen was arrested on 21 August 1936. He was interrogated and tortured seventeen times before

[18]Freudenthal and McLaughlin (2009), Graham (1985), Chilvers (2003).

[19]He had even taken lessons in philosophy from Jan Sten for 3 years in the 1920s (Tucker 1990).

being brought before a secret military tribunal, sentenced to death for membership in an invented "counter revolutionary TrotskyistZinovievist terrorist organization", and shot on the same day.[20]

It was not just the philosophers themselves that were targeted, of course. "Philosophy", directed by the Bolshevizers, became a standard part of the frame-up techniques used to extract false confessions and then to torture, execute or send to the gulags, especially after 1936. Philosophy itself was transformed and debased during the 1930s, as Yakhot shows in some detail. In his concluding chapter, Yakhot writes:

> Under the guise of intensifying the ideological struggle, philosophers actively intervened in the various fields of scientific knowledge – genetics, physics, statistics, sociology and so forth. And everywhere this produced dramatic, and sometimes tragic consequences.

Yakhot describes how the Bolshevizer Mitin operated in 1939 against Nikolai Vavilov, the great Soviet geneticist. Mitin ridiculed Vavilov for his use of probability theory in biology, alleging it was like using fortune-telling or using a horoscope:

> . . . What is this: two aces, two deuces, two fours, and so forth – if not fortune-telling, a horoscope! This has nothing to do with science! It is monstrous when people try to hand out such nonsense as scientific popularization, or even worse as scientific discoveries![21]

As Yakhot explains, Mitin would not have dared to speak like this if plans had not already been made for the arrest of Vavilov, which took place four months later. He died in prison in 1943. His fate, as Yakhot states, was a harbinger of the fate of Soviet genetics.

Yakhot only gives details of the 1920s and 30s, but after the Second World War, as noted in the comment we take from Loren Graham in Chap. 6, "the most intense ideological campaign" was stepped up, continuing until Stalin's death in 1953. Bolshevisers like Mitin remained in leading positions in the Soviet academia long after Stalin's death, and even though there was something of a "thaw" under Khruschev, no serious discussion of either the reductionism debate of the 1920s and of the "infinite levels" question that Bohm had raised, or of the relation between causality and chance could take place. Yakhot tried to get the history of the 1920s and 1930s in philosophy reconsidered but made no headway.[22] Though still a convinced Marxist, Yakhot left the USSR for Israel in 1975. His book was published in Russian in the US in 1981, and only appeared in Russia (to his surprise) as the USSR was collapsing, in 1991.

To my knowledge, although I am not a Russian language expert, although there was some interest in Bohm's scientific work, no support was given by Soviet philosophers to *Causality and Chance*.[23] One has only to consider the sensitive nature of

---

[20]Chilvers (2003).

[21]Yakhot (2012), pp. 223–224.

[22]The difficulties that the brilliant and creative Soviet philosopher Evald Ilyenkov faced are explained in Bakhurst (1991), Introduction.

[23]Since this was written I have noted that there are passing references in Svechnikov (1971), pp. 175 and 198. These relate only to the causal interpretation of quantum mechanics and not to Bohm's more general philosophy.

the topics involved. Trotsky's speech in 1925, for example, a key part of the reductionism debate, could not have been mentioned. He was only rehabilitated in Russia in 2001, but never in the USSR, where his books were banned until 1987. The issue of probability and chance would have also been a very difficult topic, as Lysenko attacked geneticists by condemning them for the use of probability and chance, as demonstrated above in Mitin's goading of Vavilov.[24]

Bohm's work would have thus been completely marginalized in the USSR, especially inasmuch as his philosophical endeavours of the 1950s are concerned. At the scientific level, Soviet physicists were understandably wary of Bohm's causal interpretation, although they were perhaps more open to discussing it in the "thaw" period. Let us consider the historical background in physics.

During the 1920s, Soviet physics made huge strides, assembling, mainly at the Leningrad Physico-Technical Institute, under the leadership of A.F. Joffe, physicists such as V.R. Bursian, Ia.I. Frenkel', V.A. Fock, L.D. Landau, and the young, but brilliant researcher, M.P. Bronshtein. This small group of physicists frequently travelled to key centres abroad. Landau, for example, spent 18 months as a graduate student visiting Cambridge, Copenhagen, Gottingen, and Berlin. Not only did they absorb the latest developments in quantum theory, translating key papers into Russian, but by the late 1920s, they also began making their own contributions, publishing in international journals. There was little concern amongst them, or amongst most Russian scientists at that time, according to Joravsky's figures, for Marxist philosophy. In quantum mechanics, they concentrated mainly on the mathematical formalism, adhering to complementarity, the probabilistic interpretation of the wave function and so on.

During the 1930s, whilst the physicists were allowed to carry out their work without direct interference in their scientific output and there was a huge expansion in the expenditure on physics, they were, nevertheless, put under considerable pressure from "Bolshevizer" philosophers. Maksimov, for example, trained in physics as well as philosophy, though apparently not as capable as his former junior Gessen, and possibly resentful about this, left the Deborinites and joined the Bolshevizers. He wrote attacks on Joffe, Frenkel', Gessen and others, presenting them as "idealists". He was even prepared to side with the crude mechanists Timiriazev and Mitkevich. Josephson explains how, in 1934, Gessen, Joffe, and other physicists were attacked in philosophical journals. At a special session of the Communist Academy in 1934, Joffe courageously defended the physicists against the attacks, utilising Gessen's work in order to demonstrate that relativity and quantum mechanics could be interpreted in terms of dialectical materialism, and showing that the anachronistic views of Timiriazev and Mitkevich, along with Maksimov's, were doing serious damage to Soviet physics.[25]

---

[24]In his speech in 1948, Lysenko stated "We must firmly remember that science is the enemy of chance." (Graham 1971, p. 236.) According to Graham, Lysenko was still influential as late as Khruschev's death in 1965.

[25]Josephson (1991), p. 270.

After the physicist George Gamov and his wife defected to the west at the Solvay conference in 1933, travel abroad was stopped, and Soviet physicists became increasingly isolated internationally. Peter Kapitza, who had worked on nuclear physics at Cambridge, UK, for 15 years, was detained in Russia in 1934. The brave defence of physics against charges of idealism by Joffe, Fock and others didn't stop several physicists and philosophers from losing their lives in the purges.

Bronshtein, known for his brilliant research and popularization of physics in Leningrad,[26] was arrested in 1936 and shot in February 1938, at the age of only 32.[27] Bursian was arrested in 1936 and died in the camps in 1945. Joravsky, in an appendix to his book*The Lysenko Affair*,[28] lists 22 physicists or philosophers of physics and 83 biologists and philosophers of biology who were repressed, though the actual numbers were much higher. ("Repression" here means arrest followed either by execution, sending to a camp, or internal exile. The official documents could cite "association with Trotskyite conspirators" and refer to their "Menshevising idealism").

By 1936, Kapitza[29] managed to ship his equipment from Cambridge to Moscow. In 1937, he was able to obtain liquid helium and then to make a great scientific discovery, the superfluidity, i.e. flow with practically no viscosity, of helium at very low temperatures. He employed Lev Landau to help develop the theory of this newly discovered process. This was at the time of the Great Purges, and Kapitza apparently had the ear of a number of top politicians, including Stalin, so he intervened on behalf of a number of scientists in order to try to obtain their release. In February 1937, he pleaded on behalf of Vladimir Fock, who had been arrested, and successfully obtained his release. Landau was arrested in April 1938, while giving out leaflets calling for socialism to be saved by "resisting the criminal Stalinist clique." A year later, Kapitza secured his release in return for a pledge that he would stop him from committing further "counterrevolutionary" acts. Landau went on to develop the theory of the superfluidity of liquid helium, receiving the Nobel Prize for this work in 1962.

After World War Two the philosophical attacks on scientists increased, as detailed by Ethan Pollock in *Stalin and the Soviet Science Wars*[30] Although the best known, and the one with the most dire consequences, was the attack on geneticists in the Lysenko issue, Pollock shows how a parallel assault on physicists was being orchestrated for their alleged "idealism". As referred to above, in Chap. 6, this was led from the top and would almost certainly have influenced Bohm in 1951, when he embarked on his causal interpretation. What was certainly not understood by Bohm was that the key players in the anti-idealist campaign, the same Maksimov and Timiriazev referred to above, were not just criticizing "idealist" interpretations of quantum

---

[26]Gorelik and Frenkel (1994).

[27]According to Josephson, Josephson (1991), p. 314, Bronshstein was well-known as a supporter of Trotsky.

[28]Joravsky (1970).

[29]The information in this paragraph is obtained from Kojevnikov (2004), pp. 116–120.

[30]Pollock (2006).

theory, but wanted quantum mechanics and relativity removed from university physics altogether and dismissed as "bourgeois idealist theories." They were demanding, in that sense, the physics version of Lysenkoism, and wanted to mount a conference in the late 1940s in which Fock, Landau and the leading physicists were to be pilloried for "idealism", which, as with the geneticists, would have led to dismissal from jobs and worse.

This conference was eventually called off in 1949. Pollock gives two conflicting versions of why this happened. The first was that Kurchatov, a leading physicist in the Soviet atomic bomb project, reported to Beria and Stalin that the conference would interfere with the development of nuclear weapons. Beria was informed by espionage in the US that their physicists had used these so-called idealist theories to develop the bomb. Stalin is reported to have said, "Leave them in peace, we can always shoot them later." According to the second version, the physicists were well prepared to defend a Marxist approach to their subject, and bureaucrats, worried that the conference would lead to damaging public disagreements, called it off.

In either case, states Pollock:

> "Without the stunning example of the power of nuclear weapons, it is almost certain that the number of Soviet philosophers and physicists willing to cross the line from criticizing idealist interpretations of quantum mechanics and relativity to dismissing the theories themselves would have been substantially larger.[31]

The attacks of Maksimov and company still rumbled on. Eventually, a letter backing Fock's reply to Maksimov was delivered to Beria by Kurchatov, backed by 11 of the most important atomic physicists, including Igor Tamm and Landau, both future Nobel laureates, and Andrei Sakharov, then unknown outside Russia. Personal comments were included in the letter by the physicist Blokhinsev, who had publicly disagreed with Fock over the "idealist" interpretations of quantum mechanics,[32] but who fully backed Fock here against Maksimov. In 1953, Fock's rebuttal of Maksimov, "Against Ignorant Criticisms of Modern Physical Theory", was published in the key journal *Questions of Philosophy*, which for the previous 6 years had only published articles from Maksimov and his supporters. They had finally been routed.

With some knowledge of this background, it is much easier to see why Fock and Landau were so concerned about a "Lysenko affair" in physics and would have feared, even in the "thaw" period, that Bohm's "mechanical" approach could be interpreted as a concession to the campaign of Maksimov, Timiriazev and others, who wanted the complete removal of quantum mechanics from Soviet physics. We have noted Blokhintsev's position in 1951, and he may, in fact, have been more sympathetic to Bohm. Fock had proceeded with what he regarded as a dialectical materialist interpretation of the standard theory, with an approach that is not dissimilar to Bohm's original one in *Quantum Theory*. In 1957, Fock flew to Copenhagen to discuss with Niels Bohr, and attempted to persuade the latter that there were weaknesses in his philosophical approach which could be improved using dialectical

[31] Pollock (2006), pp. 92–93.

[32] Graham (1966).

materialism. He criticised Loren Graham's exposition of his philosophy, insisting that he *was* a dialectical materialist.[33] However, Fock, writing in 1959, dismissed de Broglie, Vigier and Bohms' work as "examples of classical interpretations". Loren Graham translates some of Fock's paper[34]: "All these interpretations, according to Fock, were extremely artificial and had no heuristic value; not only did they not permit the solution of problems which were previously unsolvable, but their authors did not even attempt such solutions." This, of course, was hardly fair, but perhaps understandable, given the assault on physics that had taken place and Fock's lack of knowledge of Bohm's work, scientific and philosophical.

I have tried to show, in this introduction to the letters, that Bohm's approach to physics in the 1950s contained much more than the causal interpretation of the 1952 papers. One can be repelled by his Stalinist politics, and he clearly was often overwhelmed by personal problems. However, in order to do him justice, his interpretation of quantum mechanics should be taken together with his work on probability and statistical mechanics, and above all, with the philosophical work that went into *Causality and Chance*. Unfortunately, the consequences of Stalin's politics, which Bohm embraced for more than a decade, had such a disastrous impact on science, and particularly on the way in which Marxist philosophy related to science, that Bohm ended up completely marginalized, not least in the USSR itself. It is hoped that the publication of these letters will help contribute to a revival of the understanding of Bohm's philosophy, parallel to that which has begun to take place in relation to his physics.

References

Bakhurst, D. (1991). *Consciousness and revolution in Soviet philosophy, from the Bolsheviks to Evald Ilyenkov*. Cambridge: Cambridge University Press.

Chilvers, C. A. J. (2003). The dilemmas of seditious men: The crowtherhessen correspondence in the 1930s. *The British Journal for the History of Science*, *36*(4), 417–435.

Fock, V. A. (1966). Quantum mechanics and dialectical materialism: Comments. *Slavic Review*, *25*(3), 410–413.

Freudenthal, G., & McLaughlin, P. (Eds.). (2009). *The social and economic roots of the scientific revolution: Texts by Boris Hessen and Henryk Grossmann*. Berlin, Heidelberg: Springer.

Gorelik, G. E., & Frenkel, V. Y. (1994). *Matvei Petrovich Bronstein and Soviet theoretical physics in the thirties*. Berlin: Springer.

Graham, L. R. (1966). Quantum mechanics and dialectical materialism. *Slavic Review*, *25*(3), 381–410.

Graham, L. R. (1971). *Science and philosophy in the Soviet Union*. London: Allen Lane.

Graham, L. R. (1985). The socio-political roots of boris hessen: Soviet marxism and the history of science. *Social Studies of Science*, *15*(4), 705–722.

Joravsky, D. (1970). *The Lysenko affair*. London and Chicago: The University of Chicago Press.

Joravsky, D. (2009). *Soviet Marxism and natural science, 1917–1932*. London and New York: Routledge, Taylor and Francis.

[33] Fock (1966).

[34] Graham (1966).

Josephson, P. R. (1991). *Physics and politics in revolutionary Russia*. Berkeley and Los Angeles, California: University of California Press.

Kojevnikov, A. B. (2004). *Stalin's great science*. London: Imperial College Press.

Pollock, E. (2006). *Stalin and the Soviet science wars*. Princeton and Oxford: Princeton University Press.

Svechnikov, G. A. (1971). *Causality and the relation of states in physics*. Moscow: Progress Publishers.

Tucker, R. C. (1990). *Stalin in power: The revolution from above, 1928–1941*. New York and London: W.W. Norton and Company.

Yakhot, Y. (2012). *The suppression of philosophy in the USSR (The 1920s & 1930s)*. Oak Park, Michigan: Mehring Books.

Chapter 13
A "Precious Being" in Adversity

In this introduction, I have tried to set out Bohm's ideas in the letters about physics, philosophy, politics, and so on, rather than attempting to examine his psychology. However, at times, his psychological state seems to dominate the letters, making some discussion of its traits and origins inevitable and necessary. Bohm was a sensitive and creative individual, who found himself in extraordinary conditions that neither his psychological make-up nor previous experience had equipped him to handle with any degree of assurance.

Let us first consider some of the key examples in the letters in which Bohm displays what might be considered to be excessive emotional responses. Admittedly, we are reading personal letters here, but most of the instances we are referring to involved people with whom Bohm was working or was in contact, other than Miriam, Melba and Hanna.

In the physics department in São Paulo, as the term began in March, 1952, we find Bohm in a huge conflict with two other professors, Marcello Damy de Souza Santos and Hans Stammreich. In a series of letters to Hanna[1] to Melba[2] and to Miriam,[3] Bohm denounces these two colleagues as "rats", "stinkers" and "jokers" with "low moral character" and "no ability". They keep things in a state of "continual turmoil", which necessitates a continual fight with them to get anything done, a fight which has been going on for several years, and now takes up 20% of the time in the department. The director of the faculty and several others oppose their "fantastic plots" to hold back development and keep things under their control, but because they have tenure, they are difficult to get rid of. One can see here some possible justification for Bohm's complaints.

---

[1](15, **21**, p. 124) and (15, **22**, p. 125).

[2](16, **29**, p. 139), (17, **36**, pp. 149–150), (17, **37**, pp. 150–151), (17, **38**, pp. 153–154), (17, **39**, pp. 156–157) and (17, **41**, pp. 159–160).

[3](22, **72**, p. 251), (23, **74**, pp. 257–259) and (23, **75**, pp. 259 and 261).

C. Talbot (ed.), *David Bohm: Causality and Chance, Letters to Three Women*,
DOI 10.1007/978-3-319-55492-1_13

The problem escalated as one of Bohm's assistants, who was very wealthy, apparently with fascist leanings and military connections, began to finance an Institute for Theoretical Physics. The German physicist Carl Friedrich von Weizsäcker was brought in to head the institute for a period of 3 months, along with some German students. The plan was that he would be replaced by Heisenberg for a further 3 months, and so on. With "so much money, political influence and big names", they were expected to exert much influence in Brazilian physics. Bohm feared that, with the support of the "rats", von Weizsäcker's prestige was being boosted against his own. Supposedly, the "rats" and the "Nazis" were pressurizing Bohm to drop Ralph Schiller from coming to the university and to take two Germans instead, otherwise they threatened to bring in the military, presumably against "communists" in the university. Bohm was "90%" convinced that von Weizsäcker was involved in the "Nazi" plan and prepared to mobilize the opinion of international physicists, including Einstein, against it. Fortunately, though no details are given in the letters, von Weizsäcker did not appear to be interested in continuing and Heisenberg never came. Despite Bohm's hostility, it seems that von Weizsäcker valued discussions with Bohm, which inspired him in his own approach to the foundations of quantum mechanics.[4]

The second example of an extraordinary emotional response from Bohm is at the beginning of 1953, in a series of six consecutive letters to Miriam[5] regarding the remarks, allegedly made by George Yevick and Eugene Gross, that we considered in Chap. 8, "Let him give us results". Bohm and Miriam eventually call the episode the "storm" and tried to minimize it. They agreed that, most probably, there had been a misunderstanding. Gross had merely made the point that physicists in general would only respond to Bohm's causal interpretation when there were new results, he had not been placing any demands on Bohm personally.

Nevertheless, based on this apparent misunderstanding, Eugene Gross came in for a series of the most vitriolic personal attacks. Bohm thought that Gross was trying to avoid persecution by the state for his Marxist beliefs: "Genes little plan to sit around for 20 years and keep his heart pure while all his friends go into concentration camps is not very practicable". The tirade continues: "Gene is much more in danger of "going to pot" than I am, unless something happens that will knock him out of the smug groove in which he moves." Eugene and his wife "take their friends to pieces", which Bohm regards as compensation for "their own dissatisfaction with their futile life in a middle class environment closed to new ideas." Gene "lacks initiative in exploring new ideas, and in applying his philosophical principles", he had reversed his "class philosophical position", and so could no longer be regarded as a friend, etc. Whether any of these denunciations ever reached Gross is not clear, but there is no indication that there was any rift between them. In fact, in Bohm's Festschrift,[6] Gross gave glowing praise to his former research supervisor.

---

[4]Freire Jr. (2005).

[5](26, **93**, pp. 308–310), (26, **94**, pp. 310–312), (26, **95**, pp. 313–315), (26, **96**, p. 316), (26, **97**, p. 319) and (26, **98**, p. 321).

[6] Hiley and Pea (1987) pp. 46–49.

Thirdly, we consider the case of a professor and colleague in the physics department, Mario Schönberg. When he returned from Europe to São Paulo in March, 1953, Bohm reacted strongly against him.[7] He was the "strangest type of Marxist", resembling a "Jewish businessman":

> He engages in all the dirty intrigues here in a very short-sighted way, and in physics, he is a pure formalist and idealist, admiring Pauli as the ideal in [a] theoretical physicist. All this goes under the name of the purest dialectical materialism.

Miriam had suggested that Bohm was over-hostile towards him,[8] but the issues were "objective", Bohm replied. Schönberg claimed to be a Marxist in politics, but this was in direct opposition to his philosophy of science. Not only did he admire Pauli, he attempted to propagate Pauli's idealist views in relation to quantum mechanics. Moreover, in the physics department, he supported the two "stinkers" referred to above. Schönberg knew of their anti-communist intrigues against Bohm, but "does not seem to think it important".[9] Bohm suspected that the two "rats" in the department were trying to get rid of him and that Schönberg was working with them.[10] There seems to be little in common between Bohm and Schönberg in physics or philosophy. Schönberg is "100% against the causal interpretation, especially against the idea of trying to form a conceptual image of what is happening," and he thinks that the "true dialectical method is to seek a new form of mathematics". Also, Schönberg was arguing for "pure chance" as opposed to Bohm's "reactionary" and "undialectical" attempt to explain chance in terms of causality. However, in spite of this apparently total lack of agreement, as we pointed out in Chaps. 8 and 9, by 1954 the letters show that Bohm is working with Schönberg on a "turbulent ether explanation of the quantum theory," and has accepted the view that there is a dialectical relationship between causality and chance.

The final case of emotional response in letters has already been discussed in some detail in Chap. 11. on Bohm's politics. We considered Bohm's obvious anger at Miriam, who, he suspects, is siding with the "anti-Communist" left, as well as his personal attack on Victor Weisskopf. In this case, we suggested that Bohm was reacting defensively, as the outburst coincided with a decline in the support for the USSR in left politics, especially after Stalin's death and as more revelations about the regime were coming out. We will return to the political issues, but will first consider the other enormous pressures that Bohm was under. He clearly realized that he tended to act irrationally. In one of the "storm" letters he wrote: "I have to guard myself

---

[7](27, **99**, p. 326).

[8](27, **102**, p. 335).

[9]In his autobiography, Bunge refers to Schönberg as "an imaginative physicist, art critic, nationalist, a communist militant, and believer in telepathy". As soon as he arrived in São Paulo, there was an "ugly fight" between him and Bohm over the running of the department, and in the seminar "Bohm put in evidence his unstable temper as well as his prodigious imagination". Bunge (2016), pp. 90–91.

[10](28, **106**, pp. 350–351) and (28, **109**, p. 356).

about getting too excited about what I am doing, or else, these excessively intense feelings will destroy me".[11]

With McCarthyism against academics rampant in the US,[12] Bohm's situation in Brazil was extremely vulnerable. Shortly after he arrived there, the US embassy confiscated his passport, preventing him from travelling abroad. He could either stay in Brazil or return to the US, where he believed he faced the prospect of imprisonment.[13] Later, Lilli and Hanna[14] suggested that he should return, presumably because the case against Joseph Weinberg had been dropped, but Bohm felt it would be unsafe for him to do so. In this particular instance, his reaction may not have been exaggerated. It is important to understand the context. As Sean Mullet has pointed out,[15] the treatment that the US government meted out to suspects under the McCarthyite campaign was far from uniform, making it difficult to judge what would be the likely outcome of any course of action. This level of uncertainty must have increased the tension for those under investigation. In any case, as Sean Mullet points out in his examination of the cases of David Bohm, Bernard Peters, Joseph Weinberg and Rossi Lomanitz, the McCarthy witch-hunt was "a sustained pressure or presence throughout much, if not all, of their adult lives".[16]

Bohm had experienced a difficult childhood, and then found himself in an environment that was socially and culturally alien to anything he had experienced in Wilkes Barre. The transition from a small industrial American town to an elite university would be challenging for anyone at any time. At Berkeley, in Oppenheimer's group, he came under the influence of Oppenheimer himself and other left-wing physicists, fundamentally shifting his views on the nature of the Soviet Union, which he had once criticized in a high-school debate. Socially awkward and prone to depression accompanied by related digestive problems, he was also subject to feelings of alienation.[17] Bohm came under the wing of Lilli and Erich Kahler and their daughter Hanna after he moved to Princeton, and seemed to have something of a support network of friends. Wrenched out of that environment by the McCarthyite campaign, he found himself in very different circumstances. Brazil was then a developing country, and Bohm was clearly overwhelmed by noise, traffic, pollution, building-sites, widespread corruption and so on, not to mention the difficulties of finding suitable food and accommodation.[18] He experienced serious stomach problems necessitating courses

[11](26, **96**, p. 316).

[12]Schrecker 1986.

[13](15, **18**, p. 120), (15, **19**, pp. 121–122), (16, **30**, pp. 139–140), (16, **31**, pp. 141–142), (16, **32**, pp. 142–143), (17, **35**, p. 147), (17, **36**, p. 149), (20, **62**, pp. 216–217) and (21, **64**, p. 221).

[14](15, **22**, pp. 126–127) and (15, **24**, pp. 128–129) and (15, **25**, pp. 130–131).

[15]Especially comparing the case of David Bohm, who was issued a passport, and that of Bernard Peters, who was denied one, Mullet (2008), Chap. 5.

[16]Mullet (2008), p. 209.

[17]For example, in Princeton, in 1950 he wrote: "I remain sick in the intestines as long as I am depressed and depressed as long as I am sick" (14, **6**, p. 105), and "worst of all is just the loneliness, which is enhanced by the fact that there are so few people with whom I can feel in sympathy these days. (14, **9**, p. 108).

[18]See the references to letters on Brazil in Chap. 11.

of antibiotics, and the bouts of depression worsened. The strict social morality in middle class Brazilian life made it impossible to develop sexual relationships.[19]

The university of São Paulo was struggling to establish itself in world class physics and Bohm found himself with the new responsibilities of a senior faculty member. At Princeton, Bohm apparently had no administrative duties. An indication of his privileged position is that an assistant was appointed to work through the galley proofs of *Quantum Theory.*[20] Bohm is obviously aware of his previous elevated position when he refers to the impossibility of working at “some place like Louisville, Kentucky, which is 10 times as bad as São Paulo”.[21] However, any rising academic at an early stage in his career, who would have to deal with obstructive staff members and the conflicts of department politics, could find the situation overwhelming, and the two “stinkers” seem to have created exceptional difficulties.

On top of all these pressures, we must consider Bohm’s politics and the tensions created by it in all the examples cited. Politics must be seen to include philosophical tensions as well, because despite Bohm’s creative development of the dialectical materialist philosophy in relation to physics, he effectively takes the approach of a Stalinist “Bolshevizer”, as outlined in Chap. 12. The materialist line in philosophy should be seen as relating to Bohms position on class and must be understood in tandem with his pro-Soviet politics. In this sense, Eugene Gross appeared to be moving away from his previous “class philosophical position”, and Schönberg, supposedly a Communist Party member, had a duty to support materialism in quantum physics. Miriam, on the other hand, was obviously raising questions that were increasingly difficult to answer.

Bohm’s response to von Weizsäcker was also a result of his Stalinist politics. It is certainly possible that individuals with fascistic views, in league with the two “stinkers”, were causing problems for a known supporter of Communism such as Bohm. The support for Nazis, however, is hardly likely in von Weizsäcker’s case. It is true that he, together with Heisenberg and others, had been involved in the efforts to construct an atomic bomb under the Nazi regime during World War Two. There has been much historical investigation into whether they were sincerely trying to build a bomb or merely pretending, not really wanting the Nazis to have such a weapon. In fact, the Nazi regime got nowhere near to producing atomic weapons. Von Weizsäcker and Heisenberg were imprisoned after the war and interrogated on the issue by the British. Whatever the truth on their involvement with the atomic bomb during the war, there is no evidence whatsoever that they had any commitment to Nazism after they were released, quite the opposite in fact. For example, Von Weizsäcker, who held a series of academic posts, actively campaigned against Germany having nuclear weapons.

Bohm was basing himself on the Stalinist position that West Germany was a potential Nazi threat, as referred to in Chap. 11. As we saw, this informed Bohms

---

[19]There are too many letters referring to these questions to list references.

[20](14, **10**, p. 110).

[21](23, **79**, p. 270). Olival Freire, on the other hand, points out the relative advantages Bohm had in Brazil, such as support for visiting researchers, etc. Freire Jr. (2015).

great concern over "German rearmament". With such a background in politics, it was easy, therefore, for Bohm to jump to the conclusion that von Weizsäcker was leading a Nazi takeover. However, those physicists, such as Einstein, that Bohm contacted with a view to waging a campaign against Von Weizsäcker, would, presumably, have raised questions about Bohm's grip on political reality.

In conclusion, Bohm's emotional instability during the Brazil period relates to exceptional conditions: insecurity over his lack of a passport (he eventually took Brazilian citizenship in order to move to Israel), the problems of academic responsibilities, difficulties with depression and sickness in a developing country, an environment for which he was ill-equipped, as well as feeling politically beleaguered as he tried to defend Stalinist ideas.

Bohm was particularly unsuited to withstand political pressure, and it is possible that he did have some contact with the Communist Party, as we suggested in Chap. 11. in relation to a sharp change in his views on Deutscher. In Bohm's Festschrift, Eugene Gross makes the following telling remarks:

> Finally, I can only use old-fashioned language to describe his impact on me and others. Dave's essential being was then, and still is, totally engaged in the calm but passionate search into the nature of things. He can only be described as a secular saint. He is totally free of guile and competitiveness, and it would be easy to take advantage of him. Indeed, his students and friends, mostly younger than he is, felt a powerful urge to protect such a precious being. Perhaps the deep affection of his many friends helped to sustain him in the difficult years of the early 1950s.

It may well be that the affection of his friends did protect him before he left for Brazil but unfortunately, his emotional outbursts also testify to the difficulties that unfolded afterwards. Once he had made the break with Stalinist politics in Israel and had gained some emotional support after marrying Saral, he apparently steered clear of politics entirely. In his obituary to David Bohm,[22] Basil Hiley found it hard to believe he had ever even had Marxist views. Such a suggestion "is a travesty of the truth and in my 30 years working with him I never heard him even mildly defending such a faith".

No reports or discussions about David Bohm living in England, from 1957 onwards, suggest that he was a man given to excessive and angry emotional outbursts. He seems to have learned a great deal from his psychological problems in Brazil and perhaps made sure that the "secular saint" was again dominant. According to David Peat[23] Bohm "found it difficult to express anger within his personal relationships", although he could end up shouting, "if someone disagreed with his ideas". Later he learnt to adopt a non-adversarial approach, telling Basil Hiley that "arguing with people assertively is not profitable". Also as Hiley explained in his obituary, Bohm was not interested in rising up the "pecking order": "Although courageous and tenacious in defending his ideas, his natural humility and gentleness were such that he did not actively seek honours, and it was this quality that people so admired".

---

[22]The Independent, October 30, 1992.

[23]Peat (1996), p. 244.

References

Bunge, M. (2016). *Between two worlds memoirs of a philosopher-scientist*. Switzerland: Springer International Publishing.

Freire Jr., O. (2005). Science and exile: David Bohm, the hot times of the cold war, and his struggle for a new interpretation of quantum mechanics. *Historical Studies on the Physical and Biological Sciences*, 36(1). http://arxiv.org/abs/physics/0508184.

Freire, O., Jr. (2015). *The quantum dissidents: Rebuilding the foundations of quantum mechanics (1950–1990)*. Heidelberg: Springer.

Hiley, B., & Peat, D. (Eds.). (1987). *Quantum implications*. London and New York: Routledge and Kegan Paul.

Mullet, S. K. (2008). Little Man: Four Junior Physicists and the Red Scare Experience. Ph.D. thesis, Harvard University.

Peat, D. (1996). *Infinite potential*. New York: Basic Books.

Schrecker, E. W. (1986). *No ivory tower: McCartyism and the Universities*. New York: Oxford Universty Press.

Part II
The Letters

Technical Note

As noted in the Introduction, Chap. 5. the letters included here are set out as follows:

(1) Letters from David Bohm to Hanna Loewy, numbered 1 to 23 in Chaps. 14 and 15. Also two letters from David Bohm to Hanna's mother Lilly Kahler, numbered 24 and 25 are included in Chap. 15.
(2) Letters from David Bohm to Melba Phillips, numbered 26 to 55 in Chaps. 16–19.
(3) Letters from David Bohm to Miriam Yevick, numbered 56 to 123 in Chaps. 20–32. An additional Chap. 33 contains single pages from letters to Miriam Yevick and a letter, numbered 124, to Miriam's husband George Yevick.

All of the 124 letters are reproduced here as accurately as possible, including Bohm's diagrams, as well as grammar and punctuation. All mathematical formulae have been reproduced as close to the originals as possible. The Birkbeck folder number containing the letter as well as the date on the letter when one has been given, are included. All additional text I have added, indicating approximate dates, missing words and pages, etc., is included in square brackets. The only exception is the addition of full stops, which Bohm rarely used, and without which some confusion can arise! The footnotes within the letters are all Bohm's, but using numbers rather than the asterisks he uses (which can cause confusion when the pages are different to the originals).

When no date is given on the letter I have attempted, where possible, by considering the contents, to give an approximate date to the letter. I have then arranged the letters as far as possible in date order within each of the three groups (1), (2) and (3).

Note that although the folders containing the letters in the Birkbeck archives contain letters to the same correspondent, not all are in date order. Some letters to Miriam have been numbered, although occasionally the same number is used for a different letter. These numbers have been included here. Any comments added to the letters have also been included.

Chapter 14
Letters to Hanna Loewy, 1950

Letter 1. Folder C37, not dated.

[The cyclotron fire was reported February 23, 1950, dating the letter between then and May 1950, from the reference to Britten's PhD.]

Dearest Hannah

I have been waiting for you to write to me, but it looks as if I'll have to write first.

I am glad that you had such a good time in San Francisco and Berkeley. These are certainly lovely places. Perhaps we can see them together sometime.

By now you are probably immersed in work. I hope you have a good job and enjoy yourself. Please write soon and let me know what is happening to you. The Pines keep on hearing about you indirectly (from the Barnets [name not clear] and from Shirley in Los Angeles). But you promised you would write me within a week or so, and didn't do it. Have you forgotten me?

I talked recently to Esther (from Roosevelt). She said she had a good cry on your shoulders, and that you promised to visit her the day after New Years, but never did. She likes you a great deal.

As for me, the news is mixed. It looks like there may be more trouble about this Weinberg business. The man Rushmore who called you up wrote an article in the Journal American asking why I had not yet been fired. About that time, Oppenheimer happened to talk to Dodds, who indicated that there would be a lot of trouble about my reappointment, and that there was no chance for it at all unless the faculty backed me very strongly.

As for the work on the theory, it is coming along. Things look a bit promising. I just got another idea on the quantum theory also. It is based on the fact that at the microscopic level, the quantum theory deals only with potentialities. For example, the quantum theory describes the probability that an electron can realise its potentiality for a given position. But to realise this potentiality, it must interact with some large scale (classical) system, such as an apparatus which measures position. It is only at

C. Talbot (ed.), *David Bohm: Causality and Chance, Letters to Three Women*,
DOI 10.1007/978-3-319-55492-1_14

the large scale that definite and well-defined events can exist. But the quantum theory has meaning only insofar as it predicts the relation between the microscopic state of the system and large scale macroscopic events that can be observed. Thus, the quantum theory presupposes the validity of classical concepts at the classical level. This means that one does not deduce the classical theory from the quantum theory, but that the two work together to describe the whole system. This is in contrast to most theories in physics, in which we analyse all large scale phenomena in terms of the small scale components. Here, we see that at the large scale level, new (classical) phenomena appear, which are not contained logically in the small scale phenomena alone. In other words, the behaviour of the whole system cannot be reduced to a description of the relationship of all its parts, since, new properties appear in a large aggregate, not contained at all in the behaviour of the microscopic systems.

Right now I am a bit depressed by the way the US is heading for war. Also, I am lonely for someone to talk to at night (the days are still very busy).

The car is beginning to act up again. Tonight, the water began to boil for no reason at all. I guess I'll have to take it to the garage.

I should also say that Helen is beginning to get in trouble over the Weinberg business. The FBI came to the president of her University and told him that she had been a Communist, etc., etc. Now the president wants to fire her. I don't know what she will do, but it isn't certain yet that she will be fired.

Well, I guess that's enough bad news. The Brittens were finally over to dinner tonight. The Pines came over later and brought a little man, a "troll", for Erich. Erich thought that the troll looked like Panofsky.

You may have read that the cyclotron burned down here in Princeton. There was about $200,000 damage and it will take at least six months to replace it. Roy was just ready to do his thesis on it. He is now working a deal whereby he will do his thesis in Berkeley instead starting in May. Maybe you can get together with the Brittens sometime this summer.

Well, there isn't much more news. I wish very much that you would write to me, especially about what you are doing, and how you are feeling.

Love
Dave

Letter 2. Folder C37, not dated.

[This letter and the next one seem to relate to a visit, probably in 1950, to a Radiation Research Lab in Florida. No date or further details are available].

Tuesday

Dearest Hanna,

Thanks very much for your letter – I was very glad to get it. It is very sad about Peter Ceike[not clear]. I hope that he recovers. Everything seems to be in a plot to prevent you from having some peace to get your work done. Let us hope that things will soon quiet down.

The weather down here is quite warm, but on the whole not unpleasant. There is usually a brisk wind blowing from the Gulf Stream (about a mile offshore) which is very pleasant and relaxing. I just love to sit on the beach in this fresh warm damp wind. It is so much nicer than the cold winds that blow off northern seas. The water is warm, and I bathe in it often. I like to float and to watch the waves break. One gets a feeling of unity with this warm sea, and sometimes I wish I could dissolve in it and spread out to its furthermost shores. It is also very beautiful at night, especially when the moon reflects off it, outlining the sea with its palm-lined beaches and the waves breaking on the shore. It would be very nice if we could watch it together. I would like for us to sleep on the beach, listening to the water.

Except for the beach, however, Florida is not very attractive to me. The people here are wealthy but quite friendly. They want me to stay, but I just couldn't put my heart into their kind of life. It just frustrates something very fundamental in me.

I tried to get in touch with the Brittens by phone, but the man at the boatyard in Melbourne didn't know where they lived. If I stay here this weekend, I'll go out there anyway, and try to find them. I may be able to leave by Saturday, but at latest, I shall leave by next Wednesday.

I'm looking forward to seeing you and everyone else at home very much. I wish we could take a vacation together on the money I earn. Could you investigate some place in New England. Perhaps Eugene has a place. It would be no vacation if I went alone.

Love
Dave

Letter 3. Folder C37, not dated

[This letter and the previous one seem to relate to a visit, probably in 1950, to a Radiation Research Lab in Florida. No date or further details are available].

Hotel Shamrock
160 Australian Avenue
Palm Beach, Florida

Dearest Hanna

Well, here I am in Palm Beach. I have not yet been to see the Radiation Research Lab, but will go there soon. (It is now 9:30 AM and I am waiting for Coleman to pick me up). The weather is warm (about like N.Y.) and rather damp. The beach looks lovely, and I hope to swim in it soon. But it would be hell to try to live here very long, even though it will probably be relaxing for a while.

I was very lonely for you last night. I felt as if I were 1000 miles away from anyone I could really talk to on a common basis. I worried a great deal about how we're going to solve our problems of getting married and each having some creative work. I certainly hope that it is possible for us to go to Brazil.

The town here has nothing but coconut palms. They're nice for a change. But the buildings are unattractive. All of the work is done by negroes. I haven't seen a white man doing any heavy work at all – the most a white man will do is to direct some negroes. It is all very bad, both for the black man and for the white man.

Well, I'll write more later.

Love
Dave

Letter 4. Folder C97, not dated.

[Early Spring 1950 (reference to first traces of spring)]

Dearest Hannah

I was so glad to hear that you are having such an interesting time in Hollywood. You really are getting to see some of the big shots! I hope very much that you can work out something with the atom, or perhaps better yet, help Benoit Levi to start a film in France, and go to work with him. Lily read (or translated) your book (2nd instalment) and it was really fascinating. It is amazing that you met up with such nice people as the Mollenhauers. Even if nothing comes of this trip (and something probably will) you will have had a wonderful time and learned a lot.

As for me, don't worry so much. There is no immediate crisis, but there will be a long range difficulty beginning next year when the time comes for my reappointment. I spoke to Shenstone, and he was very encouraging, although he could not guarantee my reappointment, because it depends on the Regents. Even if I don't get reappointed, I'll make out somehow.

The theory is still in a tough stage, but every day, I see new aspects of it. I oscillate between being a bit tired and depressed but as soon as I stop working on it, I feel fine. It's hard to describe, but I think I have a feeling as to how to get some of the psychological causes of this tension which leads to tiredness and poor digestion. It is somehow connected with my feelings toward other people and to the world generally. But by means of an orientation that I can best be describe [as in original] as "not feeling separate", I can come to a state of peculiar good feeling and intensity, a feeling of being at one with the whole world and everything in it. Yet, it is very difficult, if not impossible, to maintain this feeling at work, because deep thinking seems to require withdrawal from the world, and this leads to strain, tiredness and depression. If I can only change my way of thinking, perhaps I can improve the situation.

The weather in Princeton is cold, windy, and rainy. Yet, the first traces of spring are in the air. Perhaps they are in me, because as I walked home tonight, I felt a sort of sense of infinite possibilities which were latent in people, something that I probably haven't felt since I was a small child. Somehow these possibilities have to be unlocked and released from the vicious nightmare which now prevents their realization. Yet one feels helpless, knowing how good life could be for everyone, and unable to take the steps needed to help bring this about. Instead, one can only watch the inexorable development of this country toward dictatorship and war. Eric and I talk every night, and we always come to the same conclusion, that the American gov't has chosen a course which has meaning only if it leads to war. The people who rule the country are getting ready to take a desperate gamble on world domination, because they know that their position is already lost, if they do nothing at all.

Perhaps I can visit Calif. for a while this summer and we can have a vacation together in the mountains. It would be nice to have a car there, but I hesitate to ask the De Soto to travel 6000 miles (it is already past middle age).

Please let me hear from you, about what you are doing, how you are feeling, etc. And don't worry about me. The Pines send you their regards.

Love
Dave

Letter 5. Folder C97, not dated.

[This letter and the next 4 undated letters are presumed to be later than Letter 4, but before the response to Hanna's letter in Letter 10]

Dearest Hannah

I just thought I would write you a brief letter. You seem to have forgotten me, since you neither wrote nor even mentioned me in your letters to Lily. Perhaps you are going around with someone else. I wish that you would write me sometime.

I am very busy. I keep working on the theory and learn more and more things, but no real results. I now know that my ideas of last Fall were much too naive, but think that I can now see how the same general objectives can be accomplished by much deeper methods. But nothing is really clear yet.

I have been thinking of going to Europe the year after next. I have an offer to go to England, but it is contingent on getting a fellowship; such as a Guggenheim.

I see that you are having a wonderful time and I certainly hope that you can either get a good job or swing this deal with Benoit-Levi. Perhaps you can get a 10% commission, and thus get a real start in the business.

I often feel lonely for you and wish that you were here. It was so nice to be able to come home to you.

Please let me hear a little bit from you

Love
Dave

Letter 6. Folder C97, not dated.

[See note with Letter 5]

Dearest Hannah

Thanks for your previous letters. It is reassuring to learn that there are so many homosexuals in Hollywood. It is ironical that the principal American efforts at glorification of heterosexual relationships are carried out mainly by homosexuals, but this just goes to show the close unity between opposites.

About your getting a job, I have mixed feelings. On the one hand, I hope that you find an opportunity to use your abilities and to work hard on something that is really worth doing. On the other hand, I am afraid that I will never see you again, after you are absorbed into the distant (from me) vortex of Hollywood. It may be true, as everyone tells you that Princeton is a good place to be from, but the same probably also goes for Hollywood. Hollywood has always tended to be insincere and excessively commercialized, but now that the gov't and the FBI are entering into the picture, the pressure towards conformity must be almost irresistible. I think that the production of an honest picture is almost impossible in Hollywood, because any such picture would contain implicit criticisms of the so-called "American Way of Life", about which the gov't is very sensitive these days. The gov't would like to spread the idea that the only problems faced by Americans are difficulties arising in making romantic adjustments. Practically no one in the movies is faced with the problem of earning a living in a none-too satisfactory way, nor do people ever suffer injustices or frustration, except in that they cannot always have the mate that they would like. Because Hollywood is so hopelessly embroiled in such problems, I think it would be much better if this deal with Benoit Levi should be successful.

I hope that your job with Marshall Field will go through successfully. I think that you will be able to do very well at it.

As for me, I have been working somewhat, but tend to become depressed. The U.S. is going Fascist very rapidly, and I am afraid that it has already gone too far to be stopped. It seems to me that at least a partial destruction of "Western Civilisation" is by now unavoidable. In fact, it's hard to see how much good can develop as long as the press and radio are able to pour out floods of misleading propaganda, which prevent people even from considering some of the steps needed for a solution of the problems before us. I am afraid that there will be a great deal of suffering here before people will consent to face some of these problems realistically.

Another problem before me is that they are reviving all this trouble about the "Weinberg case". Nothing has happened yet, but there is a grand jury in San Francisco looking into this case. With the rise in hysteria since last year, the problems will be much more serious, if I am called as a witness. Because the country is coming so close to dictatorship, I feel very uneasy about everything.

I got a sort of offer to go to England in Fall of 1951, to work with a man whom I knew while in Berkeley (Massey). However, passports are becoming more difficult. Also, I wonder even whether it's wise to wait that long.

So much for my problems. Every now and then, I talk to people who are not in science about all the new developments such as the hydrogen bomb, etc. People look on scientists in a somewhat hostile way as the creators of so many horrible weapons. They don't stop to think that science can be either destructive or constructive, depending on the people who use it. It occurred to me that a good motion picture dealing with this subject, such as the one you are planning, would help create a better frame of mind in people, so that they would want to utilize science constructively, rather than just to kill people. Only I am afraid that in Hollywood, you would have to distort the picture quite a bit, so as to make sure it contained no implied criticisms of the U.S. gov't or the "American Way of Life".

I hope that my letter was not too depressing. Actually I am not always this way, but right now I just don't feel too well. It's a vicious circle, because I remain sick in the intestines as long as I am depressed and depressed as long as I am sick. When I work on something hopeful, then I can get over both the sickness and the depression, but it's hard to do that these days. Perhaps if I stopped reading the papers, I'd feel better. Anyway, I think I'll let myself be examined, as suggested by Dr. Brunn.

I have been thinking of coming out West later this summer. I would very much like to see you.

Love Dave

Letter 7. Folder C97, not dated.

[See note with Letter 5]

Dearest Hanna

I was glad to hear that you are doing so well in Hollywood. I was away when your last book came, and it has since then been sent to N.Y., but I did get a summary. It is very good that you are in a likely position to get a job soon. A person as good as you can get a job even when the rest of Hollywood is unemployed!

As for me, I fear that there will soon be trouble with the Unamerican Activities Committee. They have already cited about 30 people for contempt. These are people who refused to answer because of self incrimination. They said that they plan to cite all those who refuse to answer for any reason at all. If I am cited, then there will be much trouble. First, as soon as I am indicted, I will be suspended from my job. Then there is all the trouble with lawyers, especially the expense. Then comes the trial. Although the law is clearly on my side, one cannot predict the outcome of a trial these days, and it is not unlikely that I could be convicted of contempt. Meanwhile, there will be no possibility of a job, and even if I am acquitted, it will be very difficult to get a job. If this happens, I confess that I don't know what to do. I have a feeling of being trapped and choked, for at best, I can see little chance to work on the things that seem interesting and important to me. Even if these troubles don't materialise immediately, however, I doubt that I can work very effectively. As soon as I finished my book, I became depressed, because I feel that the problems I tried to work on are too difficult to be achieved in the time that is probably available – if at all. Anyway, I wonder whether I really have the ability to do these things.

Then I become very lonely, because in Princeton there are so few people to talk to. Perhaps I need a vacation, but to go off somewhere by myself would just depress me still more, especially now that everything is beginning to become so dangerous.

I am sorry to have nothing but bad news. There is however nothing to worry about immediately as all of these legal affairs take a long time.

I have been very lonely for you the past few weeks. It isn't entirely this latest trouble that is the cause of it, because I have felt that way before it developed. I now realise that from my point of view, I should not have let you go, although it is certainly the best thing for you. I love you would like to marry you. But it would be foolish, I suppose, to do it now that you are just getting started. Only I don't know how I can stand another year at Princeton all by myself. This is what really gets me down. These troubles with the Unamerican Committee don't bother me so much; in fact, by giving me something definite to work on and fight for, they probably raise my morale. (Ever since they started, my digestion has improved a great deal). Perhaps my temperament is not suited to working on a thing like research which is both uncertain and something that keeps you away from people. Besides I am afraid that I no longer really believe in the possibility of success in what I am doing, although on this point I think a vacation would help a lot. I want very much to see you this summer, but I don't quite know what to do. First, I wonder whether you haven't changed in the way that you feel about me. Secondly, if you would like me to visit you, is there anywhere near your place that I could stay? Several possibilities exist:
(1) I could come out in the middle of July or toward the end. If you go to Aspen, I could drive with you.
(2) I could come to Aspen with Lily and Lotty. If you drive out, then perhaps I could accompany you back and return East by air.
Please let me hear soon what you think of these possibilities.

With all my love,
Dave

Letter 8. Folder C97, not dated.

[See note with Letter 5]

Dearest Hanna

I hope you will excuse me for writing you two such pessimistic letters as the last ones that I wrote you. But I'm now over this pessimism. I have the feeling that things will work out well in the end. There will be a long period of trouble and suffering which will reach its culmination in the next four or five years but if one can only survive then one will see all sorts of wonderful things develop. It seems clear to me that the forces which try to hold back humanity will soon be broken, partly because they are trying to do the impossible and partly because of the stupidity of the people who try to carry out such an insane policy. I don't know what changed me to this fundamental optimism, but it was the realization perhaps that the whole system is so shaky, as one can see in this Korea business.

Your letter to me was very important as it was the first thing that turned this tide of depression. The acceptance of one's small but non-zero role in the rest of the world is really a wonderful aid to sanity in troubled times like these. I can see how I can contribute something in the sense that my actions affect quite a few people around me. If they stick me in jail, quite a few people will be led to see the hypocrisy of the claim that the U.S. is part of the so-called "free world". Moreover, it's not only true that no man can be totally free while another is in jail, but also that no man is totally in jail while there are people outside to fight for his freedom and that of other people as well.

Meanwhile I am beginning to work again with enthusiasm because I can see a purpose in it after all. This work is part of a fight against the insanity that has taken over this nation both indirectly and directly. It is indirectly so in the sense that all efforts towards a rational understanding of nature are also directed at removing the feeling of helplessness with which man confronts his problems. But even more directly, I feel that the more I can accomplish in the time that is left, the more of a blow that I deliver to the people who may put me in jail. For such an action on the part of the government will then be seen more clearly by people everywhere and especially in other nations as an exposure of the actual nature of our system, which is now trying to deny freedom to anyone who will not give up the most elementary concepts of decency.

Well, I am hoping that I can see you this summer. I'll tell you frankly that I do need a woman very badly, and that woman is you. I think very often how wonderful it would be for me if you were still here, but then I think it's better for you where you are. Perhaps we can eventually get together permanently (or as permanently as the present world situation permits people to stay together). All kidding aside, I really do long to see you

Love
Dave

Letter 9. Folder C97, not dated.

[See note with Letter 5]

Dearest Hanna,

Thank you for your very wonderful letter. It cheered me up a great deal. It is hard to realise the wider aspects of this problem in a time when all the old way of life is collapsing. The thing that bothers me so much is that while people may be sympathetic with my troubles as an individual, they are at the same time solidly supporting general policies which will increase not only my own troubles but also those of everyone else. As I see it, the people of America have been in part captured by the false ideal of the so-called "American standard of living". Every person dreams of a pleasant home and a sleek car, forgetting that it is the decency of human relations that is needed to make such things at all enjoyable. Every time I see the millions of unnecessarily sleek cars parked in front of so many homes, I have the feeling that these cars are my enemies. In order to hold on to such a car, a person must be ready to conform and to do anything that the government wants him to. If I had any say, cars would not be produced in large numbers until the problem of food shelter and clothing had been solved for most of the people of the world. These modern cars are really a sort of a drug, designed to lull people into forgetting how they are pushed around. While a person controls one of these tremendous engines, he momentarily feels that he really has some power, but actually to keep this powerful engine moving at his command, he must sell himself out.

As for my plans this summer, I am still indefinite, because of the possibility of an indictment at any time. However the most practicable idea seems to be that I should drive with Lili and Lotti to Aspen and meet you there. I could possibly drive back to L.A. with you and visit Berkeley later, returning by air to Princeton in September.

I am really very tired, both physically and mentally. It isn't only this trouble with the Unamerican Committee, but the whole way of life gets me down. I am very lonely and my work gets nowhere. Then there is the fact that I feel a hostility to everything that motivates our society, arising from the knowledge that the so-called "American Way of Life" is a menace to our civilisation. I don't feel that I am contributing to anything worthwhile in my work at all, because everything is sucked up these days into the preparation for war and Fascism. But worst of all is just the loneliness, which is enhanced by the fact that there are so few people with whom I can feel in sympathy these days. I feel that the cumulative impact of so many contradictions in my life and in the life around me has shattered my feelings. How to heal them up is the problem. Working on physics only makes everything worse, because it is at best a severe strain and now almost impossible. I have spent a few weeks just in the open air, with moderate physical effort and a lot of rest, or else I'll never have the strength to face the future. But it's no use going by myself, as this will probably just depress me all the more.

I had better stop writing in this vein, as it depresses me just to write such things. Your letter revived my spirits a great deal, especially the general idea that people have not all gone bad. Nevertheless I don't see how a great tragedy can be averted in America and it is difficult to see how one can hold out through the 10 to 20 years that will be needed before the present insanity runs its course. The question always is why one should as an individual go through this horrible mess, when it is clear that there is no hope until at least another generation has grown up. I have seen this coming since 1946 when that unspeakable Truman came into power, but everyone said that I was just being too gloomy. In a sense, I have been afraid of just the present madness which is taking over the country ever since I can remember. As a child, I sensed it in people, (especially in teachers) – a tendency to act without reason, just because everyone was doing it, because it was the thing to do. When people act this way, they are capable of unlimited cruelty and destructiveness.

I am very glad that you are doing so well in meeting so many interesting people, and I hope that you will soon get a job. Does Charlie Chaplin still come to your house, to eat cakes baked by you?

I hope that my letters don't give you the impression that I am always hopelessly depressed. This is not true. But I must find some new motive in life. Why should I work and what should I try to do? Right now it seems as if everything I have done is wrong and silly, and that in any case it is futile because it is slated for destruction. But I do wish that I could be with you at this time. I often imagine that you are by my side and then it seems as if I could forget this mess for a while.

Perhaps I should explain the legal situation to you. Now to begin with, the Hollywood people were trying to establish the right of a person not to talk about his political affiliations. This right is very important because in times like these, a congressional committee can create a regular Inquisition against all opposition. My refusal to talk was based however on the 5th amendment to the Constitution which says that a person cannot be compelled to testify against himself. At present, Communism is regarded by the government as a crime. Thus, under the Smith act, 11 people are already convicted, and the Attorney General has published plans for prosecuting 12000 Communists if this is upheld by the Supreme Court. I was asked by the Committee whether I had been a Communist and whether I had gone to meetings with certain other people including Weinberg. I refused to answer on the grounds that if I did, the answer might incriminate me; i.e., might lead to prosecution, for example, under the Smith Act. Now this right against self-incrimination is an old one, and ordinarily it would be respected. The Committee is now trying to take advantage of the present hysteria to destroy this right. My legal case is good, but the present temper of the Courts is to go along with the hysteria.

Now as yet, there are several steps ahead. (1) Thus far, the Committee has only resolved to cite me for contempt. They have not yet recommended it to Congress, which they must do to make it legal. Congress will take a recess starting July 31. If nothing happens until then I am safe until Autumn, perhaps November. (2) If it goes thru Congress in July, then there will be an indictment and I'll have to be in Washington when it occurs. (3) If indicted there will be a trial. (4) If convicted an appeal, (5) Last resort is Supreme Court.

I have talked with the Executive Secretary of the Lawyer's Guild, Mr Silberstein, and he thinks he can get me a lawyer (at a reasonable cost). I have also talked with a few "liberal" lawyers, but they are afraid that if they take such a case, they will be "labelled" as "Communistic". I can probably get a lawyer soon, but several of the suggested people are now on vacation.

I hope that this gives you an idea of the legal situation. Now to answer some more questions. The Pines are going to California this summer by way of Texas, and will arrive perhaps Aug 1. David is getting his degree soon but has no job yet.

I certainly hope that I can get to see you in California or in Aspen this summer. If not, then it's really going to be hard for me to find the strength for what is coming. Whether I come depends on what develops in the next few weeks, so I can't say what I will do yet. I hope you will forgive me for writing such a depressing letter in which I worry so much about my own fate. Perhaps next time I can do better. It was very nice of you to telephone and I was glad to hear your voice.

With all my love
Dave

Letter 10. Folder C97, not dated.

[After June 7, 1950, apparently a reply to Hanna's letter of that date. Hanna's letter, not reproduced here, is in folder C37. In it she tells Bohm she had decided she doesn't want to marry him.]

Dearest Hanna

I was terribly glad to receive your letter; you are evidently living very intensely and having a wonderful time. I would so very much like to share it with you. I know what you mean about the sun and the wind and water becoming a part of you. Here in Princeton, one grows tired. The weather is hot and sticky, and the world is so discouraging. Sometimes I wonder if I am perhaps not already too old for the kind of life you describe. I cannot get up any real enthusiasm these days and I have (for the time at least) lost belief in the possibility of success. I still work on things in a desultory way, and objectively speaking, I feel that some of my ideas are good, but I just don't seem to care about them as I used to.

I am very glad for your sake that you are having the chance to find yourself. It is ironical that just when I want to marry you, you are not interested in "losing your freedom". I hope that you can get started in something very soon. The bookselling sounds like hard work, interesting for a while, but not anything to remain in for a long time.

My plans are to apply for a Guggenheim to go to either England or Denmark in Summer 1951. I can work with Massey in England, or with Niels Bohr in Denmark. The latter work sounds more interesting as, it would involve writing a book on the philosophy of the quantum theory. As for my present book, that has finally been edited for the last time and sent to the printer. Galleys should start coming in July, and the book should be ready in September. I have an assistant who will read the galleys for me this summer, so I won't have to be tied down by this job.

This summer I am driving to California with friends. I should arrive late in July. They are going to Berkeley. I would very much like to visit you in August. Also, would it be possible to go on a trip somewhere for a week or two, in the mountains or elsewhere?

To answer the rest of your questions, the Brittens are still in Berkeley, but Barbara leaves early in July, Roy a little later.

The Weinberg case seems to have quieted down there is a fair chance that it will be dropped for quite a while, at least.

I didn't get to see your last book because I was out of town, and Lili took it to N.Y. and left it there. Nevertheless I heard you were having a wonderful time. I certainly hope that there are, as you hope, some possibilities for doing good work in movies and television. You are right in saying that it does no good just to denounce what is happening. Nevertheless, I am very discouraged because I feel that the whole system is like a web of evil, in which the people who take part tend to play the dual role of being both the creators of this web and its victims. In the past, the pressure on the individual was largely economic, and could be resisted by outstanding people of exceptional integrity. Nowadays, there is the added police pressure, which is used to remove and terrorise those who object to the direction in which things are going. While resistance is by no means impossible under even the present conditions, it is much more difficult than it used to be. On top of this, we have an unprecedented splitting up and demoralising of all progressive forces, on the basis of the totally irrelevant issue of the so-called charges of communist penetration. Unless something unforeseen comes up, the country will drift into Fascism and war, in the course of which such terrible destruction will occur that the precise way in which the progressive forces could win is now unforeseeable. I hope that I am wrong, but I am afraid that people will start to work together for a common good only after this web of evil is shattered by the very conditions that it creates. I am sure that there are wonderful people in Hollywood, just as there are in Princeton and everywhere else. But under present conditions, even the very best of human possibilities are being twisted into bad ends. Thus, what is better than the wonderful achievements of the physical and biological sciences, which are now supported primarily because they provide effective means of killing more people? Similarly, the very wonderful tendency of parents to take care of their children is used to scare people into submission, for if people are responsible for children, they cannot afford to lose their jobs. The loyalty of people to their country is exploited so as to make them so suspicious and hysterical that they will consent to anything. I could go on like this for hours, but it's hardly necessary. What I mean to say is that the web in which we are caught uses the best that is in humanity and transforms it into the worst. As long as this web is so powerful, it's hard to see how isolated actions can do much good.

Of course, I can see that I may be painting too bad a picture. Perhaps the seeds of hope and confidence in humanity which are planted in a single good film will later sprout when the environment is more favourable. If enough of these seeds sprout, the web that traps us all will be destroyed.[1] As you say, no one can foresee the end of what he is doing. But sometime I must have an argument with you. I don't think education is the only solution. More people are being educated to learn the wrong things than the right things. I think that the most important thing is that people shall learn to trust each other and work together. Insofar as education helps this, it is good. But much of what passes for education these days leads to the opposite conclusions. To this extent, people who learn to work together and fight together for a common end by experience are perhaps more reliably trained.

I hope to be able to watch the sun set with you on "this blue glittering mass". I have seen it many times before, and I want to see it again (The same goes for you, only double).

Love
Dave

P.S. I have read your latest book last night, German edition. Very interesting. Would you like to drive from L.A. to Aspen with me?

[1] This is known as a "mixed metaphor" – very bad English writing.

Letter 11. Folder C97, not dated.

[Seems to follow 10 (proofs have arrived for first quarter of book)]

Dearest Hanna,

I heard today from Lili that you are still depressed and cannot get started on your job. I am worried about you. I had hoped that maybe after you got started, you could get over this depression, but apparently, it isn't so easy. Are you able to do anything at all? I think that you should still give it a try for a month or two, and then if it doesn't work, come back East. There will surely be some job you can get back here. I want very much for you to be happy, to live in a useful and satisfying way. But things are so hard these days. One must keep up one's spirit even though one can find no real outlet, for the time, because when one person is depressed this causes all his friends or associates to lose some of their courage too, and we can get through this mess only if we all keep our courage up. Each person has a responsibility to all decent people, and that is to keep going and to live his life in the best way that is possible.

So please try to work on this job to the best of your ability, even if it is so vague. Just start to do something, and you'll find that your depression will begin to leave you. You are a wonderful person and it would be a terrible tragedy for everyone who knows you if you allow this depression to overcome you. I know that you have great potentialities which will be developed when the situation is more favourable, and many other people, in an even better position to judge these things, have expressed much confidence in you. As far as I am concerned, I love you whether you will ever accomplish anything later or not, because you are the person that you are. I hope that you will be able to do the things that you want so much to do and I believe completely that when the opportunity arises, you will ultimately do these things. After all, you are only 25 years old. As you are so fond of telling me, I am only 33 years old, which is 8 years old still. So your life is hardly at an end. Anyway, accomplishments have little to do with a time scale. Under good conditions, when your whole person can be focused on a given job, more can be accomplished in one year than in ten years of disjointed and half-hearted efforts, made necessary by the bad state the world is now in.

As for me, nothing new has happened. The man who works on my case as a prosecuting attorney just got back from his vacation, so trouble should start soon. I have just been taking it easy, but am getting interested in the problem of the electron again. Proofs have appeared for about the first quarter of the book. The Brittens will remain on the farm. They send you their love, as do the Pines.

Please try to be of good cheer (and also write me soon)

With all my love
Dave

Letter 12. Folder C97, not dated.

[Bohm was cited for contempt August 11, 1950].

Saturday

Dearest Hanna,

Well, as you may perhaps have read in the papers, I have finally been cited for contempt of Congress. Now there will be plenty of trouble. The first step is that they will indict me, perhaps in a few weeks. Then there will be a trial, and perhaps an appeal and finally the Supreme Court. I cannot afford to pay the regular lawyer's fee, so am trying to get help. Thus far I have contacted the Civil Liberties Union and they are going to talk with me on Monday or Wednesday. I also saw Mr Imbrie who is going to arrange for me to talk with the National Lawyer's Guild. I am going to ask you to do me a favour. Could you find out which lawyers defended the Hollywood Ten and try to arrange for me to contact them? I should remind you that my case is different from that of the Hollywood Ten. They based their case on the First Amendment to the Constitution whereas mine is on the Fifth, i.e., the right not to incriminate myself. In ordinary times, my case would not even be considered, but with the present hysteria the future is unpredictable. By the way, I should mention that my former lawyer, Cliff Durr has moved to Denver and therefore cannot represent me.

Even though I anticipated this trouble, it came as a shock. I am trying to hold up my courage, but I cannot help feeling that the future looks black, because of the madness which is spreading over the country. This latest Korea business makes everything much more difficult. Eric and Lili have been very wonderful to me, as at least I know I have friends. But I cannot see how a good life will be possible for me for many many years. I am not really so young and would like to get married, have children etc. I have decided that my abilities are not really so great, doubtless above average in physics but not really adequate to the problems that I have been trying. Therefore to work on them makes me unhappy. Perhaps I am too easily discouraged but that is how I feel right now. Perhaps someday better people will really understand these things.

Meanwhile I am lonely for you. Your letters really mean a lot to me. Especially the last one you wrote when you showed such optimism and courage in trying to do something about taking advantage of the media provided by movies and television. I hope that your knowing me will not cause the FBI to prevent you from doing the things that I know you can do. Sometimes I think that I may be like the carrier of a plague, but then I realise that a person who would put a career before human relationships of friendship does not deserve that I should worry about his fate. So it is just one of the unpleasant aspects of the present situation. We are all in it together. The fall of one person drags the next down and ultimately no decent person will be able to have a happy life if the present trend continues.

Although the future looks dark, I have an unshakeable feeling that in the long run we are on the verge of a most wonderful era. I only hope that I can live to see it. Even if I don't, it is of some comfort to know that it is coming. The present troubles are a reflection of the breakdown of the old order.

I wish very much that you would write me a nice letter, warm and affectionate but courageous.

Love
Dave

Chapter 15
Letters to Hanna Loewy and Lilly Kahler, 1951-53

Letters to Hanna Loewy

Letter 13. Folder C38, not dated.

[October 1951, en route to Brazil].

An Airmail Letter from
Pan American Airways System

Dearest Hanna

We are now waiting in San Juan, Porto Rico, for my plane to leave. This trip has been a long series of delays. Just as the plane was ready to rise in the air in NY, it slowed down and came back to the airport building. It turned out that they wanted to remove a man from the plane whose passport was not in order. You can imagine that I was a little worried.

We should be leaving here now. What I have seen of the island is very nice, with medium size mountains coming down to the seashore. The weather is rather warm. Most of the people speak Spanish

I will write more later when I get to Brazil

Love
Dave

Letter 14. Folder C38, not dated.

[Shortly after October 10, 1951 when Bohm arrived in São Paulo].

Avenida Angelica 160
São Paulo, Brazil

C. Talbot (ed.), *David Bohm: Causality and Chance, Letters to Three Women*,
DOI 10.1007/978-3-319-55492-1_15

Dearest Hanna

At last after many long delays, I am in São Paulo. The city (which has about 2 million people) resembles the U.S. of 30 years ago. There are many small stores, and few of the large department stores or chains characteristic of the U.S. The large modern apartment buildings are however very beautiful. I am told that architects refuse to design buildings unless they are permitted to make them beautiful. The general impression is that the country is considerably poorer than in the U.S. The middle class section is somewhat shabby and tends toward being dilapidated. However, the people are very energetic and much construction is going on. The weather is cool and resembles N.Y. in late April. I stopped briefly in Rio, which is really a very beautiful city. There the emphasis is said to be on good living, whereas here it is on working and "getting ahead". Even in my short visit at the Rio airport, I could see that people had an easy going attitude towards life. For example, I was met at the airport by a cute little girl (about 18) who was supposed to escort me to the waiting room. She knew only a little English, but asked me if I knew any Portuguese. I answered "Not a word". She said "But surely you must know the word for love – amar".

I am now living at a sort of pension where I have a room and meals. The cost is about $125 a month. The food is very good, but the room is rather small. I may look for a better room, but a good room costs from $75 to $100 a month. It is convenient to stay here, however, because several other physicists are here, and many of the people speak English. You have no idea of how helpless one is in most places here where nobody speaks English (for example in the stores). Even in the hotel where I stayed the first night, only one man spoke English and that very poorly. You can't use the hotel phone until you can pronounce their numbers, and even then it doesn't help unless the party you want speaks English. However at the university, many people speak at least a little English. Meanwhile I am learning a little Portuguese, but it will be a long hard job. The country is very new to me and yet when you look at it, it is not very different from the USA. The people are somewhat darker (although there are many blonde Germans), and they look considerably less strained and worried than in N.Y. But in general everything is similar, but differs chiefly in being on a smaller scale and less lavish. As I said, the stores are all small. The food stores and restaurants are not as clean as in USA, and tend to have odors of decaying food about them. All through the streets are signs of people running for city council. 60 councilmen are to be elected and 1000 are running.

The University is rather smaller than I expected. It has about 5000 students, but the various "Faculdade" are distributed in different buildings throughout the city (a single university campus "Cidad Universidad" is now being built in another part of the city). Tuition here is practically free. There are a surprising number of women students, even in chemistry, math. and physics. Most of the students will teach in high schools, a few in the universities, and a few will do research in industry.

The Departmento de Fisica is in bad shape now and apparently they expect me to help them do something about it. The various departments (Physics, Math etc.) work almost independently of each other, as do the various professors. The net result is a sort of chaos of courses. After the war, the physics dep't was in a bad shape, having few people in it to teach. As a result, even the physics students were taking mostly math. courses. By now, this is very hard to change, even though there are more physics teachers available. Many of the physics professors like this arrangement because it gives them more time to do their own research.

Another difficulty that has arisen is that gradually by custom the students have come to beginning classes about a month late. This is also very difficult to change, since by common agreement the students just do not show up. This illustrates the state of affairs here.

The library and other equipment here is not particularly good. All in all, this would not be a good place to stay indefinitely. But for a while it should be interesting. I'll do what I can to help them, but doubt that I can do so very much.

I'm not lonely for the USA but I am lonely for my friends there. It is a little frightening to realise that the language barrier will for a long time make really close contacts with most people rather difficult. For example, in explaining physics to people who do not understand English very well, the imagination is not stimulated nearly so much. I shall have to guard against a tendency toward stagnation here. However, I hope to be able to take a trip to Europe during the winter vacation (June and July).

I am very lonely for you. Please write me what you are doing, how you are feeling. How is it going with Dr Hale? What about the translation? Also what about the N.Y. Times? Is there any other interesting news from home?

Love
Dave

Letter 15. Folder C38, not dated.

[October/November, 1951].

Avenida Angelica 160
São Paulo, Brazil

Universidade de São Paulo
Faculdade de Filosofia, Ciências e Letras
Caixa Postal: 8105
São Paulo

Dearest Hanna

Since you still haven't answered my letters, I am beginning to get worried about you. Are you well, or perhaps you are simply angry at me? I think of you quite often – really you made a deeper impression on me than I thought at first. When I got here, I was so busy with new experiences that the past was more or less submerged for the time being, but now it comes up again. I get worried about you not writing and have trouble sleeping because I wonder what is happening to you. Please write me and please let us be good friends forever.

As might be expected, I am having some trouble getting used to things. The noise is the worst thing, as I can't get enough sleep. However, Philip Smith and I are looking for an apartment in a quiet sector. Apartments are expensive – a minimum of $150 a month (unfurnished) for two bedrooms,– kitchen, bath, and a small guest room. I found a nice one (with the aid of some friends in the Physics Dep't) but the owner refuses to rent to bachelors! In any case, Smith is now laid up because of an accident. In a city where it is dangerous even to cross the street, he bought a motorcycle. Within a week, he collided with a truck, breaking several ribs, and his shoulder bone. He is out of the hospital, but will not fully recover for another month or two.

Physics is coming along nicely. I have had several new ideas and Tiomno has made some very interesting suggestions. I think we are on the track of a great unification between quantum theory relativity and electromagnetic theory, which will do what Einstein has always been aiming to do, but in a different way.

I am so tired that I can hardly keep my eyes open. I hope that I get a letter from you. Please write, and use my home address as it is better for quick delivery.

Love
Dave

P.S. Regards to Lily and Eve and Ben.
P.P.S. I am slowly learning Portuguese. It will be at least a month before I can hope to ask directions in the street.

Letter 16. Folder C40, not dated.

[November, 1951].

Avenida Angelica 160
São Paulo, Brazil

Universidade de São Paulo
Faculdade de Filosofia, Ciências e Letras
Caixa Postal: 8105
São Paulo

Dearest Hanna

Thanks very much for your letter. I enjoyed it very much, and was glad to hear that you are feeling well, and working on your book. I have just been sick with a combination of diarrhoea and fever, and spent a few days in the hospital, where they filled me full of penicillin, sulfa, etc. It wasn't serious, but it is annoying. The doctor made tests, but was unable to find the cause. The only lesson is to be more careful with my food.

Tomorrow I leave for the city of Belo Horizonte (means beautiful horizon) located north of here and further inland, where there will be a meeting of the Brazilian Scientific Society. I shall give a talk on the Quantum Theory there. Incidentally, I have given some talks on my interpretation of the quantum theory here, and have aroused considerable enthusiasm among the theoretical physics people.

Avenida Angelica is far from angelic. It is right near Avenida São João, where there is one streetcar a minute, and where the autos go at furious speeds, clearing a path ahead of them by continually blowing their horns. I haven't been able to sleep very well. This section was once a wealthy section, called "Campos Eliseos", but now the old mansions have been converted to middle class pensions, such as the place where I live.
Smith and I are planning to move into an apartment soon. We had a good one in a quiet place, but the owner didn't want to rent to "bachelors". This shows the kind of town São Paulo is. São Paulo has been called "the Chicago of South America". It is extremely bourgeois, and extremely stuffy in its morals (very unlike Rio). However, if worst comes to worst, we are assured that we can get an apartment near what used to be the prostitution district. They used to have a legalised district of prostitution, but two years ago, it was made illegal, with the result that the girls took apartments in the neighbourhood. In such a neighbourhood, the landlords can hardly object to "bachelors".
I haven't met many people as yet, but expect to start to do more of this when I get back from Belo Horizonte next week. I have almost finished the book on Portuguese, (will finish in two weeks at most), and can now read the papers and books fairly well, and can carry out very elementary conversations. The trouble is that there is no pressure to learn the language on me, because I am surrounded by people who speak English.
Well, I'll let you hear more from me later. Please write and give my regards to Erich, Lily, etc. I am very glad that you were able to detect the note of phoniness that is developing in the Shahn's. I have felt uneasy about them for over a year, because I feel they are doing two things: (1) Switching over to a very wealthy and distinguished circle of friends (2) Rationalising their political beliefs so as to fit into this new circle, entrance into which has been made possible by Ben's fame as an artist. The fact that Ben did not object to a superficial book on a level of purely personal observations is related to the whole change in Ben's position.

Love
Dave

Letter 17. Folder C38, not dated.

[November/December 1951].

Universidade de São Paulo
Faculdade de Filosofia, Ciências e Letras
Caixa Postal: 8105
São Paulo

Dearest Hanna

I am getting worried about you, partly because you didn't write for so long, and partly because I got a letter from Miriam saying that you had visited George and that George thought you were having a hard time. I wish that you would write and let me know how things are with you. Please try to keep on working, even if things are hard. I know that you can do well. I have always had a strong feeling that you have great potentialities, which could be realised if you learnt to discipline yourself. Please let me know how your work is going, how is the car, how you are doing with Dr Hale (or how you were doing). I am not going to buy a car here, so Mr Pim is all yours.

As for me, I have been sick with diarrhoea and fever quite a bit, but am gradually bringing it under control. By now I am normal except for a slight tendency toward diarrhoea. I am beginning to meet more people, but haven't done much work as yet. However, at the scientific conference at Belo Horizonte, I gave a talk on the quantum theory, which was well received. Feynman was convinced that it is a logical possibility, and that it may lead to something new. Also I got a letter from Pauli, in which he practically concedes that the idea is logical, and in which he raises mostly philosophical objections. I answered these in another letter, and am now awaiting a reply from Pauli.

I am moving into an apartment with Smith in two weeks, so for the time being, I suggest that you write to the address printed on this letterhead. Please let me know how things are. It was quite a shock to hear from Miriam that you are so unhappy, and I can't get over feeling badly about it. If necessary, I could let you have another $100 to see Dr Hale.

Please give my regards to Erich & Lily, and to Ben (are you still in the same apartment).

Love
Dave

Letter 18. Folder C38, not dated.

[December, 1951].

Miramar Palace Hotel
Rio de Janeiro

Dearest Hanna

I got a letter from Lily and learnt that you are OK but do not write because you do not know what to write. I understand this, but I shall write to you anyway to let you know how things are. As you know, they took away my passport a few days ago. I can stay in Brazil but can get back my passport only for the purpose of returning to the U.S. This worries me, first, because they may be planning to ask me to come back, and second, because I cannot travel to England this summer. I really was counting on this trip to help stimulate me, and to help me get out of the rut that comes from lack of contact with people to whom I can talk in English and who can understand my problems. Right now, I feel sort of "choked" in my scientific and personal development, in being forced to stay in Brazil, and I don't know that it would be wise for me to stay too long. Even to return to the U.S. for a vacation might be an advisable risk.

Right now, I am in Rio giving a talk on the quantum theory. About the only person here who really understands is Feynman, and I am gradually winning him over. He already concedes that it is a logical possibility. Also, I am trying to get him out of his depressing trap of doing long and dreary calculations on a theory that is known to be of no use. Instead, maybe he can be gotten interested in speculating about new ideas, as he used to do, before Bethe and the rest of the calculators got hold of him.

Rio is in a magnificent natural setting, that far surpasses San Francisco in its beauty. Steep mountains and rocks fall right into the sea, covered with tropical foliage. However, the city itself is somewhat shabby and does not measure up to its natural setting. I am staying in a hotel on Copacabana beach. The beach is in the shape of a half-moon, ringed by mountains and rocks. But the hotel is full of Americans of the less attractive kind that one associates with wealthy American tourists. Also I don't feel well, as I caught another intestinal infection. So I am not enjoying this trip much, although the people are friendly enough. Rio is a gay city, but it's gaiety is a little empty. I prefer São Paulo's energetic character and cool climate, in spite of its bourgeois nature. I am looking forward to returning to São Paulo already, where I feel a little bit "at home".

As you told me, Brazil is too "American" for me. I feel the need of finding a country not spoilt by American money, and the "American way of life". The people of Brazil have no defence against the American invasion at all, not only because they are so poor, but also because they don't realize what is happening to them. I feel an increasing dissatisfaction with life here, which can be remedied only if I can find the right kind of friends. Thus far I have found nice people, but I do not feel the possibility of a really close understanding with them.

When you feel like writing, please let me hear from you.

Love
Dave

Letter 19. Folder C40, not dated.

[December, 1951]

Universidade de São Paulo
Faculdade de Filosofia, Ciências e Letras
Caixa Postal: 8105
São Paulo

Dearest Hanna

I received your letter of Nov.15. I hope that you received my more optimistic letter, designed to counteract the very pessimistic letter that I first sent.

Let me tell you now that the situation looks a little better here. There is some chance to improve the physics dep't. Also, I have gotten a permanent identity card, which means that I can stay here indefinitely, even without a passport. Also, if I want, I can apply for citizenship. This would have some advantages; as with it, I could travel. But the disadvantage is that I could not return to the US, at least for a long time. For according to the McCarran act, they can exclude any non-citizen from the U.S. who in their opinion, was ever connected with Communism. So it's a tough decision, isn't it?

Another problem is that someone here has started a rumour that I am a communist. This rumour has reached the University authorities. It may make it difficult to renew my contract (in October).

I hesitate very much to return home. As long as McCarthy controls the State Dep't, I think there is absolutely no chance that I could get a passport. Not even what Lili suggested would be possible – namely – to return home and then to come back to Brazil. In the present semi-fascist atmosphere in the States, I would be lucky to get even an industrial job. For the Committees are now beginning to investigate people in industry. Also, if as seems likely, they should pass the immunity statute, then I should soon be in jail.

My health is a little better now. As for your idea of "getting a girl to cook for me" you should know that this is impossible in Brazil. Brazilian customs would make even my aunt in Wilkes-Barre look very broad-minded. Did you know that women are examined for virginity in the Brazilian middle class, before they are married? If she is not a virgin, the man doesn't have to take her. And if a woman is alone on the street after 8 PM everyone assumes that she is a prostitute. The men stare at her, whistle at her, and try to pick her up. The whole attitude toward women here is like the worst of mid-Victorian society. Women are regarded as cooks and bearers of children – and most of them have developed so poorly as people that there is little more that they can do.

Thanks a lot for writing me. Please don't discuss my plans with anyone but Lily and Erich, and ask them not to spread it any further. I haven't yet decided what I shall do. Perhaps I could make another effort to get my passport back, while remaining in Brazil.

Love
Dave

Letter 20. Folder C40, not dated.

[Early 1952].

Universidade de São Paulo
Faculdade de Filosofia, Ciências e Letras
Caixa Postal: 8105
São Paulo

Dearest Hanna

I was very happy to get your letter, and enjoyed reading it very much. I must apologise for not answering your beautiful and clever New Year's greetings. I always get a strong feeling of pleasure whenever I receive a letter from you, and we are certainly as you say, "still friends". You are probably right in supposing that Lili's letter tended to make me put off writing to you. I always felt "O, this is so complicated, maybe I'd better do it later". I didn't, as you think, however, really get angry. As for your friendly assurance that you wouldn't marry me if I were the last man on earth, I receive this with mixed emotions. Let's just leave it at your previous statement that neither of us was really ready for this particular marriage at that time.

I am very pleased to hear that you are working as a draftswoman, and hope that you can soon get a suitable job in the movies. It is also really very good to hear the tone of happiness in your letter – It gave me a big "lift", as I have been depressed lately by the eternal rain (which should end soon) and by the steady movement toward war. The disgraceful action of the French Socialists in deciding to ally with the Nazis and the US really is amazing. It shows that 9 times out of 10, middle-class "good wishes" come to nothing, when it is really time for them to come out for their principles. The fate of the world literally hung on their decision, and they decided to help re-arm the Nazis.

I was very interested to hear about all the young physicists at Columbia, so deeply immersed in my book. You can't seem to keep away from physicists; perhaps they have a fatal fascination for you. My article is scheduled to appear in the Jan 1 issue of the Physical Review, but this issue seems to be very far behind schedule. All I can say is that it should be out any day now.

I have been doing more work on my theory, and have shown that there seems to be a connection between quantum theory and plasma theory (on which I was working with Pines and Gross). I now think that relativity as well as quantum theory will have to be seriously altered, before further progress can be made. Briefly, I am led to the notion that all space is filled with a substratum (very very dense) made up of particles millions of times smaller than an electron or a proton, and that electrons & protons, etc. are structures in this substratum. These structures are modified by moving through space, in such a way that Einstein's laws of relativity provide a good approximate description of their motion. But in connection with processes taking place at very small distances (the size of an electron or less), or at certain energies higher than have yet been attained, the theory of relativity should fail; and "absolute space" in the form of this substratum should manifest itself. Of course, the substratum should be only "relatively absolute" since its particles are in turn made of something still smaller, etc. ad infinitum. Thus all matter contains an infinity of qualitatively different levels, all interconnected. Moreover, there is another interesting point. The so called "particles" of any given level are made up of structures in the "particles" of the lower level, etc. ad infinitum. I find that both the qualitative nature of these particles and the number in existence can be collectively conditioned. Thus, not only is the collective behaviour determined by the infinitely complex individuals into which it can be analyzed, but the individuals are themselves determined in part by the collective in which they are participating. Because of the infinity of levels, you cannot say that there are any ultimate "individuals", which are "fundamental" in the sense that their character is unalterable, and their existence eternal. At any level, any particular form of matter can always come into existence & go out of existence as a result of a transformation in the components existing at a lower level, but only matter as a whole, in its infinity of properties and possibilities, is eternal.

Unfortunately, I cannot yet put these ideas into a precise & quantitative form; as yet they remain vague and qualitative. I should also add that I believe that no law is absolute or final, but that each law provides a successively better approximation to an absolute truth, that we can never possess in a finite time, because it is infinite in all its aspects, both qualitative and quantitative. But as we work and study, we can steadily uncover level after level, and thus move in the direction of an absolute understanding of the laws and properties of matter. Humanity as a whole (combined with all other forms of intelligent life) has the possibility of expanding its knowledge without limit; and thus, by understanding the causal laws, to go beyond all conceivable limitations. (Thus through the understanding of causality and necessity is the way to freedom). On the other hand, if we fail to understand causality and necessity, both external nature + our own natures will enslave us).

Well, so much for what I am doing. Send my regards to Erich, Lili, and others in Princeton. And please write me soon.

Love
Dave

Letter 21. Folder C40, not dated.

[April/May 1952].

Universidade de São Paulo
Faculdade de Filosofia, Ciências e Letras
Caixa Postal: 8105
São Paulo

Dearest Hanna

I have had a lot of things to do, and therefore have delayed writing to you. There has been some trouble here with some very vicious characters, but it looks as if it is under control. I explained it to Eric, but perhaps I may have seemed somewhat alarmed in my first letter. Please let Eric know that the situation is not alarming, but that the letters I requested from Einstein would still be very helpful in the long run. Also tell him that I plan to answer his letter soon. Tell Lily that the statements about the Income Tax arrived a month too late. They should have been sent air mail. However, I independently filled out my tax, without the slip from Prentice-Hall.

I am glad to hear that your situation is improving, and will always be anxious to hear how you are doing. As for me, I am very busy, teaching classes (in Portuguese), engaging in intrigues with the two rats who are causing all the trouble, and working on my theory.

I was very glad to see that you understood the importance of the infinity of levels immediately. The substratum is another word for the ether, but I prefer "substratum", because "ether" implies something "ethereal" while the substratum is apparently very dense. You are right in saying that it leads to a commonsense view of the world, which at the same time can treat problems of the subtlest type. Erich is also very much in favour of this kind of idea. I have some more ideas, and as soon as they crystallise a little, I'll let you hear some more about them. Meanwhile, please let me hear some more about you.

Some reactions to my theory are coming in. The older people like Pauli, Rosenfeld, + Bohr, object to it, but their objections are almost childish. I expect to publish some papers soon answering these objections. Among the younger people, especially the experimental physicists, there is much interest in it. Of course, the Institute says that it is "of no importance". Thus my theory has passed two important tests in its favor.

(1) It has been opposed by Pauli (Pauli has opposed every new idea that was ever suggested).

(2) The Institute people think it "unimportant".

Give my regards to Lily, and to Erich, and to Ben

Love
Dave

Letter 22. Folder C39, dated: Oct 6, 1953.

Universidade de São Paulo
Faculdade de Filosofia, Ciências e Letras
Caixa Postal: 8105
São Paulo

Dear Hanna

I shall have to apologise for not answering your last letter, but I have been tired and depressed, and always tended to put off answering you for this reason. I am glad to hear that you are well, and as happy as anyone can be expected to be in this age. I have also heard from Lily that you are in good health and spirits.

As for me, I am afraid that Brazil and I can never agree. First, there is the climate which is too warm for me (even in São Paulo and especially in Rio) and which makes it very hard for me to work. I could get used to the climate, if there were some compensating factor which was attractive, but there isn't. Then there is my health, which is poor here. I suffer from a very weakening form of diarrhoea 50% of the time. Doctors can't seem to find the cause, but I suspect that the cause is in the rather dirty conditions of the country, combined with the high temperatures, which is conducive to the growth of bacteria.

Now if there was something else worthwhile here, I could accept both the high temperature and the bacteria. However the University has nothing to offer – there are no good students, and the whole set up is (like the rest of Brazil) so disorganised that there is no hope to do much. The physics dep't contains a number of professors of very low moral character and practically no ability in physics, who are able to keep things in a continual state of turmoil, in which nothing constructive can be accomplished

Now I could get used to all of this if there were some people in São Paulo with whom I could have a close relationship. But the Brazilians are first of all a very individualistic lot – they rarely make friends even among themselves. Then their way of thinking seems to be different enough from mine so that there is little satisfaction in talking to them. I have had a few friends here but these consisted of people from the U.S., Argentina, Chile, and France, who were here temporarily. It seems that of all the countries I could have chosen, Brazil is most alien to my own spirit. It is difficult to define the cause. But the problem manifests itself as a sort of complete lack of interest in trying to penetrate deeply into things, a sort of intellectual deadness, combined with a very superficial point of view, and a tendency to take things very easy. Even when they work hard, they take no passionate interest in their work, but merely a sort of utilitarian interest. I don't know whether I have analyzed the difference correctly, but it is really there. Other people have noted it too.

Finally, there is the city of São Paulo, which is practically unliveable. Although it looks nice from a distance, it is easily seen to be a disorganised, chaotic, poverty-ridden city from a closer inspection. Everything is falling apart. The traffic is horrible and the buses emit dense clouds of foul-smelling smoke into the hot sunshine, while the trolley cars make a never-ending din in which you can't even think, much less talk to someone. Although there is a terrific shortage of housing for poor people, no housing is built for them. Instead one sees nothing but expensive apartments going up, about two every block, making a terrific racket, so that it is almost impossible to find a quiet place to live (They start six o'clock every morning, many of them even on Sundays, and some have a night shift). Moreover, there is absolutely nothing to do in São Paulo. A typical idea of Sunday amusement is to stand in line for two hours and wait to enter a motion picture theatre. There are very few places near here worth visiting, and the hot sun makes hiking or walking out of the question. Moreover, travelling is impossible, as there is no place to visit. The roads and public transportation are very bad, while good food, water and lodging are unobtainable outside the 2 main cities (Rio & São Paulo). Even in Rio and São Paulo it is not safe to drink the water, and I cannot eat in restaurants without getting sick, while in the interior, everybody gets sick unless he carries his own food and water.

So you can see that I am practically alone (except for an American couple that I have managed to have invited down here, the Schillers), my health is bad, and there is little possibility for constructive work. I am good and sick of it, and I'm becoming depressed. In fact, in the past six months, I have made little progress on my work. I am becoming discouraged also because I lack contact with other people, and feel that there is a general lack of interest in new ideas among physicists throughout the world. Everybody pays lip service to the necessity of new ideas, but nobody really wants to bother, unless they are presented with some magic formula that solves all problems.

Lily has written me a letter in which she mentions Phil Morrison's suggestions that I return to the U.S. I cannot accept these suggestions. First, there is the problem of the immunity statutes. I saw recently in the N.Y.Times that the administration is going to press next year for the passage of such laws, which will compel people to testify in Congressional investigations. Hence, if I returned, I would be in jail in a year or two. Actually, for the present, my position would be much more dangerous than that of Weinberg. For they would not offer him immunity from prosecution in order to get him to testify, since he is the man that they are trying to get. Hence, it wouldn't make sense to promise him that he will not be prosecuted. Hence, he will have the right to refuse to testify. But people whose main role is that of witnesses can be offered immunity, and then they either have to testify or go to jail.

Secondly, I think the situation in the US will remain bad and get worse for the next 10 to 20 years. Not only could I not get a job (except perhaps in industry) but a depression is very likely, so that the numbers of jobs will be low for everybody. Also, they are already cutting support of non-military research by not renewing Army, Navy, and A.E.C. contracts, so that even if the general economic level remains high for a while, there will be plenty of unemployed physicists, who cannot or do not wish to work on military research.

So I think the only hope for a solution to my problems is to go to Europe. How to do this I do not know, for the present. But maybe something can be worked out.

Please let me hear from you again soon

As ever
Dave

PS. Did Lili get my answer to her letter [?]

Letter 23. Folder C39, dated: Nov 8, 1953.

Universidade de São Paulo
Faculdade de Filosofia, Ciências e Letras
Caixa Postal: 8105
São Paulo

Dear Hanna

Yesterday I sent you a rather depressing letter. Please don't take it too seriously, as it was the result of a combination of particularly difficult circumstances that was making me feel bad. I am not always in such a low mood. I apologise for sending you that letter, and promise not to send such letters to you or to other people hereafter. What I said in there is true enough – I do want intensely to leave Brazil, but I shall nevertheless make an effort to fight the feeling of hopelessness that occasionally overwhelms me. I also have a difficult problem of decision, as in some ways I would like to return to the US (to see all my old friends) but very much doubt the wisdom of doing this for the next 5 to 10 years. But to stay here for 5 to 10 years is out of the question.

I am glad to hear that you have "found" yourself to some extent and have begun to develop a new level of consciousness. However, the problem is never really solved finally, and each of us needs continually to try to reach higher levels of understanding, as the situation develops, especially because life is so difficult these days and conditions change so rapidly. So you must not rest on your laurels, but should instead keep on working on the problem.

How is Erich doing? Has he made progress on his numerous books? Please send him my love. When I think of him now, I appreciate what a magnificent spirit he has, to keep struggling for his ideas, and to keep a youthful attitude. For that matter, the same should be said about Einstein (even though I am in partial disagreement with him on scientific questions). Please give him my best regards.

I have made some progress on my ideas, but now I seem to be facing an impasse. Perhaps I must try some totally new track. Perhaps in a few days I shall write Erich a letter in which I shall try to summarize the situation. Naturally, the letter will really be meant for you too and for Lily.

As for starting on a new idea, I feel I could do this better in a more favorable environment. Here it is difficult. However, I have a great many things to write up on the existing state of my ideas, and this should keep me busy while I am still in Brazil.

As ever – Dave

Letters to Lilly Kahler

Letter 24. Folder C96, dated: Dec 3, 1953.

Dear Lili

Thanks a lot for your second letter. I am glad to hear that Erich is better and that everyone is doing well.

As I said in my 3rd letter to Hanna, the situation looks a little better here than it did before. I have just managed to get a permanent identity card, which means that I can stay here indefinitely if I should wish to, even without the passport. But to travel, I should have to take Brazilian citizenship, to get a Brazilian passport. This I cannot do in less than about 18 months from now; it may take two years.

There is a serious disadvantage of taking Brazilian citizenship. It will mean that I cannot return to the US for a long time. For according to the McCarran Act, the US will not give visas for non-citizens of the US to enter if they have ever had any so-called "Communistic" connections.

A number of possible plans present themselves. First of all, I am not very happy about staying in Brazil. Einstein may be able to work independently of his surroundings, but I find that I can't. When I remain too isolated from people with whom I can have some relation, my thinking seems to dry up. I reach a stalemate, from which I can get out, only after discussions with other people. Even if these other people aren't experts, it seems to help me to formulate the problem by talking with them and to get their reactions. New ideas suggest themselves, which do not appear when I am just thinking by myself, as if I just discuss my ideas always with the same small group of people. Also, there is the relaxation which comes from having some relation with people. Without it, I tend to develop a very tense mood that makes work difficult. None of these things are available in Brazil, and I have little hope that there is any possibility for such things here. I think I know the country well enough to feel fairly sure of this.

Now a possibility is to try to get a job in England. Unfortunately all the British people that I know suggest that I first come there, and then something may develop. Unfortunately, I cannot get there without a passport. Perhaps if I already had a definite offer, I could get the passport back, but without an offer, my chance is zero.

A second possibility is Israel. I have been in touch with friends in Israel (Americans who went there). They find that they like the place. Scientifically, it is perhaps poorer than Brazil in experimental physics, but probably the chance of finding good people with whom to do theoretical work is better. I have been in touch with Rosen in Haifa (indirectly) and have learned that it would probably be possible to get me a job there. Israel would have the disadvantage that life is more difficult, but the advantage that there would probably be more people with whom I could have some kind of relationship.

There are two possible ways of getting to Israel. One is to arrange the job offer and apply to the State Dep't for a passport on the basis of this offer. Perhaps the Israeli gov't would be willing to make some representation to the State Dep't.

The second is to go to the Israeli embassy here and ask for a special document to travel to Israel. This method would imply that I intend to become a citizen there.

With either of these methods, I should need some help, especially in the form of letters of recommendation. Einstein's help would be especially good in this connection. I'm not asking him to do anything against his principles but would he:

(a) Write a letter of recommendation for me to get a job in a place like Haifa?

(b) In case of necessity, write a letter to the Israeli embassy in Brazil, suggesting that they give me a special document to travel to Israel?

I have not definitely made up my mind on this question; I probably have a few months to decide. If I could get an offer from England, I would leave Brazil immediately. Israel, however, is a more difficult question to decide.

Please give my regards to Einstein and to Mrs Dukas.

Love
Dave

P.S. The idea of my returning to the US is for the time absolutely out of the question. It seems clear that the US is coming very close to Fascism. In any case, McCarthy is running the State Dep't these days, so that I cannot imagine how I could ever get out of the country, if I once returned. Also, even industrial jobs are being investigated by Congress these days. It is the gov't's intention to make it impossible for anyone who opposes its policy to make a living. It is difficult to believe that they will leave open the chance for an industrial job.

Letter 25. Folder C96 (also C28), not dated.

[December 1953 (references to Christmas and last summer's visit of George Yevick and Jean-Pierre Vigier)].

Dear Lili

Thanks for your latest letter. I am glad to hear that everyone is well. Please give my best regards to Erich and to Hanna. Also best wishes to all for Christmas and New Year.

I thought that I would like to correct some wrong impressions that my letters seem to have conveyed. I am by no means satisfied with staying in Brazil. For a while, I thought there would be an improvement in the physics dep't, but now it looks as if this won't happen. Let me explain that ever since I got here, there has been a vicious fight going on in the physics dep't. This fight started several years before I got here. It makes things very unpleasant here, and makes it almost impossible to do anything here to correct the bad situation. Let me also say that physics students are almost disappearing. A typical class may have 3 or 4 students, and the number who enter is decreasing every year. The students are not a very good, as they get bad training in high school, and equally bad training in the University. Also, there is a general atmosphere here, in which people don't feel any desire to work. Besides, after the students graduate, there are no jobs for them, except to teach at very low salaries in the high schools.

So you can see, the general morale here is low. I felt that if at least the fight in the physics dep't could be resolved, some improvement could be made. We spend 20% of our time here in meetings to discuss this fight fruitlessly, and in thinking & talking about it.

But even if the physics dep't could be improved, I could not stand living in Brazil. For I have no one to talk to, and nothing to do in my spare time. I cannot work efficiently if I have no relaxation from time to time. Life is so tedious here that each morning, I wonder how I am going to get through the day. Under these conditions, my ideas tend to dry up, and my work just simply does not get anywhere.

As for my health, I am already sick of going to doctors and taking tests, etc. There is simply nothing that doctors can do about it. The cause of my diarrhoea is clear. The food is bad, usually cooked in heavy fat, and not very clean. The weather is hot, and conducive to the growth of bacteria. I am sensitive to these bacteria. Some people are sensitive, some aren't. We had a visitor from France last summer, a very tough man, but he was sick every day with diarrhoea. Other visitors just didn't get sick at all. The best that you can do is to try to eat the cleanest, simplest food that you can find, and not to overeat. This helps, but at best, I can reduce the fraction of the time I have diarrhoea to about 1/2. Psychological depression tends to make the diarrhoea worse. And with my situation here, I alternate between 2 states. In one I feel depressed and hopeless. In the other I just feel hopeless without feeling depressed.

So you see, I must eventually get out of here. The idea of this staying here as a home, or of staying 10 to 20 years simply horrifies me. I don't see how I could possibly do it. It is like spending the rest of your life half-dead. On the other hand, for the next 2 or 3 years, at least, return to the US will also be impossible. I cannot agree with Phil Morrison's Utopian predictions. It is true that McCarthy's personal prestige may have dropped a bit, as a result of his attack on Truman. But as far as the issues are concerned, McCarthy won. For McCarthy's basic issue is that the most important problem facing the US is Communist espionage. This is really ridiculous. But Truman didn't answer it and indeed couldn't answer it. For Truman is the one who started all this anti-Communist hysteria. Anyway, the basic point is that the Republicans have decided to base the next election campaign on the issue of "Communism" and the Democrats will not answer back. So we can expect some big spy trials in the near future, and a lot of excitement generally. Let us see next November how the American people respond to this.

So in general, my feeling is that I must eventually get out of Brazil, but I will not return to the US until I saw clear evidence of a change in policy. I might be able to stand Brazil for another year or so, but I am afraid that a much longer period would be very bad indeed.

Well, I hope that makes my situation clear. Thanks very much for your efforts on my behalf thus far.

Love
Dave

Chapter 16
Letters to Melba Phillips, October-December, 1951

Letter 26. Folder C46, not dated.

[Shortly after October 10, 1951 when Bohm arrived in São Paulo].

Avenida Angelica 160
São Paulo, Brazil

Dear Melba

After many delays, I am finally in Brasil. It is hard to write down all my impressions. São Paulo resembles an American town of 25 years ago or more. The large department stores and chains of stores are absent. Everything is on a sort of small - scale chaos. The new buildings however are very beautiful. Apparently architects refuse to design buildings unless they are given scope to make them beautiful. The entire city has an air of great energy with much construction etc. The people move rapidly. The air is cool, resembling New York in late April. The comforts of living are definitely far below these of the US however. I stopped in one of the best hotels in town (about $7.50 at the official exchange rate) but it resembles a second rate hotel in N.Y. The people at the physics dep't got me a room in a sort of pension where we also eat. The food is excellent, but the room is small and not very nice looking. Good rooms are however very expensive. Room and board here is about 2500 cruzeiros a month (about $120) which isn't too bad.

The language problem seems tough right now. A lot of people near the University understand enough English so that I can get along. But only one person in the hotel understood it, and very few of the store keepers, etc. One feels really helpless without the language.

The University is rather smaller than I thought, about 5000 students in all, of whom 1000 are in the Faculdade de Ciências, Letras e Filosofia. The different "faculties" are separate (being in different parts of the city) although a new "Ciudad Universidade" (university campus) is now being built in a separate part of the city. Not only is

C. Talbot (ed.), *David Bohm: Causality and Chance, Letters to Three Women*,
DOI 10.1007/978-3-319-55492-1_16

each faculty independent, but each chair (i.e. professor) can work independently. The result is a general chaos. Thus, it turns out that because there was a shortage of physics professors right after the war, most of the science students (including physics students) are taking 2/3 of their time in math. Now it is very hard to change this. Many of the physics professors prefer it as it gives them more time for research. Also, recently a custom has arisen in which students report to class about a month after the term begins. This is also very hard to change.

The library is not in very good shape, and they are far behind on foreign journals. On the whole, conditions are far from those in the States. I am not really unhappy, however. I just don't take these problems too seriously. I'll do what I can, but one cannot expect miracles here. They are evidently hoping that I can bring the department into better order.

Summer vacation begins soon, but they are expecting me to help get things going here now. I can probably wangle a vacation in winter for a trip to Europe (this June). This suits me quite well.

I do not really regret leaving the U.S. All that I really long for are the good friends that I had there, especially you and the rest of the house at 298. These are among the memories that stand out most vividly. Much of the rest now seems hazy and almost unreal. Here in South America, we seem to have turned the clock back 25 years or more. The horrors of war and the uniformity of opinion characteristic of US have not reached here. Yet, one can see that the people here are vulnerable to the same things, because behind this concern for the every - day things of life there is no understanding of the long range social processes that are taking place. It is hard to express what I mean, but here we seem to have a lot of individuals each trying to make his way and get ahead. They haven't yet got to the point where this game forces them to combine in great organizations that smother all independent thinking. Yet one can see that this is the inevitable end of what is happening here, just as it was in the USA, unless there is a fundamental social change. Meanwhile, however, the people seem to be less worried and friendlier than in the U.S. In Rio, they are even more different, because the prevailing atmosphere in São Paulo is that of getting ahead, whereas in Rio, the emphasis is on enjoying life. I could see this even in the short time I spent at the Rio airport. To illustrate one point in this direction, I might tell about getting off the plane. A cute little girl (about 18) came to take me to the waiting room. She asked me if I spoke Portuguese and I said "Not a word". She said "Surely you must know some words. [For example – words crossed out CT], the word for love - amor".

There is a great deal of poverty here. You can see the people on the streets especially the workmen, who look much poorer than in the US. They compare with the workmen I used to know when I was a child but are even poorer. Even the middle-class homes have a certain shabbiness about them, a tendency to fall into disrepair. However it isn't too bad. Perhaps it is good preparation for living in Europe. Only the new apartments are really in good shape, but an apartment of this kind is $200 a month, clearly not worth it for me.

Well, I shall write more later. Give my regards to all and let me hear news from home before I lose touch altogether.

Love
Dave

Letter 27. Folder C46, dated: Oct 22, 1951.

Oct. 22, 1951
Avenida Angelica, 160

Dear Melba

Thanks very much for your letter. It is the first one from the States. I advise you to write to my home address, or to the Faculdade de Filosofia, Ciencias e Letras. The Physics Dep't address is not very good as there are 3 or 4 physics dep'ts in the university.

Thus far I have been studying the language and getting used to things a little. On the whole, Sao Paulo is not very different from the U.S. except that the poor people are poorer. I live in a very noisy place, but hope to move to an apartment with Smith as soon as he recovers from his accident. He was crazy enough to buy a motorcycle in a city where you take your life in your hands when you cross the street. As a result, he collided with a truck, and suffered cracked ribs and a broken clavicle. We are now trying to convince him to sell his motorcycle.

The language is rather ugly compared even with English. Every now and then you run across a pleasant or musical word, but you can always be sure that it is Indian. For the rest it sounds like a mixture of Spanish and French from which half the syllables have either been dropped or replaced by the nasal sound "aũ", which should be said "ah-oong". We live near Avenida "São João", which is "Sah-oong Joh-ah-oong", but this must be said rapidly and comes out roughly "Sahn Jo-ahn". The word for "to drink" is "beber", and the conjugation is <u>pronounced</u> as

present		past	
eu beboo	nos bebaymos	eu bebee	nos bebemos
tu bebes	vós bebays	tu bebastee	vos bebasta
ayle bebe	ayles bebayn	ayle bebee	ayles bebarem

I am gradually getting used to it, and can make the sense of newspaper stories, signs on the street, and subtitles in English movies. But it will be a long time before I can understand the speech.

During the first week or so, I was not lonely, but now I think quite often of the people back in the States. Also, scientifically speaking, I miss the stimulation of talking to people. One feels far away from things. Even though there was a lot of opposition, still it was sort of stimulating. I guess I'll have to get used to working without such stimulation. Also, I'll have to get out and meet more people. Smith has quite a few friends, but they seem rather naive to me. Smith himself is rather naive, but I like him because he is so full of life and because he has at least good intentions.

I had another idea about the quantum theory recently, tying it up with the plasma theory. This idea is based on a remarkable formal similarity between quantum theory (in my interpretation) and the plasma theory. In the plasma theory, one has the alternative of a description in terms of individual particles in a large assembly, or of a collective description of these particles in terms of an effective field, which is constituted by the density of particles at each point. Now, I propose to regard the negative energy electrons in a vacuum as forming a sort of plasma, while the ψ field (and also the electromagnetic field) represent the collective aspects of this plasma. The individual particle and collective aspects interact, but each provides a different description of the same system. This idea promises to give a more rational interpretation of the fact that the particle velocity is restricted to $v = \frac{\nabla S(\vec{X})}{m}$.

Well, I hope that I can get some sleep tonight. The street cars and autos are very noisy, there are roosters near here that start at 2 AM, and at 7 AM, they start to clean up the house.

It's hard to believe that we are 5000 miles apart. The idea hasn't sunk in fully yet.

Give my regards to all

Love
Dave

Letter 28. Folder C46, not dated.

[November, 1951].

Avenida Angelica 160
São Paulo, Brasil

Universidade de São Paulo
Faculdade de Filosofia, Ciências e Letras
Caixa Postal: 8105
São Paulo

Dear Melba

Thanks for your several letters. I am falling behind in answering my correspondence because I got sick last week with a combination of fever and diarrhea (3°C of fever). We called a doctor, and he took me to a hospital, where they filled me so full of penicillin and sulfasoxidine that the fever disappeared in a few hours. The diarrhea went away in a few days. The doctor made many tests, but couldn't find the cause. (Perhaps because I had so much sulfasoxidine that no bacteria could possibly survive within a foot of me.) Anyway, I am now back to normal, a little weak, and set back about 2500 cruzeiros (≅\$125) for a few days in the hospital plus the tests and the cost of the medication. The only moral is to be more careful with what I eat and drink.

Tomorrow I leave for the city of Belo Horizonte (means "beautiful horizon") which is located to the north and some distance in the interior of Brasil. Here there is a meeting of the Brazilian Scientific Society. Everybody will be there, and give talks. I shall talk on the Causal Interpretation of the Quantum Theory. I gave two talks on the subject here, and aroused considerable enthusiasm among the people like Tiomno, Schutzer, and Leal - Ferreira, who are assistants (Roughly assistant professors). Tiomno has been trying to extend the results to the Dirac equation, and has shown some analogy with Einstein's field equations. In other words, one not only needs a quantum - mechanical vector potential, but also a tensor potential, $g^{\mu\nu}$, analogous to the gravitational tensor. Thus, there may be some connection between Dirac's equ. and the unification of electricity and gravitation, which may perhaps occur only at the quantum level.

Everybody here has been very nice to me, especially O. Sala, who is Smith's boss, and who is trying to set up a few - MeV van de Graff machine, for doing very accurate work in the low energy range. People here suffer from lack of money, lack of help on routine things (such as painting, cleaning the machinery, etc.) and lack of skilled machinists and technically advanced machine shops in the neighborhood. In order to make his van de Graff tank, Sala had to train the machinists in a local shop to do precision work. Things therefore move very slowly. He has a very beautiful building on the outskirts of the city, where a new university campus is being built.

Brazilian science is in for some atomic trouble. The gov't and some rich men and the army would like to build an "atomic city" at Belo Horizonte. Lattes supports them by giving interviews in which he says that Brazil could make an atom bomb, and that Argentine is getting ahead of Brazil. Of course, he later qualifies his statements by saying that they aren't ready yet, and must do other things first. But the harm is done, because nobody reads to the end of the interview. The Head of the Brazilian Research Council, (an admiral) came out with a statement that they would develop atomic energy, and would expect the help of every Brazilian physicist. This is of course ridiculous, since Brazilian physics would be destroyed in the attempt, and since they do not have enough physicists. But the "big - shot" fever seems to be spreading, despite the opposition of most of the Brasilian physicists. Every time they get a machine (like the 23 MeV synchrotron at São Paulo) they make a big spread in the papers. Once the papers got mixed up and said that São Paulo was going to build the biggest cyclotron in the world (This happened in 1948 while Sala was working in Illinois as a student). The American papers picked up this item and printed it. The FBI then called on Sala and asked him to explain. He of course said he knew nothing about it, but didn't believe it. I asked Sala what right the FBI had to object to their building the biggest cyclotron in the world, and he told me that one must accept the fact that there is much American control in Brazilian affairs.

I haven't done much science in the past week nor have I seen many people. Perhaps in a week or two I'll be able to get further ahead. I am almost finished with my Portuguese grammar, and can read newspapers, and books, and carry out very elementary conversations. The words often have Latin roots that can be recognized in print but not in speech.

Give my love to everyone.

Dave

Letter 29. Folder C47, not dated.

[November, 1951].

Universidade de São Paulo
Faculdade de Filosofia, Ciências e Letras
Caixa Postal: 8105
São Paulo

Dear Melba

Well, I no sooner got back from Belo Horizonte than I caught another infection, which incapacitated me for a week, with diarrhoea, weakness, and fever. I finally called a doctor, who prescribed a new drug from USA "chloromycetin", which is weight by weight, three times as expensive as gold. It seemed to help, but I am still worn out, and have a trace of the infection, as evidenced by the presence of some diarrhea. I am getting sick of this, and don't know what to do about my susceptibility to these infections. It takes away all one's excess energy and spirit, so that even a few blocks' walk is tiring.

The conference was interesting. I talked about the quantum theory and created a favorable impression, even convincing Feynman that there might be something to it. Incidentally, Feynman has begun to get a sort of "social conscience" in Brasil, and is very much impressed with the extreme poverty of the people, which was especially evident in the interior, in such places as Belo Horizonte. In fact, when the city of Belo Horizonte gave us a dinner at a modern country club (designed by Niemeyer), Feynman became angry and when the Mayor came in with some pomp & ceremony, Feynman walked out angrily, because he felt that so much money should not be wasted in the presence of so much misery.

Lattes was not at the conference, because he was busy talking to Gordon Dean in Rio. When I arrived in Belo Horizonte, the papers were full of the story of a deal between Brazil + US, in which Brazil agreed to sell the U.S. Thorium, and in return, US would build Brasil an "atomic city" in Belo Horizonte, with a pile, etc. Some of the "military physicists" and military men want this, as in the secrecy of an atomic city, nobody will ever know that they aren't accomplishing anything. I feel that Lattes is opportunistically backing them up. There is a general feeling, however, that such a city would absorb the few Brazilian physicists away from real physics, and thus prevent the growth of Brazilian physics. Incidentally, Feynman shares this view. Dean was in São Paulo, and had dinner with Souza - Santos, the most influential physicist at the University. Smith tells me that Souza - Santos is a ruthless opportunist who drives for power, and that although Souza - Santos has in conversations with many people come out against this pile, he may easily change his mind. So you can see trouble ahead here.

I was amused by Charlotte's remark that "there isn't even a machine at São Paulo, and Lattes isn't even there." Is Lattes equivalent to a machine, and if so, what is his voltage? It is true that one J R Oppenheimer remarked that Lattes was too much of a "high-pressure salesman", and events seem to bear him out. Anyway, our problem in Brasil is less to get more machines than to keep them out, lest all of the efforts of Brazilian physicists be absorbed in servicing machines. Also, the time may come when Charlotte will be happy, if Bob is able to get a job, even at a "machine - less" university, although this idea has probably not yet fully reached her consciousness.

Well, I hope that I can approach a more human status, by getting rid of these bacteria. I hope that it will not be necessary to eat my weight in precious chloromycetin.

Give my regards to all

Love
Dave

Letter 30. Folder C46, not dated.

[November, 1951].

Universidade de São Paulo
Faculdade de Filosofia, Ciências e Letras

Dear Melba

I am writing to let you know that something alarming happened today. While I was in the physics dep't office, a representative of the Consulate came and told me that I should go to the consulate for "registration and inspection of passports" as is required of all American citizens. When I went there, and registered, they informed me that they were keeping my passport, and that I could stay in Brazil, but that I could get my passport back only if I returned to the U.S. He would give me no more information.

What this means I don't know. At least, it means that they only want to be sure that I stay in Brazil, but at worst, it may mean that something more serious is cooking. Frankly, I am worried. I would appreciate it if you would watch the papers to see whether anything is developing in connection with the "Weinberg case".

Well, it looks as if my brief interlude of peace is coming to an end. I wish I had used it better. This uncertainty is a bit hard on the nerves but I guess I had better get used to this uncertainty again. After all, I think I did my best work under such conditions.

When I got out of the consulate, I suddenly felt as if the U.S. had come down into Brazil. The people moved around with the comparatively placid and unworried expressions of Brazilians, but it seemed as if the sun were being shaded by a faint cold haze that foretold the coming of the same kind of fear and tension that one sees everywhere in N.Y. It was really a shock to discover that I am not really out of the U.S. I knew it intellectually all the time, but now I really felt it. The traffic jams in São Paulo began to resemble the traffic jams in N.Y. with all their nightmarish associations (for me). And as I looked into the faces of the people, I felt that they were well meaning and good natured, but totally incapable of coping with the corrupting and destroying influences that were coming from the U.S., already visible in the form of American type advertising, American products of all kinds, American movies, etc. I had the feeling that they would not understand what was happening to them (just like the American people) and would helplessly fall into the same trap. But perhaps I was hasty in my judgement. On talking with Tiomno, I began to feel that I can get some support here, perhaps a surprising amount. People are quite different here in some ways, because the same overwhelming pressure toward uniformity is not present (although as I said, it is foreshadowed). Also, it is not as dangerous to support a person in trouble here as it is in the U.S. Time will tell.

In the midst of all this trouble, I might as well tell you of one bright ray. We have just had a 30% salary increase, so that I now earn the magnificent sum of 19,000 cruzeiros monthly, 12 months a year, worth $1000 a month at the official exchange rate, and about $800 in purchasing power. What I shall do with all of this wealth I don't know (Now that I can't travel). So at the last moment, I may become a wealthy man, just before wealth loses all its meaning.

Well, I expect to go to Rio in about a week to give a long talk on the quantum theory. Rio is probably hot as hell these days. São Paulo alternates. Last night was an uncomfortably cold rain. Tonight is a cold clear windy night. A few days ago, it was very hot. It is said that in Rio, there are only two seasons, Spring + Summer, while in São Paulo, there is only one, which is a mixture of all seasons in rapid alternation.

As to how to solve all of these problems, I don't know, but I am going to begin to talk to various people here, to see what their attitude is. Right now I feel almost care-free in a sort of intoxicated way, but I wonder how I will feel when I wake up in the middle of the night.

I miss very much having someone to talk to about these things. People like Tiomno and Phil Smith are very nice, but one cannot relax completely with them, because there are certain regions where there is no mutual understanding. Incidentally, I am moving in to an apartment with Smith on Dec. 15. I used to jokingly tell him that it was risky for me to share an apartment with him because (a) He may be drafted and (b) He may be killed on his motorcycle. Now he says that the risk is even bigger for him, because I may be called back to the States, leaving him responsible for all the costs. So goes life.

Please let me hear from you soon. Regards to all.

Love
Dave

Letter 31. Folder C46, not dated.

[December, 1951].

Miramar Palace Hotel
Rio de Janeiro

Dear Melba

I am now in Rio, and as was to be expected, it is hot. The first thing that happened to me in Rio is that I caught an intestinal infection which left me tired, so that the trip is not too enjoyable. Rio itself is in an incomparable setting of bay, ocean, mountains and rocks, which is even more spectacular than San Francisco. It is particularly impressive from the air, from which you can see tall modern apartment buildings sitting at the bottoms of steep valleys, with the tops of the buildings almost in contact with the tops of the hills. Alongside the hills are homes, set in rather dense tropical vegetation. When ones sees the city at close quarters, however, it seems a bit shabby and dirty, compared with expectations aroused by the air view. There are many incredibly poor hovels only a short distance from the most magnificent apartment buildings. I am staying at a rather expensive hotel to which I was sent by the people at the Center. This hotel faces the Copacabana beach, which is in the shape of a half-moon ringed by mountains and very steep rocks (resembling the sugar loaf). It is very spectacular and wealthy, but most of the guests are American (Feynman stays here as a regular home). Most of the Americans are not particularly attractive specimens of our compatriots. On the whole, I don't like this location, despite its beauty, and I wish I had gone to a cheaper place where Brazilians stay. Only when I got here, I didn't know where to go, since it is my first time in Rio.

Mrs. Yevick wrote me about an idea of hers to get in touch with I. F. Stone, in connection with an article on the causal interpretation of the quantum theory. Stone is organizing a socialistically inclined periodical, the "Monthly Review" and Mrs. Yevick wondered whether one shouldn't suggest to him that he write an article on the causal interpretation (after he is briefed by a letter from me and by discussions with other people). I have told her to get in touch with you and to discuss the idea with you. My own idea at present is that it would perhaps be a good thing, but that there is danger that it would come to some Congressman's eye and get me called home. Without a passport, I am a sitting duck for such an attack. What do you think of the idea?

This loss of the passport makes me feel uncomfortable. It probably does not represent an immediate danger, but it certainly does not look good from the long-run point of view. Besides I feel sort of "hemmed - in" or "choked" because I am stuck here way off in Brazil, with no one with whom I can discuss things, etc. There is much danger that I will lose interest in the whole business. I can see that at this distance, there is no hope of convincing anyone of something new. My experience with Feynman and other people proves that nobody will read a paper carefully enough to convince himself that it is correct, if it is very far from the usual run of things. Also, there may be various wrong trends, blind alleys, etc. that could be corrected if I had enough people to talk with. And also, I find that much of my thinking was connected with talking. Very often, an idea will lie dormant until I talk about it and make it more definite. Partly because of language difficulties and partly because of a lack of people who understand, there is little opportunity for that here. I note that already much of the impetus for thinking of these problems has been reduced, and there is danger that if I have to stay here indefinitely, I may give it up because of lack of interest. Either I will have to overcome these difficulties or else solve the problems of where else to go. Neither of these alternatives is particularly easy. Ever since I lost the passport, I have been depressed and uneasy, particularly since I was counting very much on this trip to Europe as an antidote to all the problems that I have mentioned.

Please give my regards to all

Dave

Letter 32. Folder C46, not dated.

[December, 1951].

Sat. morning

Universidade de São Paulo
Faculdade de Filosofia, Ciências e Letras
Caixa Postal: 8105
São Paulo

Dear Melba

Thanks very much for your letter. It looks as if they didn't plan any immediate trouble in this passport business, as nothing more has happened yet. This business about a safe - deposit box for the passport would just not have worked. You have no idea how powerful the U.S. gov't is here. Although there is a lot more freedom here than in U.S., this is not due to the relative good intentions of the Brazilian gov't but rather to their comparative ineffectiveness. Thus, it is said that when the local police really want to know about anyone's political ideas here, they get in touch with the U.S. Consul; this is much easier than it would be to find out for themselves. The police are very ineffective, and for this reason, very few Communists, for example, are actually bothered, although Communism is against the law. But a prominent Communist (such as Catunda, who is a Math prof. here), is likely to be arrested about once every six months, beaten up a bit, and then released. The Brazilian police are notoriously brutal. A hint by the American embassy would make them very interested in any individual. Moreover, the University people here are almost as scared of a red label, as their counterparts in U.S., because it interferes with the supply of money. If half of the big shots in Brazilian science knew my background, they would be very unenthusiastic about me indeed. In fact, in situations like this, experience has shown that the only possible protection is to know a powerful, courageous, liberally - inclined person, high in the gov't. It is not out of the question, that I can ultimately establish such a connection, but it is unwise to do so now, because nothing has actually happened, and before he is called on to do something, he may have many months to think it over, discuss it with "wiser" heads, etc., and cool down quite a bit. So until we are ready to ask for something definite, it is better to say nothing to anyone. It will just scare people to hear about it.

The apartment is much nicer than the gloomy pension in which I lived. It overlooks the city from a hill - side. The rooms are moderately large, bright, and cheery. Unfurnished, the cost is about \$125 a month (purchasing power equivalent) for 2 bedrooms, bath, kitchen, living room + maids room. As Smith says, the maid's room contains just enough room for a short maid. We use it now for storage. We have a garrulous German woman, a vegetarian and a demon of industry, to clean 3 mornings a week. We cannot force her to take more than 30 cruzeiros a morning (about a dollar). In return she forces Smith to listen to lectures on vegetarianism, etc. She believes coffee and tea are very bad for us. Smith is hiding his liquor supply, for if she discovers it, she will surely refuse to work for us.

The address is Rua Brasilis Machado 380, Apt. 703, but for the time I suggest you use the address printed on this letterhead.

The furniture here is terribly expensive. To take care of kitchen + bedrooms, we have already spent about 12,000 cruzeiros, or about \$400, another \$300 is needed to finish the job.

You will be glad to know that Pauli admits the consistency of the causal interpretation of the quantum theory, but he still objects on philosophical grounds. He refuses to believe a theory which even permits us to conceive of a separation between the observer's thought processes (taking place in the brain) and the rest of the world. Also, the French physicists, Vigier, and Regnier write that they are getting marvellous generalizations of relativity and connections of relativity with qu. mechs, causally interpreted, but as yet, they have sent no details that I can understand. I have urgently requested such details and am waiting for their answer.

As for the British, I am not sure how helpful they are likely to be. I sent an article on quantum theory to Massey, suggesting that he sent it to Nature. $1\frac{1}{2}$ months have passed and no reply. I suspect that they are leery of such ideas until the ideas are accepted. Last year, I got a fellowship offer of £600 from Mott, provided that I was willing to work on the collective theory of electron gases. I suspect that he knew about my work on the quantum theory and didn't want any work on such notions going on in his university. This sort of thing makes me a bit angry. Various people, when I ask them about quantum theory, say they are interested in this collective theory (in a rather pointed way), saying that it has promise of various applications, never failing to mention nuclear physics. Every time I hear this sort of thing, I cannot avoid a reaction of loss of interest in this collective theory, because every time I hear the word "nuclear physics", it calls up to my mind an image of the most boring possible subject in the world. The surest way to discourage me from working on the quantum theory would be to continually remind me that it might be useful in nuclear physics.

Well, regards to all at 298 and on the farm

Dave

Letter 33. Folder C46, not dated.

[December, 1951].

Dear Melba

I am feeling much better now that I have recovered from my second illness, and life looks much better. I am eating a diet in which fried foods and fresh vegetables, milk and cheese are excluded (these are possible sources of infection) and supplementing it with a horrible tasting vitamin compound. I am even getting back to work. The only bad point is that the best theoretical man here, Tiomno, is going to Rio to join his wife, who is working there in the Physics dep't, teaching and doing research. I can't blame him and yet it puts a heavy load on me, not only in removing one of the few people with whom I can discuss physics, but also because he knows the ropes here. I need such a person, at least until my Portuguese is good, or else I can hardly deal with the administration and get the necessary things done. There are only 4 professors in the dep't, and of these, Schönberg is on leave for another year in Brussels. Thus I am the only professor in the theoretical group. We are trying to improve the courses, making them less high flown, and trying to get the students to do a few problems. This will be quite difficult, and Tiomno's help would have been a great aid. Also, he is the only person who can really operate at an original level in suggesting problems for graduate students, and in working closely with them. He doesn't leave till March, but to take his place, I'll have to develop in a number of directions and do a lot of work.

It turns out that I can hire an instructor, on a contract of from 1 to 2 years, at a salary of about $4000 a year (which will be raised to $5500 in a few months when everyone gets a 40% raise to cover the effect of the steady inflation that has been going on for 10 years). Now it also turns out that Peter Bergmann has a student, Ralph Schiller, who has read my Mss. on the quantum theory, and wants to come down here to work with me on these problems. Peter says that you can tell me about his undergraduate work at Brooklyn. If you think a great deal of him, I can probably have him hired, starting Sept. 1 1952. What do you think? Peter has a high opinion of him.

Please let me hear how things are going. Regards to all.

Love
Dave

P.S. Do you know of anyone else who would be better than Ralph Schiller? (Not that I know anything about Schiller, but I am just looking for information).

Chapter 17
Letters to Melba Phillips, 1952

Letter 34. Folder C48, dated: Jan 15 [1952]

Universidade de São Paulo
Faculdade de Filosofia, Ciências e Letras
Caixa Postal: 8105
São Paulo

Dear Melba

I've been waiting to answer your last letter until my ideas are more definite on certain subjects. First, the article that I sent you will be published in "Nature". Massey wrote me at last, saying that he had sent it in. Nevertheless, it might be worthwhile to publish something a little more popular and more extensive in Scientific American. Perhaps you can sound out the editors, using the articles I sent you as a base.

I have made much progress on plasma theory and on the causal interpretation of qu. mech. Interestingly enough, the two are now very closely related. I have become convinced that the time has come to reconsider the concept of an "ether" that fills all space. For in the causal interpretation of the quantum-mechanics, the negative energy electrons, protons, and neutrons are all supposed to be really there, forming a medium of incredible density. For this reason, the word "ether" is bad, as the medium is anything but "ethereal". A better word would be "substratum".

As a matter of fact, there never has been a proof of the non - existence of the "ether". The exp'ts leading to relativity merely showed that if the laws of relativity are strictly true, our velocity relative to the ether cannot be observed. But from this we cannot deduce that the ether does not exist. All we can deduce is that at a certain level, it should be possible to formulate the laws of physics abstractly without reference to the "ether". But since it is always conceivable that in some domain, the laws of relativity may fail, it is always possible that the ether not only exists, but is also observable, by methods not yet known to us (note the similarity to the problem of interpreting the quantum theory).

C. Talbot (ed.), *David Bohm: Causality and Chance, Letters to Three Women*,
DOI 10.1007/978-3-319-55492-1_17

The ideas I have been led to are as follows: Matter at our level is an inhomogeneity in the substratum (a stable and highly localized one). Light waves, etc., are other possible types of inhomogeneity. The "disappearance" of matter at our level is simply a transformation in the substratum, in which energy changes its form from localized to a spreading wave (and vice versa). Of course, the substratum particles are not indestructible either, but are structures in a sub-substratum, etc., and so on ad infinitum. But for the present, we can regard them as approximately permanent. When we look at space, it appears to be empty, because a uniform background does not scatter light, or electrons, protons, or planets, etc. A similar situation arises in the theory of metals, where a perfectly regular array of ions does not scatter an electron. Thus, a positivist might be tempted to conclude that a metal consists of empty space, if he had only evidence from the scattering of low energy electrons.

In studying plasma theory, I have found that each particle is surrounded by a cloud of charge of elliptical shape, which shortens in the direction of motion, as the particle approaches the velocity of waves in that medium. When the particle exceeds the speed of the waves, it throws off a "wake" (as in the Cerenkov effect) and loses energy rapidly, until it falls bellow the speed of the wave. Now, I can conceive of a substratum theory along these lines, in which a "particle" accelerated slowly (adiabatically) would shorten in the direction of motion according to the formula of Einstein, and would get heavier, because its associated "cloud" became more and more intense. Thus, in processes of this kind, the particle could not be accelerated faster than light, as it would get infinitely heavy. But if one of the substratum particles were given a violent acceleration that pulled it right out of its "cloud", it could go at theoretically unlimited speeds, except that, because of the "Cerenkov effect" it would throw off energy and slow down very rapidly until it fell below the speed of light. Let us imagine that processes at our level and at the atomic level are too slow for such phenomena to appear, but that at the level of "creation" of elementary particles, processes are fast enough for this to happen. We should then have a theory that is "relativistic" in every domain where the theory of relativity has been tested, but non-relativistic in those domains where the present theory is known to fail.

I should also add that it is possible to put a purely classical plasma theory into a form which is remarkably close to that of the quantum theory, as causally interpreted. I therefore think that it is not impossible that by means of suitable assumptions about the laws prevailing in the substratum, we might obtain both relativity and quantum-mechanics as approximations holding at atomic and higher levels. Also, there is some possibility of bringing in gravitation, but I prefer not to discuss that yet. Roughly speaking, the idea is that "matter" is (as in general relativity) a disturbance that is localized in a small region, but actually spreads out over all space, in a steadily weakening form (by the inverse square law). Thus, in a sense, everything that we call "matter" (i.e., the inhomogeneities in the substratum) are really in contact, and "interaction" of the central points is our approximate way of describing the effects of this contact. We might make a picture of a forest, where each tree seems to be separate, whereas in the soil underneath, the roots intertwine, so that the forest is in a sense a single thing. But the interconnections are (as in the substratum) at a level that is not "observable" by readily available methods. Thus, a naive observer could see one tree affecting another, and he could postulate an "interaction", inversely proportional to some power of the distance, because as the roots of a given tree spread out, they grow less dense.

Well, as you can see, I am enjoying myself, scientifically speaking. It's nice to be off by oneself for a while, away from people who prevent the free use of one's imagination. Actually, I find that the pieces are slowly fitting together, and each day, some aspect of the problem becomes clearer.

Regards to all,

Dave

Letter 35. Folder C47, not dated.

[Early 1952].

Universidade de São Paulo
Faculdade de Filosofia, Ciências e Letras
Caixa Postal: 8105
São Paulo

Dear Melba

I have just heard from Bernard that O. is going to be called before some committee again. Perhaps he'll be able to avoid it again this time - perhaps not. Let's hope for the best. But I am curious about the status of the immunity statutes. Have you heard anything about them yet, or other significant items of information? I am a little worried because no one in the States has written for over two weeks. You might try writing your next letter for a change to Brasilio Machado 380, Apto. 703, and see of this helps speed up the delivery. I am also curious to know how people react to my paper. (This should be published in Jan Phys. Rev.) I have a few hints so far. Von Neumann thinks my work correct, and even "elegant", but he expects difficulties in extending it to spin. The older Bohr didn't say much to Art Wightman but told him he thought it "very foolish". The younger Bohr thinks it logically consistent, but "something like an absolute frame in relativity", but thinks if "I could find some modification significant in domain of 10^{-13} cm, it would be wonderful". De Broglie is now fairly friendly to me, saying in a letter that I have carried the pilot wave theory much further than he did in 1927. You know of course, what Pauli's reactions are. Rosenfeld thinks the theory very "ingenious", but "basically wrong". However, he seems, according to my informant, to be remarkably vague on just what is wrong with it.

I must say that events worry me, not only with regard to my own special problems, but because it seems hard to see how war can be avoided. Of course, my information here is bad, because the papers are atrociously prejudiced, and I hope that my present very depressed feelings are unfounded. Sometimes I feel as if "the game is up" for everyone living in this generation, and this largely because of atomic power, which seemed such a bright dream when I was a child. Now I have dreams of new levels of reality more fundamental than the nuclear particles, which may help lead to concepts and ways of thinking that will compensate somewhat for the damage done by the atomic bomb. But this is so far in the future that it no longer seems so significant to me, now that the threat of catastrophe is really here. I have a great confidence in the ultimate future, but no confidence whatever in the next 20 years. This will be a disheartening time, when the decay prepared during the last 50 years will finally manifest itself openly, in the form of everything going to pieces. Personal relations, careers, children, or whatever else people live for, will become meaningless in a disintegrating society. It seems hard now to hold on to the idea that a concept of things not yet known, such as I am working on, is worth pursuing, but I suppose that it is.

I hope I haven't bored you with all this depressing talk. I'll probably get out of this mood soon. By the way, did you ever give that talk to your students on causality in the quantum theory, and if so, how did it go? With regard to the talk on the crisis in physics, I now feel that it would be very difficult to present a good case, without a really exhaustive and rather frank analysis, connecting with all sorts of factors, political, social, etc. What I could do would be to present a much more limited case, in favour of a "realistic" point of view in science, as the most fruitful in the long run. But even this could be strongly attacked, because as Pauli said "I am cashing in on future dividends" when I maintain that this method is likely to be helpful in the domain of 10^{-13} cm. The whole thesis is right, but positivist confusion has spread so far in our science that it is remarkably difficult to carry the argument to that large majority, which without realizing it, has come to regard the positivist point of view inevitable. I can make an impression on a person after 3 or 4 hours of discussion, but a short paper would have little effect. There are an infinite number of twists to the positivist argument, and each person must be reached through the peculiar twist that he has absorbed.

I don't like to repeat it, but I wish I could shake off this awful feeling of foreboding and depression. I am hoping for some letters from old friends in the States, which are really more helpful than you would think, not only because it is good to hear from them, but also because it gives me an idea of what is going on behind the curtain of the "free press".

Incidentally, I have a job offer from Mott in England, and could probably obtain offers in other countries if I really wanted them. The passport situation doesn't look so good, however. I am looking into methods of getting the passport back, by being invited to England, but if they spread O. over the front pages, I'll be lucky if I am not called back myself. I am praying that my unimportance + inconspicuousness will come to my rescue. To feel unimportant & ineffective is an annoying thing, but it sometimes has its advantages (I hope, this is one of the times).

Coming back to the ubiquitous positivist point of view, I find that Brazil is full of positivists. Relativity is the culprit here, as no one here understands quantum theory well enough to be confused by it.[1] People have the idea that by reducing everything to a study of the role of the observer, Einstein got the whole theory of relativity, and that this is the best way to proceed in science from now on. This also ties in with the disapproving attitude to manual labour characteristic of Latin countries; for people feel that math equations deduced by an analysis of the role of the observer, are the basis of physics. The flexibility of positivism is amazing, for among exp'tal physicists in U.S., there is a belief that physics flows solely from empirically observed data, or "operations", which is also combined with a belief that theorists take these numbers, and with the aid of a few geniuses like Dirac, produce equations that fit these numbers.

Give my regards to all

Love
Dave

Letter 36. Folder C46, not dated.

[April, 1952].

Tuesday

Dear Melba

Well - all hell is ready to break loose. The two dep't stinkers (Prof. Marcelo Damy de Souza Santos and Prof. Stammreich) have joined forces with the local Nazi group. I am dismissing my "assistant" whose father organized the Institute of Theoretical Physics. V. Wiseacre [Bohm's name for von Weizsäcker – CT] has discovered my troubles in the States, and knows I don't have a passport. Now, the skunks are trying to blackmail me, saying that if I don't take Schiller, but take two German boys instead, all will be well. Otherwise, they will ask for army intervention. Everyone here says it is a bluff, and we are going to call it. Meanwhile, a long and bitter fight is in the offing. It will be a fight to the finish, and if we lose, science in Brazil is done for. The rest of the dep't and many people in Rio realize it, and are getting ready to stake their careers on this fight. Also, the director of the Faculdade + a powerful faction are on my side, as they hate the 2 dep't stinkers and the Nazis. So we'll see what happens.

I would like to find some way to let the world know what a skunk von Weissacre is, or at least, let all physicists know. If you have ideas, let me know. We are only 90% sure of v. W's role in this, not 100%, but we'll soon be sure. Perhaps Phil Morrison has some ideas. In any case, you should let Phil know what is happening. I have already let Einstein know, by way of Shenstone. Who else should hear of it? It might make a good story "Nazis taking over Brazilian Physics".

[1]For example, Tiomno expressed surprise when I pointed out to him that q. theory as usually interpreted, implies a subjective point of view, since all is finally referred to the consciousness of an observer. However, he had often wondered how q. t. treated this problem and readily agreed that the ultimate implications of usual point of view are nonsensical.

Schiller's case comes up in the Congregação of the Faculdade Saturday. This is a crucial test, but we can probably win it. But this will be hardly the beginning. Our only solution is to discredit the 2 stinkers (Everybody knows that they are incompetent, and that at least one steals dep't money) and to discredit the Nazis. Either they will be smashed, or us. There is no choice. Meanwhile, they go around saying that "v. Weissacre is a genius" every time he says "$F = ma$", and they are clearly getting ready to say that the causal interpretation of the quantum theory is nonsense. I have asked Einstein for some statements on this, which we can publish, if necessary. I will have to refuse to hire the German boys on with a public statement that I don't think they are suitable. Then the 2 stinkers will bring in their great Nazi genius, v. W. (or Heisenberg), and ask how I can disagree with him. I will just have to say "He is a competent physicist, but no genius, and that is that". Well, here we go. Where we will end up, who can tell?

Please, let me know what is happening at your end. I hear that you are under attack too. I hope it is nothing serious.

Regards to all

Love
Dave

Letter 37. Folder C47, not dated.

[April, 1952].

Universidade de São Paulo
Faculdade de Filosofia, Ciências e Letras

Dear Melba

If I haven't been writing much to you lately, it's because I've been very busy. First, there is the chaotically complicated situation developing here. There are two jokers, Prof. Souza Santos, and Prof. Stammreich, who are throwing monkey wrenches into everything that they can. Two years ago, they nearly wrecked the physics dep't, and they were put in their places by the Director of the Faculdade, but now they are up to their old tricks again. Souza Santos is almost on the verge of paranoia, suspecting everyone of being against him, and as a result opposes almost every move that anyone makes, seeing in it only a move against him. Stammreich is a professional pessimist, believing that nothing can ever be done in Brasil. Nevertheless, there is some pretty good evidence that he made a good thing out of his job, for he arrived in Brasil penniless (excuse my tendency to Portuguese spelling) and is now quite rich. But nobody has ever been able to really prove anything against him. In any case dep't meetings with these two characters are a big waste of time, as nothing can be accomplished. I am in a tough spot, as the dep't has only 4 professors, and one is on leave, leaving behind an assistant, who is on my side.

A typical sort of trick engaged in by these characters is to get various functionaries to lose contracts and important documents, thus delaying things for many months.

The number of mislaid contracts is really impressive. That is why I'll be sure of Schiller's contract only when I see it.

But these difficulties pale into insignificance compared with those that seem to be looming on the horizon. Recently, an assistant of mine, whom I inherited, who is very wealthy and has Fascist leanings, succeeded in getting about $200,000 from the state gov't and $800,000 from private sources to finance an Institute for Theoretical Physics. This fellow's military connections seem to have been responsible for his success in raising money. To give you an idea of his history, when he was a student here, he was able to go to the gov't, and ask the army to take over the physics dep't here, because "it had so many communists in it". He created a serious threat that was stopped only by the action of the professors. Anyway, now they have gotten hold of von Weissacker, (who is here with a few German students right now) to head the Institute, alternating every 3 months with Heisenberg. The two will commute apparently from here to Germany or Chicago, or wherever it is that they now are. The physics dep't generally resents them, but they have so much money, political influence and big names that they will inevitably exert much power in physics here. And now I have to decide whether or not to fire my assistant, since he is not supposed to have two jobs at once. As yet, we are on overtly friendly terms, but the tension in our relations is quite obvious.

In Rio, meanwhile, your friend, the Admiral Alvaro Roberto has come back from the U.S. with plans to build a pile. This will put the military in control of a very large section of Brasilian physics research.

On the top of all this, I am waiting until one of these characters discovers my history in the States. Tiomno warned me against talking about it, as there are too many people around here who would seize on it to cause trouble. But if I get into the inevitable fight that is brewing on two fronts at once, it is quite possible that the whole situation will become really ugly.

Meanwhile, I am working on many things. The ether theory is quite a bit more developed, but I have set it aside for a while to write a paper showing that if, in the causal interpretation of quantum theory, you start with an arbitrary probability distribution, $P(X)$, you will ultimately get $P(X) = |\psi(X)|^2$ as a result of chaotic collisions with atoms, molecules, etc., undergoing random thermal motion. Thus, quantum probabilities have the same origin as those of classical statistical mechanics. This will nail down the last point in the theory.

I have received reports of a seminar given by Pauli + de Broglie, in which they make some rather childish objections to my theory, calling it "metaphysical". I hope they publish this nonsense, because if they do, I can really do a good job of tearing them to pieces, I can hardly wait!

I heard from Schiller that Phil Morrison agrees with my philosophy, but prefers the Bohr interpretation, "because it is simpler". This inconsistency, (which Schiller also remarked on) amazes me, because Phil ought to know that Bohr's interpretation is "simpler" only because it does not attempt to explain what is happening at the quantum level. In any case, it is not really "simpler" as I can vouch for after teaching it for several years. In practice, it is an exceptionally complicated and difficult point of view, compared with that of the causal interpretation. This type of inconsistency in

Phil disturbs me. He should be helping me, instead of raising irrelevant obstacles. Or at worst, he should do nothing. But to aid in the creation of confusion is something he ought not to do. Sometimes I have a feeling (knowing him only a little, however) that he does this sort of thing to create a greater feeling of friendliness with his colleagues. Of course, I realize that he is making serious enemies in the political field, and I admire his courage in this regard, but from my observations of his speech, mannerisms, etc., I often got the idea that he tries to compensate for this a little in the scientific field, by being very moderate in all of his opinions, and tending to agree with the majority. I have found that such impressions are often fairly accurate, because quite often after speaking with a person, I have a vague feeling of distrust - that something is not quite what it seems to be - and subsequent experience shows that it is justified. Some of your remarks about his having swallowed "Oppenheimeristic" ideas of a fellowship of scientists who would be responsible for the wise use of atomic power, would tend to confirm these impressions. In any case, I find it difficult to believe that a person of Phil's calibre can honestly be that confused.

I have received an encouraging number of favourable comments on my papers (about 15, at least) as well as 30 requests for reprints. However, since both Pauli + Phil Morrison make the argument that my interpretation is more "complicated" than it needs to be, I'd like to summarize my objections to their arguments.

The crux of the problem is in the statement of the usual interpretation that two systems having the same wave function are "physically identical". This follows because the w. f. is said to define all physically significant properties. Thus two uranium atoms in the same quantum state are said to be "physically identical" even though one of them may explode tomorrow, and the other in two billion years. But these atoms might be each connected to an atom bomb in such a way that the bomb would go off when the uranium atom disintegrated. The usual interpretation then asserts that the two atom bombs are "physically identical".

How does the usual interpretation then explain the obvious fact that they behave differently. Different physicists adopt one of 2 possible points of view:

(a) They can assume that the relation of past and future is arbitrary. Then two "physically identical" things need not behave in the same way. But in my point of view, the two uranium atoms are not "physically identical" because each of them has a particle in it in a position that will determine when it will disintegrate. Thus, in my point of view, I make a hypothesis, namely that there exists an as yet unobserved particle, but with this hypothesis, I can unite things that were previously arbitrary. But this is one of the criteria for a good hypothesis; viz., "Does it relate things that were previously unrelated?" The usual interpretation may be "simpler" in the sense that it makes less hypotheses, but it achieves "simplicity" only by leaving a large number of things arbitrary. We could always achieve simplicity in this trivial way, but only by giving up one of the most fundamental methods of making progress in science.

(b) One can take Bohr's point of view that if two systems with the same w.f.'s are "physically identical", then the obvious difference between them must be associated with something "unphysical" (i.e., not subject to physical investigation). This difference is what Bohr has called the "identity" of an event, a property not subject to physical investigation, because in the usual interpretation, only statistical properties can be studied in physics. The property of "identity" is therefore intrinsically undefinable - it is connected with real physical events, but it is itself forever beyond the domain of physics.

To this point of view, one can raise the objection that it is certainly the purest possible form of metaphysics. A person who holds it cannot with consistency maintain that he objects to the causal interpretation because it is "metaphysical". For in the causal interpretation, the hidden variables are at least observable in principle, whereas Bohr's "individuality" is unobservable as a matter of principle.

I would like to suggest criteria for theories to replace the positivist criterion that only those elements that are already in principle observable can have physical significance. Instead I would suggest the following:

(1) In order for a hypothesis to be admissible, it is necessary that there be no inconsistency in supposing that every element appearing in it has been observed. However, the hypothesis itself does not have to include within it a method that permits the observation of every element. Thus, it is permissible to suppose that some of the elements may later be observed with the aid of interactions or processes that are not yet known. This is exactly what was done with the atomic hypothesis, and it is what is done in the causal interpretation of q. t. To refuse to admit such hypothesis is to cripple the imagination, when the time comes to investigate new fields, and to confine ourselves to the existing domain of concepts. (It is clear that Bohr's point of view does not satisfy this criterion, for Bohr is compelled to admit the existence of systems, having an "individuality" whose observation would be inconsistent.)

(2) Now, in order to prevent the multiplication of hypotheses which might satisfy (1), we also require that every hypothesis whose elements are not yet observable be required to unite facts or domains of facts which had hitherto been arbitrary. (This, as I have shown, the causal interpretation does.) Experience shows that such a hypothesis has a high probability of being correct, and therefore of being fruitful.

The application of these two criteria is almost instinctive for most people who have not been confused by a positivist philosophy, but the positivists have created so much confusion that all sorts of nonsense can get by. The principal job in science now is to remove this confusion.

Well, I guess that's all I have time for now. Regards to all.

Love
Dave

P.S. If you see Phil could you please discuss these points with him. As for the cosmic rays job, I got answers from a number of people, but the only one who wants to come now is John Tinlot of Rochester. Do you know anything about him?

Letter 38. Folder C47, dated: June 28, 1952.

Dear Melba,

I must have failed to get one of your letters (at least one, at any rate). But I was glad to hear that all is well in N.Y. Things here are still quiet. The Institute got started at an official ceremony full of generals, colonels, etc. but it isn't certain that they can keep von Weissacker, while Heisenberg probably won't come. As for me, I am getting tireder & tireder of Brazil, but maybe things will brighten after Schiller gets here. I really have no one to talk to at all. The students are so young that they seem like children, while most of the older people just don't know anything. Smith is OK but limited. I was going to go on a vacation in the mountains next week with Andrea Wataghin (the son of Gleb Wataghin), but he is too busy to go. So I shall go instead with a German immigrant – a former member of the Wehrmacht. So you see how desperate I am!

As for the article, let's give it up as a hopeless case. I did write Rosenfeld but all he said was that there isn't any controversy because there is nothing to argue about. He is writing an article for a de Broglie commemorative journal in which he seems to argue that as long as my point of view leads to the same results as complementarity, the whole idea of mine is just silly. He seems to have missed the point completely that the similarity of the two points of view is only in a limited domain, and that my point of view may lead to something very different in the domain of 10^{-13} cm. When the atomic theory was first proposed, positivists like Mach argued that after all it led to the same results as the macroscopic theories (such as the perfect gas laws) so that the extra assumption of atoms was just a pretty little piece of imaginative fiction. Mach would have said that new results of the atomic theory, not yet observed, did not have any meaning, particularly since no one knew just what to predict at that early stage. I suspect that like most complementarians Rosenfeld is a confirmed positivist - so deeply confirmed that like M. Jourdain, who spoke prose without knowing it, they speak positivism without knowing it (and call it dialectics).

I have come to the conclusion that barring a lucky accidental discovery, it will take at least 20 years before theoretical physicists will alter their general point of view very much (This is the time needed to replace the present generation of physicists (especially the big shots) by a new generation). So what's the hurry? I may as well take it easy and try to enjoy life. But this cannot be done by me in Brazil. However, there is a small chance that I may get my passport back, in which case I'll take a trip to England in November. I have asked my lawyer to apply on the basis of the argument that Massey + Mott have invited me to give a few talks.

I have just finished a new article that removes one source of confusion. It proves that as a result of random collisions, an arbitrary probability density $P(X)$ approaches $|\psi(X)|^2$ with the passage of time. One of the boys (Keller), at NYU in Courant's group has just written an article for the Phys Rev (reviewed by me) in which he asserts that behind my theory is an implicit assumption that a "deeper" kind of probability is needed in my point of view than the usual kind (so that as a result, I didn't really succeed in establishing a new interpretation of qu. theory, he says). For if P must equal $|\psi|^2$, it cannot take on an arbitrary value. He said this, despite the fact that I pointed out several times in my articles that a reasonable conjecture was that with collisions, an arbitrary P approaches $|\psi|^2$ with time. I hope he will get the point at last. I sent him a copy of my paper. (I am trying to have the paper mimeographed, but with typical Brazilian confusion, this will take 2 months, plus a considerable amount of effort). But anyway, people seem to be able to find every possible source of confusion in the theory, and even some impossible one's, like Phil Morrison agreeing with the philosophy, but saying that complementarity is "simpler" because you don't have to make so many hypotheses.

I can't help becoming discouraged, because if I suggest some new idea, it will surely be misunderstood, not only because people are so mixed up, but also because they want to protect themselves from understanding something that could require them to revamp their ideas. I know this "protective confusion" because I used to use it almost consciously to avoid embarrassing arguments (especially political arguments with reactionaries). At this distance, it's very difficult to do anything about it. By talking with people, I can clear it up. For example, Feynman said that he first thought my ideas were crazy, then when I explained them he saw they were at least consistent and later with this proof that $P \rightarrow |\psi|^2$, he even began to prefer this point of view. But I am afraid that Feynman's readiness to consider new ideas is very unusual. (Did you know that he just got married, after returning to the U.S.?)

I have some new ideas on the theory of probability now. I am trying to reduce the subjective element in classical statistics by means of developing the concept of chaotic fluctuations which follow from the causal laws. I am writing up an informal summary, and in the unlikely event that I live long enough to get the article past the endless series of hurdles needed to get it mimeographed, I'll send you a copy. These new ideas on probability fit in integrally into the further development of the causal interpretation of qu. th. & into the ether theory (which is now in the background for a while). If I can succeed in my general plan, physics can be put back on a basis much nearer to common sense than it has been for a long time, and therefore much easier for everyone to appreciate (at least in its over-all aspects). But the present generation of bright young physicists will resist such changes to the bitter end, so I have little confidence that such ideas will be taken seriously for a long time. It's a little hard, but I have to get used to working with the new time scale of 20 years or more, now that I have seen what life is really like in the dep't of initiating (or fomenting as we would say in Portuguese) new theories.

I hear that you have been honored by getting on to various lists of citizens of doubtful character. I hope that your job is still safe. I suppose that you could always retire to the farm and raise vegetables + wild flowers instead of physicists. Perhaps the results would be more satisfactory in the long run.

Give my regards to all the others. Perhaps some day we'll get together again.

Dave

Letter 39. Folder C48, not dated.

[July, 1952].

Universidade de São Paulo
Faculdade de Filosofia, Ciências e Letras
Caixa Postal: 8105
São Paulo

Dear Melba

Thanks for your letter. It looks as if the arrangements for Schiller are complete, but you can never trust the people here to do what they say they are going to do. In any case, if all goes according to plan, Schiller should be here before the end of the month.

I had a pleasant vacation in the mountains, and after that we had a Congress in Physics that was a bit dull. 8 physicists from the States (including Wigner, Rabi, Herb [Anderson], [Donald] Kerst, and others), 10 from Mexico, Argentina, and Bolivia, and a few from Europe, were brought here by the Unesco and by the Brazilian National Research Council. Most of the meeting was in Rio, where it is even now quite hot. As usual when I visit Rio, I got sick, and this with the heat & the insane traffic made life quite unpleasant. I gave a talk on my hidden variables, but ran into much opposition, especially from Rabi. Most of it made no real sense, but it boiled down to this: "As yet, your theory is just based on hopes, so why bother us with it until it produces results. The hidden variables are at present analogous to 'angels' which people introduced in the Middle Ages to explain things". People have a peculiar double standard. When someone tries to make a meson theory or a non-local theory and says he hopes for results in the future, even though it is not even self - consistent today, everyone says - fine. But when I present a new point of view which has hopes of solving the problem people say "We are interested only in results".

I am still somewhat discouraged about getting anywhere with the hidden variables, since it is a difficult problem, and since there is little stimulation or help to be gotten from others. But things should improve somewhat when Schiller gets down here.

I am also discouraged about Brasilian physics. I think that any serious efforts made to help in this direction are just being poured down the drain. The set up here is just plain crazy. Part of the insanity stems from the fact that physics here has no solid base - it is being transplanted en-masse from other countries. Industry has not developed enough yet to need much physics. As a result, physics is at the mercy of various asses who have gotten influence and control of the money, like the admiral Alvaro Alberto, and like Professor Marcello Damy de Souza Santos of São Paulo. There is so little consciousness of the nature of physics, however, that people like these can easily fool everybody, pose as geniuses among non - physicists, (there are only about 6 or 7 physicists in the country) and have their way. The latest insanity of the admiral is his plan to import a 450 MeV synchro-cyclotron, plus the people to build it and run it. It is told to me that Lattes arranged this as a "compromise", since originally the admiral planned to build an "atomic city" like Oak Ridge or Los Alamos. (He still plans to do it.)

It is really impossible to describe the poor education and the confusion shown by most of the physics students here. Part of it is rooted in the generally bad situation in Brasilian education, part of it in the contempt endemic in the culture for manual labor, but a large part of it could be remedied in São Paulo if we could get rid of this paranoically insane Marcello Souza Santos, who controls the physics teaching in the first two years, and not only does not teach the students anything, but ruins them by giving them the impression that physics is done only by "geniuses" like him. Incidentally, I have never seen such a contemptible individual as this man. He presents a good impression at first, but in a few months, one gradually becomes aware of what he really is.

The general picture of the organization of physics is very bad, both here and in Rio. I have little confidence that anything will come out of it, except disappointed and soured individuals who have tried to do some physics. In general, I feel that I ought to try to get out within a year or so, perhaps taking advantage of the time to do some work with Schiller.

As for the manuscript, it has been typed and mimeographed finally. I sent you a copy, and it will be interesting to see whether it arrives, because with printed matter, the post office employees have a habit of robbing the stamps and throwing the mail away.

Well, regards to all. Let me hear from you soon.

Love
Dave

Letter 40. Folder C46, not dated.

[Fall, 1952].

Universidade de São Paulo
Faculdade de Filosofia, Ciências e Letras
Caixa Postal: 8105
São Paulo

Dear Melba

I was very sorry to hear about your troubles - as yet no word of them has reached here in the press or otherwise. I hope that you can weather the storm and retain your job, but apparently you don't believe that this is likely. I am wondering what you will do if the "sword falls". Do you have any idea? I know that the situation must be very unpleasant for you, but I also know what kind of a person you are, so have complete confidence that you will come through O.K. in some way or other. My strongest hope is that at least they don't bother you with trials & things like that. It's one thing to lose your job, and another to face that sort of thing. Incidentally, could you let me have some idea of the source of the trouble? If you would rather not, then of course, just say nothing. In any case, let me welcome you to the fraternity. I shall always remember a very true statement of yours to the effect that the earlier the blow falls, the more fortunate you are, since the first blows are just the fore - warnings of the storm, while the later ones are really murderous.

Incidentally, if you ever need money, I have quite a bit. Of course, it's in cruzeiros, but it could be changed to dollars in a pinch. Perhaps you could come to Brazil. Have you ever considered that possibility? I don't know whether something could be arranged, but it could be looked into if you are interested. Whether it's worth while for you depends on your relative devotion to physics or to accomplishing other things in a place where as they say, you have "roots". In Brazil, little can be done outside of physics by a foreigner, especially by an American.

You once asked about snakes in São Paulo. There are quite a few of them here - so many that it's dangerous to walk in the woods (more accurately, the jungle). But I spent my vacation at Campos de Jardão, where there are relatively few snakes, because of the altitude of 6000 ft.

As for my work with Schiller, it looks promising. I wouldn't be surprised if something resembling Einstein's dream of deducing qu. mechs from general relativity could eventually be achieved. Einstein's general orientation appears to be correct, but his weak point was in not paying enough attention to the specific clues coming from a careful study of quantum theory, especially the Dirac equation. I have vowed to understand the Dirac equation, even if it kills me to do it. As I told Phil Smith, the day that we defeat the Dirac equation, we are going to have a special victory party, with a case of champagne.

In general, I admire Einstein more and more as time goes on for the general correctness of his line of thinking, and for his tenacity in sticking to it. The only place he goes wrong is in his lack of a dialectical orientation. But in general, his ideas are basically clear, simple, and well-founded. Eventually they will probably be superseded by the substratum point of view, but much progress can and must be made, before we are ready to seriously study the substratum.

I want to say that I still miss you very often, and hope that we will eventually come to see each other again. I also miss the house at 298 W 11th St. That is really quite a family that you've got there. I heard from the Schillers that Charley has fixed up a fancy new amplifier system. Please give my love to all of them. How is Paul doing?

As for this business of a trip to England, it looks unlikely, according to my lawyer, who is handling things in Washington. But it may come though. If so, it would be wonderful. I am very much looking forward to such a trip, even though I really don't expect to go. Incidentally, I hear that the causal interp. of qu. theory is being extensively discussed in England + France.

Once again, I send you my best wishes for a satisfactory finish to this encounter with the Senator.

Love
Dave

P.S. I would be interested at least in seeing clippings about your encounter.

Letter 41. Folder C46, not dated.

[December, 1952].

Universidade de São Paulo
Faculdade de Filosofia, Ciências e Letras
Caixa Postal: 8105
São Paulo

Dear Melba

I was glad to hear that all is still (relatively) well with you, and that you are still holding your job. Please let me know how things are going with you now. The Schiller's are also anxious to know.

The Eisenhower victory, with such a smashing majority, was quite a blow. It means trouble of a higher order of intensity. Everyone here was in favour of Stevenson, even the conservatives. Eisenhower's victory has provoked some anti - U.S. reaction, but in general, most people in Brasil do not realize what it really means. This is probably so in other countries too.

There has recently been much trouble in the Unesco. The Mexican and Brazilian delegates resigned, because the U.S. and Britain have been trying to prevent the Unesco from expanding and playing a real role. Instead, they want to keep it under their control as a pretty plaything which exhibits "Western" generosity to sub - developed countries, etc. In addition, the Mexican delegate has been under American attack because he is a "neutralist" and therefore does not actually support the "cold" war.

All of this has significance for us, as it is no longer certain whether contracts, such as we have in mind for a professor of Cosmic Rays, will continue to be renewed. Things are further complicated by the fact that Tinlot, who is officially indicated for this job by the Unesco at present, just told us that he can only get leave from Rochester for a year. This is not enough to get anything done. Therefore, we would like to change to Kurt Sitte who can stay longer. But now, in the present confusion, it isn't clear whether we can do this, as we have been working through the Brazilian delegate who just resigned. Also, we hesitate to ask Kurt to come down here for several years under conditions in which he gets a contract year by year, subject to continual renewal. In the past, such contracts have been renewed, but the future looks doubtful. So we don't know quite what to do.

I just went on a vacation in the mountains again for a week. It was very restful, as the mountains of Brasil are really peaceful (so unlike São Paulo). They give the impression of being timeless - i.e., nothing seems to have happened there for millions of years.

Progress on research in the quantum theory is at present a bit slow. In general, I tend to feel a bit discouraged these days, because there has been very little reaction to my papers. Many people have asked for reprints, but the few communications I received showed (except of course, for those from Peter and a few others) a general lack of understanding of the nature of the problem. Reports that I get from people who travel in Europe say that the reaction there is "So what". In U.S., we have a similar reaction.

The situation in Brazil doesn't look too good. The gov't faces a severe economic crisis due to lack of dollars, and is doing nothing about it, except to sell out to the U.S. Even the National Research Council here has done this, as there is now an employee of the U.S. Atomic Energy Commission who goes around inspecting all laboratories to which this Council has given money.

The situation in the University here is basically bad. There are a few potentially good students, but their basic education is very inadequate. The University runs by personal politics, and we'll never change that. One of the two jokers here, Marcello

Damy de Souza Santos, is thoroughly entrenched, because he grew up with the University, and has the reputation of being an internationally famous genius, even though he is really utterly incompetent. But to get such a reputation here, you need only be a Brazilian physicist. The other character, Stammreich, whom people generally characterize by the word "Nauseous", we may be able to hope to get rid of, as he may soon have to take the competitive examination for his chair, and he knows nothing but spectroscopy, so he might fail. But I don't like the situation here at all. It isn't just the physics dep't that is bad, but the same basically rotten system extends throughout all departments.

The Schillers are gradually getting used to things here. Bunny [Berenice] was very sick for a while, because of difficulties with food, but she is recovering and has begun to work on a thesis problem. Ralph and I are working on the qu. mechs., but things look difficult right now.

Otherwise I have nothing much to say. Please let me know how things are with [missing]. Regards to all (and have a good Christmas!).

Love
Dave

Chapter 18
Letters to Melba Phillips, 1953-4

Letter 42. Folder C47, not dated.

[April/May, 1953].

Universidade de São Paulo
Faculdade de Filosofia, Ciências e Letras
Caixa Postal: 8105
São Paulo

Dear Melba

The reason for my delay in answering you is largely that I have been busy, and sick part of the time. So please accept my apologies.

I am making some progress on the quantum theory, having found a relation between the Pauli spin theory (causally interpreted) and Dirac's new electrodynamics, which introduces a theory highly reminiscent of the old ether theories. The spin appears definitely to be related to vortex motion in the "ether". I am now trying to extend this theory to cover the Dirac equation. If this could be done, then it would be an important step, (as well as a victory in my life - long struggle to understand the Dirac equation). I'll keep you posted about any further progress.

I am working on another line, which can perhaps be called "The Causal Interpretation of the Theory of Probability". As you may know, the theory of probability is a sort of a mess, especially in regard to defining probability and explaining why probability is generally close to the relative frequency of real objects or events (in certain types of problems, of course). To attack this problem, I have defined the concept of "chaos", which is qualitatively speaking, a lack of order or regularity in the objects or events under consideration. A chaotic distribution of objects or events generally results from an unstable process, such as collisions of molecules, in which the final orbit is very sensitively dependent on the initial one. Thus chaos comes from the very form of the causal laws. Only where chaos exists can we correctly use the

C. Talbot (ed.), *David Bohm: Causality and Chance, Letters to Three Women*,
DOI 10.1007/978-3-319-55492-1_18

theory of probability. Thus, probability reflects, not incompleteness of information, but instead, chaos in motion or process. Of course, when a process is chaotic, it will in practice be very hard to get precise information about it. Thus, in practice, our information will be incomplete in precisely those cases in which the theory of probability applies. But in principle, chaos and complete information could exist side by side. Thus, in principle, if we knew the initial positions + velocities of all molecules of a gas precisely we could predict the precise future behaviour of the system perfectly. But the precise future behaviour would be chaotic; and as a result, the relative frequency with which molecules entered any given region ΔV would be in general nearly equal to $\Delta V/V$, where V is the total volume. In other words, the relative frequency would be close to the probability, whether we had precise information or not.

With the aid of the concept of chaos, I can throw new light on the problem of irreversibility. Thus, I can show that causal changes in a system can be made only in the direction of increasingly chaotic motion of its parts (which is also the direction of increasing entropy). Since the future is by definition that which we can in principle change by action in the present, we see that the future must also be the direction of increasing entropy.

I am becoming very discouraged about Brazil. I definitely do not like the place. It is in a hopeless mess in every way. The latest mess is a rapidly developing economic crisis, brought on in part by a scarcity of foreign exchange, but mostly by sheer mismanagement and corruption, carried out to a degree that even an American finds difficult to imagine. It is for this reason that I am in Rio now. I am trying to see if we can get Sitte's research budget paid. Because of the recent devaluation of the cruzeiro from 18.6 to the dollar to 45 to the dollar (which incidentally turns my former opulence into a very modest income indeed) it may be very difficult to get enough money for Sitte to work here. It is this crisis which makes Bonner's coming difficult too, as the fantastic salary only holds if it is paid in dollars at the old rate. This, the Bank of Brazil has, as yet, refused to do. This is the principal problem connected with Bonner's coming here. If he would take his salary in cruzeiros, it would be easy, but the salary wouldn't amount to very much. Will you please tell him about this.

The way things are going, I'd even consider a return to the US, if there were any possibility of some kind of employment in the NY area (presumably very unlikely). Living in Brazil is like walking on a mud - flat, the top of which has hardened enough to be stepped on, while the bottom is still liquid. You can only pray that it won't rain. In addition, I find it hard to get to know people here, so that life is pretty dull. São Paulo is an intrinsically bad city to live in, with poor public transportation, horrible traffic problems, failure of light + water, lack of parks, and nothing for anybody to do in his spare time. Besides it is dirty, and difficult to obtain good food. I am beginning to get very tired of it. In physics, the situation is a hopeless mess for many reasons. There are very few students, and the support for physics may disappear overnight (as already seems to have begun to happen).

I am glad to hear that you have began to do some writing. Please let me hear how it goes. I am also going to write a book on Kinetic Theory + Atomic Physics for 3rd or 4th year level; and after it has made more progress, I shall get in touch with some publishers.

Please give my regards to all

Love
Dave

PS What you say about the superstition that "there is no future in science" is very cogent. In fact, the only justification that I can see for my remaining in Brazil is to have a chance to try to demonstrate the falsity of this superstition, partly by my work, and partly in terms of a book which is gradually taking shape. Thus far I have given a series of lectures on the concept of causality, including a history of this concept, leading up to modern problems, showing how modern "superstitions" lead people to believe that there is no causality, and suggesting new points of view in terms of an infinite number of levels of reality. This is now being mimeographed (in English) and I'll send you a copy when it is ready.

Letter 43. Folder C48, dated: Oct 13, 1953.

Universidade de São Paulo
Faculdade de Filosofia, Ciências e Letras
Caixa Postal: 8105
São Paulo

Dear Melba

This is just a note to thank you for sending me Paul's book, which I found very interesting, and also for taking care of looking into the publication of the article on Causality. What do you think about the idea of cutting out the historical part (which is after all very sketchy) and stressing the other parts, perhaps expanding the last section on the infinity of levels, chance + necessity, etc.? The chapter on classical physics could be retained with more stress on a precise definition of mechanism or "Laplacean determinism", and a discussion of what is wrong with it. (This latter discussion might perhaps better take place in the last chapter.) In general, I also think that the article could be tightened up – the amount of repetition decreased, etc. Also an important additional aspect of causality needs to be discussed in more detail – namely – causality as a means of determining the mode of being of qualitatively new things, which grow out of the old things. The basic aspect of mechanism is that (as in an idealized machine) the universe is conceived of as made of basic elements (particles, fields, or what have you) which simply interact according to fixed roles, and which themselves never change as a result of the processes in which they take part. Naturally, every physical theory has some non-mechanistic aspects. For example, in the field theory, new entities (waves + particle - like singularities) can arise out of the interconnections of the basic field elements through the field equations (especially if the latter are non-linear). Also in a particle theory, new entities can arise out of interactions. For example, the atoms in a crystal can oscillate collectively as a wave and the quantization of this wave can give rise to new types of "particles" which are called "phonons". Nevertheless, the basic elements in such theories are usually conceived of as fixed and eternal. However, the concept of the infinity of levels shows that there need exist in nature no such thing as a basic element which never changes. Thus, causal laws not only determine the future in a mechanical sense; i.e., in the sense of determining quantitative changes in the arrangements of entities whose intrinsic character is fixed. The causal laws also tell when qualitative changes will occur and may define the characteristics of the new entities that can come into being. Thus, causality is a broader concept than that of mechanical determinism. It contains limited mechanical determinism as a special case. Indeed, the concept of causality is continually evolving with the development of science and other aspects of human activity, so that the potential richness of this concept has no limit. In other words, we may expect future generations to discover more and more aspects of the concept of causality, thus transforming this concept in a way that we have at present no inkling of. Yet these changes will not be arbitrary, but will instead grow in a definite way out of the efforts to solve real problems presented by the successive levels of reality that we shall be able to reach. A "mechanistic" attitude toward science however, tends to limit the growth of our concepts in an arbitrary and dogmatically conceived way. Such a mechanistic attitude refers not only, however, to the mechanistic determinists, but also to the "mechanistic indeterminists", who insist that in the quantum of action, we have reached an ultimate, indivisible, and unanalyzable entity, which will never be found to have a structure understandable in terms of a deeper level. In fact, the quantum of action presents many aspects of the ultimate particles of the atomists, so that the insistence that the quantum will never be analyzed is as mechanistic as a theory of point particles following determined orbits. Similarly, the insistence that chance + probability are not subject to a causal analysis at a deeper level constitutes a mechanistic attitude toward these things, since chance + probability are conceived of as existing in themselves and functioning under all possible circumstances according to fixed rules.

Another important problem is that of irreversibility in time. In any mechanistic point of view, nothing basically new ever happens. The future is conceived of as some rearrangement of fixed types of already existent possibilities. The mechanistic determinists say that the precise future events are already determined by the existing state of things, while the mechanistic indeterminists say that only the probability is determined. But they both agree on an essential point. The general character of the kinds of events that can happen, the kinds of things that can exist, is already determined by some fixed set of rules. The choice of which of the particular allowed combinations will actually happen is fixed by some mechanical factors existing in the system, according to the mechanistic determinists. According to the mechanistic indeterminists, it is fixed by an equally mechanical "chance" which is conceived of as absolute and not itself capable of change or development. We may make an analogy of a man who is offered the possibility of 100 different ways of being executed. The deterministic school of executioners would choose the way according to certain definite factors, e.g., the chemical concentration of the blood, the wave - length of the light emitted from his skin, etc. The indeterministic school would chose the way by spinning a roulette wheel. The non-mechanistic school would seek a qualitative change - i.e., to find a way to escape execution, taking advantage of all factors, both "determinate" and "chance". So the essential point is that because of the infinite complexity and depth of the laws governing the nature of matter, no pre-assigned scheme of things can remain adequate forever, not even if it is restricted to being a general framework or outline. But this is just what most people find it difficult to accept – perhaps because our society requires us to accept the idea that a certain general form of social organization is inevitable, although within this general framework, we may make various quantitative changes, either by chance, or by determinate rule, as we please, as long as nothing essential is ever changed.

Please, let me know if you think these ideas should be included in the article. There is also another important criticism of the usual interpretation. In connection with the uncertainty principle, you may recall that this principle is deduced by using both the wave and the particle models for light, in different connections. Thus; from the particle concept, we get for the change of momentum

$$\Delta p = \hbar \Delta k$$

where Δk is the change in wave vector of the light quantum. From the wave theory, we get for the resolving power $\Delta x \cong \lambda \cong \dfrac{1}{\Delta k}$
Thus we get

$$\Delta x \Delta p \sim \hbar$$

But we may ask "How are we justified in using two different models for the same thing?" To justify this, we may point out that wave + particle models are consistent to a limited extent. For to the extent that we can build a wave packet, we can localize the electron as of it were a particle, and still treat it as if it were a wave. This extent is given by $\Delta x \Delta k \sim 1$. But if $p = \hbar k$; we get $\Delta p \Delta x \sim \hbar$. Now it is clear that both models are needed, the wave to describe interference and the particle to describe localization of energy + momentum. Yet it is clear that each is in the last analysis, inconsistent with the other. From this fact, the Bohr school concludes that a unified conceptual model is possible only in the limited domain in which both are consistent. Hence, the principle of complementarity in a more general domain. But this conclusion is based on the explicit or implicit assumption that no other models can be conceived of than classical particle or classical wave. A more natural point of view would be to seek a new kind of model which would contain classical particle + wave as limiting cases. The Bohr school insists that no such model be sought, without giving proof that such a model cannot be found. Indeed, the causal interpretation gives such a model in terms of the synthesis of a particle plus a field in interaction. It is true that this model is somewhat crude, and that a deeper synthesis should be sought. One possibility is that particle + field are different aspects of a more fundamental non-linear field, the particle representing a region of relative concentration of field, and the ψ wave a dispersed region of field. This would correspond more or less to Einstein's point of view. My own opinion is that the synthesis will eventually have to be on a still deeper level and will have to introduce new kinds of entities that are neither particles nor fields, of which we have only a vague idea at present. These entities will not exist purely in a very small region of space + time, as is necessary for particles and field variables of the types to which we are accustomed. But in any case, the Bohr point of view seems intentionally to lead to an acceptance of a permanent limitation to our concepts.

So much for these points. I have been feeling very sick lately, but feel better in the last few days, probably because the weather has turned colder. The warm weather not only tires me, but predisposes me to digestive disturbances, which trouble me 80% of the time here. If for this reason only, I hope to leave Brazil. I am investigating ways, which would however involve giving up US citizenship. I hesitate to do this, as I might want to come back in a few years, if things quiet down. What do you think? Leaving Brazil would probably mean going to England or France. There I could be happier, but the situation is very unstable there, both politically and economically.

Do you have any idea of a suitable position for Ralph Schiller next year? He wants to avoid being involved in war work, but wants a job where he would have time to do some research.

I hope that you enjoyed your vacation (and I am sure that you did). As you say, your requirements seem to be less than mine. Perhaps I should lower my requirements. But Brazil is a bit too low. One important point is your "family" on 298 W 11th ST, which must provide a lot of compensation for other aspects of life.

Give my regards to all

Love
Dave

Letter 44. Folder C48, dated: Dec 22, 1953.

Universidade de São Paulo
Faculdade de Filosofia, Ciências e Letras
Caixa Postal: 8105
São Paulo

Dear Melba

It has been a long time since I wrote you; I must apologize for the delay.

I am glad that you are working out some way of making a living. However, things don't look so good for the next year in the US. It is true that the foreign policy of Dulles isn't doing so well. But internally, the Republicans will try to cover this up (as well as domestic difficulties) by a strong anti - Communist offensive. In this regard, the projected immunity statute is clearly an important part.

I am enclosing a check for $100 for Christmas (or perhaps New Year). It is the only practicable way of sending a present from Brazil. I only hope that there is enough money in the bank to cover it up.

During the past few months, I have not been feeling well. I guess that I am just plain sick of Brazil. First of all, these people here are very difficult to make friends with; even among themselves they do not really make friends. Moreover, they tend to be very shallow, especially in the middle class. Secondly, I am disgusted with the University and the physics dep't. You have never seen such chaos (which is only a reflection of the general chaos in the whole country). The students are badly trained and don't have the habit of work. Most of the courses are bad, and nothing can be done about it. There is a vicious fight going on in the physics dep't, which I fear will eventually reach the stage of one group throwing the other out. It is almost impossible to avoid being dragged into this fight. We spend 25% of our time on questions concerned with it, and it is very nerve - wracking and depressing. Thirdly, I have diarrhoea much of the time. The food here is not very good, and is not very clean. Nervous tension and general boredom make the problem worse. Fourth, the climate is too hot, although I could take it if [...missing...] compensating factors. Fifth, the city is so noisy that I can't find a room quiet enough to sleep in - no matter what I pay, unless I move an absurd distance away from my work. Sixth, there is absolutely no way to relax here. If it were not for the Schillers (and the Sittes who have just arrived, but who probably will not stay more than a year) I would have no one really to talk to. There is nowhere to go at night or on Sunday. The average Brazilian idea of relaxation is to stand in a line 2 hours to wait to see a lousy movie. You can't even travel much, as there are few good roads, while hotels, food and water are bad, and even dangerous once you get off a very narrow beaten track.

Scientifically, I have been able to do little for the past few months, partly because of sickness and depression, partly because of time wasted in dep't quarrels. However, I have been critically analyzing my ideas and trying to find a new way to attack the problem. I now have seen some possibilities. As soon as there is more peace (and as soon as I can find a room where I can sleep) I hope to be able to make progress on new lines which appear to have some promise. Meanwhile I hope to be able also to explore possible ways of moving to some new country to work in.

Best wishes to all for Christmas and New Year

Love
Dave

Letter 45. Folder C46, not dated.

[Early 1954 (Oppenheimer's security clearance was suspended on December 21, 1953, Ralph Schiller will return August, 1954).]

Universidade de São Paulo
Faculdade de Filosofia, Ciências e Letras
Caixa Postal: 8105
São Paulo

Dear Melba

Thanks a lot for your long letter. I am glad to hear that your book with Pief [Wolfgang Panofsky] is at last approaching completion. I shall be very interested in seeing it when it comes out. Could you manage to have a copy sent to me?

I cannot agree with some of the things you said in your letter doubting that it is worthwhile to consider philosophical questions in science. Your example of Bacon and Descartes is not valid here. For obviously, I do not wish to claim that no "idea" is necessary, no application of the specific experimental facts and theory that applies in the special conditions we have now in physics. Pure and general philosophy is admittedly inadequate. My claim is that the two must be united, that a person can search for a specific idea more effectively if he is guided by a conscious philosophy that has a lot of truth in it (and even sometimes if it has only a little bit of truth in it, but just the right little bit needed in that particular situation). We may take as an example Einstein, who was guided by a conscious philosophy, not entirely correct, but good enough at the time to be of a lot of help. And Planck had a philosophy (mechanistic materialism) as did de Broglie (Cartesianism) and Schrödinger (idealism) and even Heisenberg (may his name be cursed!) who had positivism (and Bohr is a Kierkegaardian Positivist). Many of these philosophies are way off the beam, but by orienting a person they keep him moving in some direction and thus keep him from wandering around at random. Almost no great physicist worked entirely without a philosophy - even Dirac has one if you will read his book. Naturally, the closer to reality a philosophy comes, the better will it serve, in general, for orientation. But modern American physicists are so decadent that they like to think they are guided by nothing at all except pure opportunism so to speak, the search for the trick "idea", the little "gimmick" that without changing our way of looking at things fundamentally, will permit us to embark on a new orgy of calculations. Just as they think that by making bigger and bigger machines with more and more ingenious mechanisms, we shall inevitably be led, without any intelligence at all (or perhaps with the aid of a flash of unexplainable and miraculous genius on the part of a new Dirac) to the ultimate laws, out of which everything comes. In any case, no ordinary physicist (who is not a new Dirac) had better be so "immodest" as to try to think about what he is doing - he's only to calculate until our new genius bestows upon us the new "idea", and then to calculate some more.

But the reality is that a new "idea" does not spring thus out of the blue. The history of science is full of examples of how long and halting efforts in many directions were needed to lay the background in which a new idea could develop. It was necessary to criticize old concepts and to raise philosophical questions (what do we mean by space, time, matter, field, etc.) and then it was possible for the "idea" to come. Much imaginative speculation entered into the foundations of the atomic theory, and even more into the foundations of the theory of relativity. (Thus Lorentz worked out models of the electron, and the very Lorentz transformation grew in an analysis of ether theories). Just because a line of research is not the final one that prevailed does not prove that it did not contribute in a fundamental way to clearing up many questions.

But modern Western theoretical physicists are no longer willing to engage in this kind of work. All they want are "results" that "pay off" immediately in higher jobs, recognition, job security, navy contracts, etc. Even those that don't want these things absorb from the general atmosphere a point of view, a set of objectives, that has grown out of this orientation of physics. You can see how it works in the research of that stinker Levy,[1] who came out with a new little "gimmick" that claimed to permit him to calculate 4 or 5 numbers correctly. He was immediately raised to the stars, he became the great white hope of physics. Then it was discovered that the reams of calculations on which these "results" were based were wrong. Now they are waiting for a new "gimmick".

What we need specifically now is to study the sub-quantum mechanical level, out of which the present quantum theory with its statistical regularities must arise. In doing this, we must be guided by imaginative speculative concepts based on an explanation of existing known facts and regularities. These concepts may give us a glimpse of what lies beyond and deeper, just as the atomic theory, originally based on an explanation of the regularities of chemical combination and the gas laws, gave a glimpse of what to look for at the microscopic level, without which it is very doubtful that anyone would even have known where to begin, even with the biggest and most ingenuous machines in the world.

I don't think that one should give in to the enemy on this point. It is very important, as it represents a point of view that is as rotten in science as it is in all other fields of human activity - namely, that we would do better not to orient our actions by a conscious philosophy, directed at a long range goal, but that rather we must do nothing but consider the immediate problems of the moment, the trick tactics that will solve all of our problems without the need for us to think deeply or to question any of our fundamental assumptions. Indeed, it is claimed that we are not even making any fundamental assumptions, but are just being guided by the situation itself. (You see the implications of this in politics, don't you?)

Well, I hope you will pardon all the heat. I am working on many questions now - and am also looking for "ideas". I think that when one understands the Dirac equ. in a rational way, one will have a new "idea"; for the picture so obtained will have many new implications. Thus far I have make a lot of progress in this direction.

The business about J.R. is very nasty. Please keep me posted about details, as we don't get the full details here.

The Schillers will soon return. Bunny [Berenice] is going to have a baby, and she leaves in about a week. Ralph will wait until August.

Please give my regards to all.

Love
Dave

P.S. I hope you will excuse the tone of some of the things I said. It is not directed at you at all, but at a point of view prevalent in America, which everybody tends finally to give in to, especially at those moments when he is not feeling very well.

Also, thanks a lot for your efforts to get the causality paper published. I quite understand the point of view of Cameron and Kahn. I am now revising the paper extensively, and will have it published (in English) as a lithographed book in São Paulo - by the University. Later I shall try Lawrence + Wishart, or Kegan and [Paul] (Who published some of Bernal's books).

---

[1] A French physicist who worked at the Institute.

Letter 46. Folder C48, dated: March 15, 1954.

Universidade de São Paulo
Faculdade de Filosofia, Ciências e Letras
Caixa Postal: 8105
São Paulo

Dear Melba

Thanks a lot for your letter. I hope that you will soon finish your book with Panofsky. In any case, you have had a very nice vacation. I too, am taking a brief vacation right now in the mountains in order to get away from the heat, smoke, and noise of São Paulo. But my vacation is only for a week, and it is not very interesting, as the hotel here is pretty dull. In the morning I generally work. In the afternoon I sleep from one to three, then hike till about six, then read a bit, then eat, write letters and go to bed. But my health appears to be improving anyway, even in the few days that I have spent here thus far.

About the causality manuscript. I feel the need for considerable revision. My concepts have been evolving since I wrote the article. First of all, it is necessary to sharpen the distinction between causality and mechanism (or deterministic mechanism). Mechanism is characterized by two fundamental aspects:

(1) Everything is made of certain basic elements which themselves never change in essence (i.e., qualitatively).

(2) All that these elements can do is to undergo some quantitative change according to some fixed laws of change. For example, if they are bodies, they can move in space. If they are fields, they can change their numerical values, etc. But the basic elements themselves never undergo qualitative change.

If we postulate an infinity of levels, then we make a step beyond mechanism. For the elements existing at each level are made of still smaller elements in motion (i.e., changing quantitatively), and the mode of being of the higher level elements arises out of the motions of the lower level elements. Thus, there are no elements that can never change.

Indeed, even if we have a finite number of levels, some qualitative change is possible within a mechanistic theory. For example, with atoms in chaotic motion, we obtain new large scale properties, such as pressure, temperature, etc., new entities, such as gas, liquid, solid, and qualitative changes between them. Now, at first sight, it may seem that we could eliminate the large-scale level by analyzing it in terms of its basic molecular motions. And if there were a finite number of levels, this would be true. But if there are an infinite number, then each level stands on a footing that is, in the long run, as basic as that of any other. For every level has below it a deeper one. Indeed, matter can be regarded as made up of the totality of all levels. Each level makes its own specific contribution to the totality. Of course, each level finds an image in others, so that one can deduce many properties of a given level by studying other levels. Yet, there may be properties that cannot so be deduced. Not only may these properties be peculiar to a given level, but they may involve "crossing" of levels. For example, the general large scale conditions, such as electric field, gravitational field, may actually change the conditions of existence of smaller particles, such as electrons, neutrons, etc., so that in strong fields, the very "elementary" particles into which we now analyze matter would change. Thus, there can be a reciprocal influence from a higher to a lower level, which by itself would make impossible a complete analysis of all properties of the higher level in terms of the lower. In addition, a given level may also have new properties not entirely deducible from those of the lower levels. Hence, if there are an infinity of levels, then we can go beyond mechanism and can have genuine qualitative change which may be conditioned and determined by quantitative changes at other levels, but which may have new properties not fully deducible from the laws of other levels.

One can formulate causality more generally thus:
Everything is a necessary consequence of something else.

Various forms of causal laws exist. Thus, the simplest case is mechanism. For when we investigate the changing world, we often discover that certain elements remain approximately permanent, and that the rules governing the quantitative changes in these elements also remain approximately permanent. But more general types of change are possible. Thus, we may have qualitative change, as in evolution or in embryology. The appropriate causal laws then govern the qualitative changes, and tell which things will change into what and under which conditions. Here, we have no permanent elements but a rule of change which is permanent, provided that all conditions are repeated. But since it is impossible to repeat conditions perfectly, no rule of change can be perfectly permanent, but instead all of them must be capable of infinite variations, both quantitative and qualitative.

Of course, beneath the qualitative change, we have a substratum of quantitative change. Thus, in embryology, the successive forms of the organism arise in molecular motions of various kinds. But we are not necessarily justified in supposing that the laws of embryology could in principle be deduced completely from those of molecular motions; for the following reasons:

(1) Some aspects of embryology may require a more precise expression of the laws of motion than are now available – The new terms in these laws may be important only when there is a specific organization of the molecules characteristic of life.

(2) The correct laws may take such a form that they could not be expressed completely in terms of entities existing at the molecular level. They might have to be expressed in part in terms of concepts appropriate to the higher level.

The correct attitude here, I think, is that both the qualitative and the quantitative aspects of the laws are needed here. Each may influence the other, each may provide a partial image of the other, so that by studying one, light may be thrown on the other. But this is only a sign of the inter-connection of all aspects of the world. It does not mean that the qualitative aspects can be eliminated in terms of the quantitative aspects, just because a great many features of the qualitative aspects are so closely connected to the quantitative aspects that they can be inferred from a careful analysis of the quantitative aspects.

Naturally, we do not wish to throw out mechanical laws. In certain limited domains (the limits of which of course, cannot be known a priori) mechanical laws will surely be good enough approximations. In general, when we study a new field, we should try to find elements that undergo only quantitative changes, and which thus may permit the setting up of a mechanical law. But this does not mean that we should thereby assume that all laws will necessarily be mechanical in form. We must also seek laws of qualitative change, as well as the relationships between the qualitative and quantitative laws.

Now, a mechanical law is characterized by the fact that it specifies a rule governing quantitative changes of elements that are fixed in nature. A more general causal law may express the conditions governing qualitative change. But if it does this, it must do something else that a mechanical law is never called upon to do. It must not only determine the mode of change, but also the mode of being of the elements when they are not changing. A mechanical law simply postulates a certain fixed and eternal mode of being of the elements, so that there is a sharp separation between the laws of change and the mode of being of the elements. A more general causal law does not make such a sharp separation. Thus, in the theory of evolution, the principle of natural selection enables us to say something about the mode of being of the various forms of life, in terms of their past history of evolution, struggle for survival, etc. Similarly, in embryology, one can in part, understand the characteristic properties of an animal at a given stage of development in terms of its past history which helped make it what it now is. Thus, a more general causal law may be historical in form. By this, I mean that the very mode of being of the elements which enter into the laws is a necessarycon-sequence of the causal laws governing the whole chain of development. It is clear that social and economic laws are often of this form, for the institutions that have arisen and the relationships between them are a product of a whole complex of historical developments. Similarly, human personality is such a product, and it is probably wrong to postulate fixed elements such as "subconscious", "ego", "id", etc., which are independent of the way in which the individual has developed in his natural and social environment.

This extended concept of causality is clearly important, and it is only partially contained in the infinity of levels. Indeed, we may say that the infinity of levels provides one of the factors which underly this more general concept of causality. But it is not necessarily the only factor. We may have to look for others too; perhaps an infinity of such factors. But, in any case, we must recognize this important point: A causal law may express the necessity of a fundamental qualitative change, so that what develops may have something new in it. This something new arise[s] as a necessary consequence of what is old, and yet it is not just a rearrangement or a quantitative change of the old elements. (Even though such a quantitative change may be the factor that sets in motion the process that produces this new element; e.g., a change of temperature may set in motion the change of a liquid to a gas, and yet the qualitative transformation is something more than a mere change of temperature.)

I have not yet got all these ideas straight, but am making slow progress. This is a very difficult question.

I am now working on many lines. I have a lot of ideas on the fire, but nothing definite has come out yet. Nevertheless, there is still some hope, although I fear that to really solve the problems that now face physics will require a gigantic step or series of steps which will be very hard to make. It seems to be like a very complicated jig-saw puzzle in which many interconnected pictures are involved, but in such a way that you can make little progress with each picture before you have guessed approximately what the entire pattern is like.

Except for work, life goes on in its very dull and unsatisfactory way here.

Incidentally, I seem to have forgotten or perhaps did not receive the name of the city in which your friend lives.

Please give my regards to all

Love
Dave

PS I am thinking of applying to Lawrence + Wishart, London, to publish my book on causality when it is ready.

Letter 47. Folder C47, not dated.

[June, 1954 (CIA operation in Guatamala).]

Dear Melba

I must apologize for this long delay in answering you, but I have been very busy. Also, I have been away on a vacation.

I am glad to hear that you are still doing O.K. Have you gotten any more definite consulting jobs? I suppose that Pief is too busy to concentrate on the electrodynamics book, but I am sure he will eventually get it done. It will be a good thing to have a good modern text on electrodynamics available. We need one here, and I am sure that it is needed elsewhere.

As for me, I have made a great deal of progress on the causal interpretation. The Pauli equ. is already finished, 100% and a paper is being written. The Dirac equ. appears about ready to yield at last, and even an explanation of 2nd quantization is appearing on the horizon. The general picture that is emerging is this: An "ether" composed of spinning body-like elements. These bodies must spin at the rate that the fluid is turning in the neighborhood, so that the bodies in a given neighborhood must spin at practically the same rate. This leads to a coordination of motion in different parts of space (both spin motion and that of space displacement). The need for a single - valued field of spin orientations explains quantization of energy, momentum, etc. There is a random motion in the ether, but it is not powerful enough to destroy the coordination, which makes the fluid have qualitatively new properties. (An analogy is that the "molecule" of the quantum fluid are executing motions more like a ballet dance than like the motions of a disorganized crowd of people). The coordination follows from the causal laws and has the property of stability; i.e., small displacements from this motion lead to stable oscillations around it as a mean.

Matter as we know it is an inhomogeneity in the ether, either an excess or a deficiency of "ether particles". Stable quantum states of motion will contain some discrete excess or deficiency of particles in a given region. But the "elementary" particles as we know them are higher level entities which come into being when the ether moves in a certain way. Because the possible stationary modes of motion are discrete and stable, they act to some extent like single entities at the atomic level, but at a lower level, they can be "created", "destroyed" and transformed.

If we can finish this general model, this will carry us over a crucial hurdle and open the way for a treatment of new problems, which the present quantum theory cannot treat.

As for my plans, I am inclined to wait here another year or so to see what happens. The political situation in Europe is immeasurably better than it was a year ago; and I even think that things may change in the US perhaps enough so that I could return in a few years. Of course this Guatemala deal is very dirty, but the US has lost a tremendous amount of prestige in Latin America by engaging in it. Even the most conservative Brazilian papers criticize the US very strongly for their actions in Guatemala.

I suppose that you will be off on a vacation soon (if not already). Give my regards to everyone.

Love
Dave

Letter 48. Folder C48, dated: Nov 22, 1954.

Dear Melba

It has been a long time since I wrote you what with all sorts of delays, being busy, etc. I am glad to hear that you have succeeded in finding a new home, and that it will be a satisfactory place; also that the book on Electrodynamics is finally coming out. I shall be very interested in seeing a copy.

As for me, I have not been feeling too well lately. I don't know the cause, but part of it could just be boredom with life in São Paulo. If I do not maintain a certain pace of activity, I tend to become slack and groggy, digestion doesn't function very well, etc. However, since I shall soon be leaving São Paulo (for Israel), perhaps this will change.

The Dirac equation still doesn't come. By now I am convinced that a correct interpretation will require a treatment of second quantization (because of pair creation). My general idea now is that one must assume a sub-quantum-mechanical level, in which the fields undergo a rapid, irregular, quasi- ergodic type of motion. The ψ function of quantum theory will then represent certain average properties of the sub quantum-mechanical level. Thus, quantum theory would be an essentially statistical theory. In very rapid processes (high frequencies, hence high energies) there would not be time for this averaging to take place, and the laws of quantum theory would cease to be valid.

Thus far, I have succeeded in expressing the quantum theory as an average of such lower level motions. But as yet, I do not understand physically very well what the theory means. Some very difficult mathematical problems seem to be involved in the effort to obtain a deeper understanding. These problems are closely related to those involved in proving the quasi-ergodic character of the motion in classical statistical mechanics.

Well, that's about all I have to say for the present. I'll let you know more about my plans as they become definite. I have my fingers crossed about Israel. Perhaps it will be better than São Paulo, perhaps not. But then, one doesn't have to stay there forever, does one?

Give my regards to all

Love
Dave

Letter 49. Folder C48, dated: Dec 30, 1954.

Dear Melba

Well, at last I'm ready to go to Haifa. I leave before Jan. 15 but will spend some time in Argentina and in Paris, probably arriving there some time in February. My address in Argentina, (from Jan 15 to about Jan 22) will be
c/o Mario Bunge
Dean Fures 1876
Florida, F.N.G.B.
Argentina

In Paris, write (to arrive not later than Feb 15)
c/o J.P. Vigier
Institut Henri Poincare
Rue Pierre Curie
Paris

In Haifa, write to
Physics Dep't
Technion, Israel Institute of Technology

You may be interested to know that I am now "um Brasilieiro" complete with p.p. An extra copy of the p.p. photo is in this letter.

I would prefer that you wouldn't discuss my trip until it is over. Also, there is no need to discuss the fact that I am a Brazilian citizen.

Of course, I am saddened a bit at not being able to return to see my old friends, at least for some time. But things should have their compensations.

I am now finishing the book on causality, also an article on probability.

Hence, have little time, and will have to write more later.

Regards and best wishes to the family

Love
Dave

P.S. Please accept the belated Xmas gift (perhaps the last I can give you for a while, as the Israeli pound is not easily changed for $). But don't cash it for a few weeks, as I am not sure that the money is in the bank yet.

Chapter 19
Letters to Melba Phillips, 1955-6

Letter 50. Folder C49, dated March 18, 1955.

Dear Melba

Well, at last I have arrived in Israel, after a month's trip in Europe. I stayed in Paris 3 weeks and in London for a week. Paris is quite beautiful, even in winter. However, I am afraid that 0.0. I didn't get much chance to enjoy it, as I spent so much time discussing physics with Vigier, giving seminars, etc. Incidentally, the seminars were well received, both in Paris and in London. We have been making further progress in understanding the Dirac equ. with 2nd quantisation. If we could do this, then I am convinced that we would get a big following, because people are getting discouraged in the usual approach.

Right now I am feeling somewhat exhausted after all of the working and talking. Haifa is a beautiful city on a hillside overlooking the bay on one side and the sea on the other. It is rather reminiscent of Berkeley, also of the port of San Francisco near the Golden Gate bridge. At present the weather is cool, but I am told that the summers are fairly hot.

The physics dep't here is rather small and in a primitive state of development, behind that of Brazil. But there is a good spirit here. People are optimistic and friendly. Sitte seems much happier than in Brazil. He is still spending most of his time learning Hebrew. The next 4 or 5 months, I shall have to do the same, going to school from 8 AM to 1 PM. It seems a terrific waste of effort, but what can one do?

Rosen is head of the Dep't here. He is a very nice person. Also, David Fox is here, whom I knew in Berkeley. In general, the atmosphere is more friendly here than in Brazil. Also, I find that the food agrees with me better. But it will definitely be harder

[the next pages are the ones attached, but they do not follow. Probably from a different letter]

C. Talbot (ed.), *David Bohm: Causality and Chance, Letters to Three Women*,
DOI 10.1007/978-3-319-55492-1_19

Additional complication is that I can renew the passport only 3 times, then must return to Brazil to get another. Who knows what the political situation will be there in five years?

The weather here is quite pleasant now I am feeling very well and I'm getting a lot of work done. Haifa is a pleasant city to live in during the winter time, if perhaps a bit dull in some ways. They are building a new University Campus on a beautiful and relatively cool place high on Mt Carmel, but this will take another two years. Meanwhile, we teach in tiny rooms in hot wooden shacks, where there is hardly space to turn around when you are writing on the tiny black board.

I hope that we can get to see each other soon. I may be able to come to Canada this year, but I am not certain yet. I don't think there is any danger of trouble from the USA. Do you?

I am glad to hear that the book on Electrodynamics is doing so well. It must be a great satisfaction to you. Besides, it will make a tidy contribution to your income, which is not to be sneezed at. By the time you write all these other books, you should approach a state of relative financial independence. I think that you will get well over $1000 a year in royalties on the Electrodynamics alone. If a person could write about ten or twelve texts, then he could face a relatively secure old age. Do you think that this scheme is better than life insurance, with annuities at 65? We could perhaps arrange a company that will guarantee payments after 60 to anyone who would write a certain number of texts. It could be called "Text Book Writers Mutual Benefit Association".

At times I feel discouraged about the state of the world. A thing that particularly strikes home to me is the report I got from Burhop (confirmed by others) on Russian physicists. Apparently, they are all busy on doing calculations on electrodynamics according to Feynman, Dyson, et al. Their orientation is determined strongly by the older men, such as Fock and Landau, who in addition to their training, are influenced by the fear of a sort of "Lysenko affair" in physics. The typical physicist appears to be uninterested in philosophical problems. He has not thought much about problems such as the re-interpretation of qu. mechs., but tends to take the word of the "big-shots" that ideas on this such as mine are "mechanistic". Actually, the standard procedure is just to label such a point of view, and then most people accept the label without even bothering to read about such questions. There are some philosophers in Moscow who criticize the usual interpretation, but they haven't had much influence on the physicists. All in all, the situation in Soviet physics doesn't look very different from that in Western physics. It is disappointing that a society that is oriented in a new direction is still unable to have any great influence on the way in which people work and think. In both societies, one finds that the typical person finds it safer and more convenient to avoid "controversial" subjects and to become an expert in some narrow technical field. I must say that there is some justification in the effort to avoid another example of the extremes to which Lysenko went in criticising backward trends in biology. But the reaction has been for people simply to avoid fundamental questions altogether, and to say that experiments will eventually indicate the new theories that are needed, while meanwhile, everyone works on the old ideas until this happy day will be reached. I have the impression that the Soviets were not able to maintain an objective and scientific attitude in their criticism of philosophical ideas implicit in scientific concepts, and as a result of the violence of their criticisms, the net reaction has been bad. First of all, most people probably feel that it is best to keep away from such dangerous questions altogether, while others reacted with equal violence to protect their life-long lines of research and the results that they have achieved. As a result, a very bad atmosphere has been created, in which certain questions simply cannot be discussed freely, because too many questions charged with emotion are involved and too many people's careers are at stake.

Well, that's about all I have time to write for the moment. Give my best regards to everyone.

Love

Dave

Letter 51. Folder C49, not dated.

[About July, 1955].

Dear Melba

This is just a brief note to let you know that Saral and I are ready to leave for Europe on July 30. This is none too soon, as it is becoming impossibly hot here. I haven't been able to do much work in the last month or so.

I am now feeling a bit better about things. I was somewhat depressed because Paul Zilsel and David Fox left for North America – Zilsel to Hamilton, Ontario and Fox to Rochester. This leaves me no one to talk to except a few students, and Rosen, who spends all his time on administrative matters. However, I am beginning to get back to to work anyway, and to feel that life will perhaps at least be liveable next year, if not, in all probability, very interesting.

We are going to spend a month in Austria, (or perhaps 3 weeks and 6 weeks later in Holland with the Vigiers). This will be a vacation. Austria has the advantage of being cheap, cool and beautiful ($\frac{1}{2}$ the price of Italy, and France is still worse). Then we spend 6 weeks in Paris, where I'll do some work with Vigier, then 10 days in England where we'll have (alas!) to visit various relatives and in addition perhaps to see if I can't line up a job there. If you want to write to me before Nov 1, please send the letter c/o J.P. Vigier, 5 Rue des Beaux Arts, Paris France.

I have sent you (by ship mail) a mimeographed set of my notes on Modern Physics. I also sent a set to Warren Blaydell of Addison Wesley, who said you mentioned the book to him. I suggest that you get in touch with them about the book (but wait a few weeks as they won't have it until early September). Maybe you could get a good idea of their reaction. You will notice that the last 4 or 5 chapters are not written 100% from the usual point of view, even though I do try to give both sides and say the issue is not yet decided. I am afraid that it is still a bit early to hope that such a book will be widely adopted. Incidentally, I expect to add a chapter on atomic structure, chemistry, X-rays, Zeeman, Raman effect, etc., an Appendix on the theory of probability and an appendix covering a more advanced treatment of statistical mechanics than is given in Chap. VI.

Please let me have your opinion of the book, when . . .[remainder of letter missing].

Letter 52. Folder C49, dated: Oct 12, 1955.

Dear Melba

Well, it's a long time since I wrote before. As you know, I've been in Europe (Holland, France and England). I got a good rest there and a lot of work done. It is evident to me now that the hot weather of Israel is bad for me. It makes it hard for me to work and is conducive to intestinal disturbances. Anyway, I have pretty well decided to go to Europe next year, either England or France.

As for the book, the publishers have asked me to shorten it. Also there is one chapter (Chap. V) in which there is a fairly technical treatment of my ideas on the causal interpretation. They felt that this should be put on the same level as the rest of the book, as it will discourage the general reader and yet not cause the book to be purchased as a technical work.

I have made the suggested revisions, also cutting out most of Chap. IV, which was a criticism of positivism, and condensing Chap. VI, which contains my own philosophical point of view. The book now contains about 80,000 words, in 5 Chapters. It should amount to about 200 pages of a medium sized book (smaller than my Quantum Theory, but with pages about as large as in Einstein's little book on relativity, or a little larger). I think the new version is quite an improvement. It is being typed now.

I asked the Yevicks to give you a copy of the book, to be passed on to Schiller, Freistadt, and Gross and later anyone else whom you think might be interested. I shall send you a copy of the new Chap. IV and the rest of the main revisions, as soon as it is ready.

I got a lot of work done on the Dirac equation with Vigier in Paris. In fact, we now have a complete causal interpretation. The basic picture is that of an ether or substratum composed of spinning bodies. In the normal vacuum state, this ether is undergoing random fluctuations in the spin orientation and in the velocity at each point in space. The fluctuations in velocity are very big, going very close to c, the velocity of light. Indeed, to make a theory of a covariant ether, you must, as Dirac has shown, have a uniform distribution of velocity fluctuations that goes all the way to c, so that no matter what frame you choose to look at them, the distribution of fluctuations looks the same. Actually, however this would not make sense, as it would imply an infinite energy. To avoid this, you can assume that the fluctuations go very close to c. Thus, for a very wide range of velocities, the edge of the distribution would be so close to c that the distribution would not change much in a Lorentz transformation. Only for particles having fantastically high energies would we then obtain significant deviations from Lorentz invariance of, for example, the mass energy relationship. But as far as we know, such deviations may exist – they should be looked for in experiment.

Now to return to the Dirac equation. What we have shown is that small systematic oscillations around this fluctuating background satisfy the Dirac equation. Furthermore, it looks as if new modes of motion are possible, in which there could be small oscillations satisfying Maxwell's equations with the correct coupling to the [missing] that satisfy the Dirac equation. Also [missing] oscillations may be possible, I think, which [missing] by the meson equations. Thus, there is hope that [missing] substratum will support all the waves corresponding [missing] various "elementary" particles.

The next step is to go on the second quantisation,[missing] treats the many body problem. Here it looks [missing] we are well on our way to solving the problem [missing] indeed, the same medium appears to be capable of supporting waves corresponding to the presence of one, two, three or more particles of a given kind. The particle like character of the oscillations then arises out of the "quantisation" of the field energy. This [missing] yet fully worked out, but what seems to happen is that particle-like concentrations of energy are always forming and dissolving. They tend to form with discrete amounts of energy, momentum, charge, etc., thus simulating the discrete characteristics we usually associate with stable particles. Naturally, if particle-like concentrations of energy dissolves at a certain place, it is very probable that it will re form nearby. Thus, on a macroscopic scale, the concentration stays in a limited region of space and moves on a fairly well-defined track, while microscopically, it does not continue to exist as a permanent entity.

Well, so much for this. I'll write more later. Please let me know how the family is, and give my best regards to all of them

Love

Dave

P.S. Do you still have the cat with the wide eyes?

Letter 53. Folder C49, not dated.

[Late 1955 or early 1956].

Dear Melba

I received the book on Electrodynamics. It strikes me as a very fine book, and one that has been needed for a long time. The treatment is systematic and clear, physical ideas are there, and it is appropriate for the graduate level. I think that it will catch on rapidly.

It was nice to hear of Willis getting the Nobel prize. By the way, why is he going to Oxford?

I have sent you by ship mail a copy of the corrections to the book on Causality. Please replace the material by the corresponding corrected material when you circulate the book.

I am now giving my lectures in Hebrew. It is clumsy and slow but it goes.

I am working on two problems which have formal relationships – the treatment of liquid helium as a field (Bose) and the problem of the substratum underlying the current quantum theory. I have made some progress, but it is slow. Now that the weather is turning cooler, I get more ideas. There is a regular gale blowing in from the sea right now, and today I saw a large number of things clearly for the first time.

In the theory to which I referred in my last letter the charge comes out as a kind of internal angular momentum of the "ether" particles. The electric field corresponds to a kind of circulation of the fluid with vorticity. The turning particles and the turning vortices interact with each other, and thus one gets the connection between charge and field. Positive charge corresponds to a positive component of the angular momentum along the principal axis of the particle, negative charge to an opposite component. Don't confuse the ether particle with the so-called "elementary" particles, which are complex meta-stable moving structures in the ether. Neutral "elementary" particles correspond to equal involvement of positive and negative "ether" particles.

Another important point is that the meaning of the exclusion principle is becoming clear. One has complex quasi-ergodic motion in the phase space of the ether particles. But this motion takes place in a sub-space of dimensionality equal to the number of "elementary" particles (This latter number is conserved in all canonical transformations). To describe the shape of this subspace, one uses a set of complex determinants, which turn out to be just the (antisymmetric) wave functions of many particles in the configuration space. The squares of these determinants represent a volume element in the subspace in which quasi-ergodic motion is taking place, and thus automatically represent a probability distribution

[Missing lines]

Things seem to be working out, but the details are still puzzling. It may be a long time before this theory is really ready.

The situation here doesn't look too good. There is a general belief that war is inevitable with Egypt. Let us see what happens.

Give my regards to everybody and let me hear from you soon

Love
Dave

Letter 54. Folder C49, not dated.

[Probably between March and July, 1956 (Khruschev's "Secret Speech" was made February 25, 1956, and presumably Bohm would leave for Europe by July, "Not by Bread Alone" appeared in Russian in 1956, in English in 1957)].

[Heading on photocopy: "Israel"]

Dear Melba

I am sorry that I did not answer you until now, but I have been very busy. I have many new ideas on the foundations of the quantum theory, and if I do say so myself, some of them are fantastically beautiful, especially in the scope of the things that they unite. I have been studying Hegel (along with some other people here in Israel). It is true that Marx and Engels stood Hegel's ideas on their feet, by making them materialistic. Nevertheless, there was a tremendous wealth of ideas that they did not use, because the science of the time did not require them. But now, with the further development of science, these ideas applied to space, time and matter are surprisingly fruitful, as well as beautiful. My ideas on the sub quantum mechanical level have made tremendous steps forward; and I wouldn't be surprised if I didn't make some real applications to physics within the year.

Meanwhile, I wish to thank you for your efforts with regard to the Modern Physics book. I am still correcting and revising it; and will also add some chapters on the structure of atoms, molecules, Zeeman effect, Raman effect, solid state, etc. I hope that a publisher can be found.

As for the job in America, my feeling is to delay it for at least a year. In any case it could not be arranged before October. Meanwhile we shall go to Europe (either England or France).

You are quite right to say that I have a talent for unhappiness. I am trying to correct this, but it isn't always easy. I did present too dark a future in my previous letter. China is definitely more promising than Russia is for the moment, and even in Russia, the future is not so bleak as all that. Many positive features exist. For example, Russian industry, technique and science have made good strides forward. The standard of living has risen considerably. Education is reaching new high levels, both qualitatively and in the number of people involved. Some of the worst excesses of Stalin are being corrected. One could go on like this for quite a while. Nevertheless, we must look the facts in the face, see both sides of the problem, and try to frame a policy, based on a correct evaluation of reality (or at least the most nearly correct one that we can make).

Before coming to this problem, however, I should like to comment on a few points that you raised in your previous letter; first of all about your statement that people began by making the mistake of worshipping the leaders of the C.P. as superhuman and then, when disappointed, jumped to the opposite extreme, thus demonstrating a considerable emotional immaturity in both kinds of reactions. This is true, but not the whole truth. For it is essential to remember here that the party systematically encouraged and organized such worship, not only by the character of the propaganda that it put out, but also by its very undemocratic rules, which made criticism of the leaders almost impossible. You just have to remember how even people like Foster did not dare to criticize Browder until the day after they had thrown him out. The situation did not change in this regard after Browder was expelled, nor was it basically different in the CPs outside the USA. Indeed, in Russia, from which the whole pattern originated, criticism of Stalin was equivalent to signing one's death sentence; and K. and B.[Khrushchev and Bulganin] managed to do this only some time <u>after</u> Stalin died (it is possible, not without a bit of help from K. and B.). It is hardly surprising then that after being required to believe that their leaders were supermen for many years, many members found it difficult so suddenly to adjust to the fact that far from being superhuman, many of them were subhuman in their cruelty and stupidity. Indeed, many of the members who might have been disposed to doubt the superhuman qualities of the leaders had already been expelled (or in Russia, they had been sent to Siberia or perhaps otherwise disposed of). Therefore I do not think that your criticisms of the reactions of so many members reached to the root of the problem. It is true that their reaction is in many ways childish, but it is even more important to note that this childishness was in a large measure organized and fostered by the structure of the party (not only by direct propaganda work, but also by threats of expulsion, etc. and also by tending to attract people having an emotional make up with a tendency to hero worship).

The second point on which I would like to comment is your statement that we must be less emotional about the whole thing. Here again you are quite right. We must look at the whole problem as coolly, calmly and objectively as we possibly can. Nevertheless, your advice leaves out a whole side of the problem. We ought to remember that people struggle for a better society not only for intellectual reasons, but even more because they feel within them the inescapable necessity for a better, fuller, warmer, and more satisfying life, which can only be gained for themselves in the common struggle for a new organization of society. In doing this one cannot help feeling a certain kind of love for a better life that will come, and for the germ of this life that is now in the hearts and minds of the people who are carrying out the struggle. Indeed this feeling is perhaps one of the most important motivations for putting the full intensity of one's being into whatever aspect of the struggle that one may be connected with, as one must in order to obtain really good results. It seems natural to suppose that people who work together and profess the same goals should really believe in them. If they really believe what they are saying then one would expect to see Communists behave at least as well toward each other and toward other people as most non-Communists do. Indeed, one would expect more, for their belief in a better society and better humanity should lead them to a higher standard of personal conduct. This does not mean that one expects a god-like perfection. Rather, as during the days of the Resistance in Europe, one expects that Communists will fight with more dedication and less selfishness than people who have not been penetrated by the vision and hope of a better life for all. The same is to be expected when the war is over and the time of construction begins.

But what actually happened? We found that Communist leaders behaved worse than most other people. They began by framing their enemies, torturing them, beating them up, murdering them, sending them to slave labor camps in Siberia, etc., and ended up by doing the same thing to their own "comrades". If only a small fraction of these things had gone on in the USA, you would certainly have heard plenty of protest. For example we know that the Negroes in the South are very badly treated. But millions of people in Russian slave labor camps have been treated as badly as the worst treated Negroes in Georgian prison camps. Many of these people were perfectly good Communist, framed by the Secret Police. And those who have been in Communist prisons (e.g. in Hungary) are quite ready to testify that they are worse even than were the Fascist prisons under Horthy. Only the Nazis have exceeded the Communists in their cruelty and brutality (Perhaps Franco's Spain has the dubious honor of being in second place, instead of Russia and Russian dominated countries). Whatever you may wish to say about it, one must agree with the article No 1 that I am enclosing (On How Stalin Wronged the Jews). As the Committee (of British

Communists) who went to investigate Anti-Semitism in Russia concluded "Rightly or wrongly, we have expected something vastly different from this".

But one may perhaps try to excuse all this cruelty as an excess of zeal for socialism. If this were the real reason it would be bad enough, but at least one would say that it was just the result of mistaken actions in a good cause. But as so many stories indicate, there was a general demoralization of the Soviet bureaucracy. Gov't officials, scientists, writers, artists, etc became scared and cynical. They began to seek fame, comfortable jobs, etc, instead of trying to build socialism. They sat still and kept quiet, while many of their comrades were framed and sent off to Siberia or murdered. In some cases they even cooperated in framing some of their colleagues in order to favor their careers. High gov't officials, bureaucrats, etc began to live in an atmosphere pushing careerism and terror, in which no man was any longer the friend of another. In this regard, you might read the second article that I sent you, which is a review of the recent Soviet book "Not by bread alone". Here you get a more detailed explanation of the corrosion of human relations that seem to have occurred in the upper crust of Soviet society. It appears that much of the Soviet bureaucracy (if not the major part) is now composed of people whose dedication to principles is much less than their dedication to pushing themselves into the best possible positions. This kind of decay and cynicism is, in my opinion, even more dangerous in the long run than the cruelty and the brutality of the Stalin era, especially inasmuch as the latter is, to some extent, on the way to being corrected.

It is true that the younger people are beginning to protest against the rottenness that they are meeting. Although these protests are to a large extent being suppressed at present, they will probably increase in volume. It is from here that one can see the principal hope of the future in Russia. If Communism is to continue as the goal of the Russian people, the present bunch of bureaucrats must go. But they will fight with every one of the many weapons at their disposal to stay. Moreover, the younger people, in their efforts to criticize the gov't, will doubtless suggest many foolish things, especially since they haven't been allowed to think or to discuss freely up till now (and even now, only a little), so that they have little knowledge or practice on which to base their ideas. Besides, many of them will surely be corrupted by the possibility of a good job, and frightened by the danger of losing the possibility of such a job, as well as by still worse possibilities that they can easily imagine. Thus, although the future development of Russia centers on the struggle between the bureaucrats and the young people (who have been raised to an educational level where they do not easily accept the old fashioned nonsense from the bureaucrats) it is by no means clear as yet how this struggle will turn out, or how long it will take before there is a favorable outcome.

In any case, however, it is clear from all of this that the old way of looking at the Communist movement cannot work. There can very rarely be personal trust of one person in another (at least not in the upper part of the bureaucracy), nor can people now have the kind of love for their leaders that people had for Lenin, F.D.R. (and on a mistaken basis) for Stalin. Moreover, because of the character of most of the present Soviet bureaucracy, it is doubtful that one can even have much respect for the leadership, which not only went along on many crimes, but which, sheep-like, followed orders until they came to power, to demand the same sheep-like obedience from their own subordinates. These men are no longer developing new ideas – their slogan is "Back to Leninism". The idea that new theories are needed for the new situations that have arisen does not seem to be considered. At least Stalin proposed some ideas of his own, even if many of them weren't very good. But one could hardly imagine the slogan "Forward with Khrushchev-Bulganinism".

So to sum up this point, I believe, that while we have no choice but to look at the present situation with a minimum of emotion, we must remember also that this need reflects a most unfortunate fact, namely that by their very bad conduct the leadership has destroyed much of the respect, trust, confidence, and even love that people tend to feel for leaders who are really leading. Even worse, they have shaken very badly, and perhaps even destroyed the belief of many people in the possibility of a better society, at least in the foreseeable future.

The third point that I want to discuss is the idea you mentioned that China and other countries in Asia and Africa would now begin to play a leading role in the development of humanity. This is quite true, but I am not sure that it is going to solve some of the problems that we have been discussing. There is no doubt that it is a good and necessary development that the Afro-Asian nations are liberating themselves from colonial rule. Yet (except for China which I shall discuss later), most of them are in an even more primitive state than Russia was during the nineteenth century. The primitiveness is not only in the domain of industry and agriculture; it is even more in the domain of ideas about politics and society. Here, even among people who call themselves Communists, you will find the strangest mixtures of ideas. Moreover, the degree of cruelty and disregard for human life is probably much greater than it was in Russia. The people are primitive, superstitious, and under the influence of hatreds of tremendous intensity.[1] There is a long tradition of cruelty in these people, as well as of corruption and cynicism. If Marxism became so badly distorted in Russia, partly because of primitive and cruel traditions, it will tend to be even worse in the Afro-Asian nations. It is true that one must be in favor of the trend against colonialism. But one must recognize that such developments may well sharpen rather than solve the problems arising from successive cruelty stupidity and brutality among primitive peoples who take up Communism.

With regard to China, the outlook is much brighter. Perhaps this is partly because China has actually been civilized for thousands of years, so that people are better able to think and have some other habits and traditions to fall back on. Yet even in China, there has been, as admitted by the Chinese gov't itself, a great deal of unnecessary cruelty. This seems to be partly due to an unfortunate tendency to copy the Russian model, a tendency that the Chinese have recently begun to outgrow. The outlook is hopeful that they will have avoided the Soviet errors before it is too late. Yet, even here, there are many disquieting signs that they still follow the Russian

[1] Think, for example of the caste problem in India, or of the Hindu-Moslem relations, where in riots, many thousands were killed a few years ago. Yet India is a country where the passive "non-violent" traditions of Ghandhi have some influence. What will you expect for Egypt or Central Africa? Even in Russia anti-semitism hasn't died out after 40 years of Communism. What do you expect from these other nations?

methods of requiring mechanical acceptance of slogans without thinking. This can lead to certain dangers, if it is not corrected in time. But on the whole, reports from China have been almost uniformly good, often reports from the same people who were unfavorably impressed by the Soviet Union.

One other point – about the "historical necessity" of Russia's developing in the way that it actually did. The exact degree to which this necessity can be traced to various causes is not yet clear. Certainly the backwardness and cruelty of life in Russia played a key role. Also the lazy habits of the Russian peasants; for because of the specific organization of peasant life before 1860 (communal life in a village, or else serfs), many Russians developed the habit of working only when they were forced to – literally when they were driven by the master's whip. To a certain extent, the cruel and brutal system of punishments adopted by the Communist gov't was a continuation of such methods – which, under the conditions, could almost have been said to be necessary, since without them, people might not have worked, and the Communist gov't would have collapsed. In China, it seems that people are basically more industrious, that they are used to working on their own initiative. Thus, the gov't does not have to have such cruel punishments for people who do not work or who do their work badly. From this point of view, the outlook for the Middle East is very grim, since the degree of laziness, hopelessness, cynicism and corruption is very high indeed in these parts. Such characteristics are not fundamentally changed in less than a few generations.

Finally, there is the question of the role that Stalin played in the whole mess. Certainly, it was very unfortunate that he came to power when he did. Even Lenin had already warned against him in his recently published last will and testament, as well as on other occasions. (There is also the famous incident when Stalin insulted Lenin's wife in a very arrogant way when Lenin was very ill). As Lenin pointed out, Stalin suffered from an excessive desire for power. This desire was perhaps natural for a man who was raised in the very bad atmosphere in which Stalin actually grew up. Moreover, even then, he was already suspicious of people, feeling that they were plotting against him. To feel more secure, and to increase his power, he put his own men into every office that he could, and plotted the downfall of his opponents, preparing traps for them, etc. At first he began to frame people like Trotsky for espionage (it is now implicitly admitted by the Soviet gov't that such charges were false) and later, he found this method irresistibly convenient for all of his opponents. No doubt he felt that in doing this, he was furthering the interests of socialism (Just as Hitler must have felt that his schemes were benefiting humanity in the long run, too). As time went on, he killed more and more people; and as a result, this time quite justifiably, felt more and more insecure, since he was making more and more enemies by his conduct. Eventually the whole thing seems to have developed into a kind of paranoia. Exactly when it became a kind of insanity is not clear – perhaps roughly around the time that his wife committed suicide, apparently because she could not accept the implications of Stalin's actions with regard to the future development of Communism. In any case, it is clear from Kruschev's descriptions of Stalin's actions that during the war years and after, he really was off balance. For example, his engineering of anti-semitic plots to frame Jewish doctors, intellectuals, etc is a clear sign of unbalance in a man who is committed to Socialist principles of equal rights for all races, nations, minorities, etc.

It is rather interesting to note that Stalin's so-called basic theoretical error, namely, that the struggle would intensify rather than diminish after the establishment of socialism, is exactly the error that Stalin needed, in order to justify the craving for power and the suspicions of other people that were [missing word] basic to his character from the earliest days. The Communists in Israel are always tracing all the troubles back to this theoretical error, implying that if Stalin only had developed the right theory, most of the subsequent excesses would not have happened. But Stalin's tendency to seek power and to be suspicious did not begin in 1932 (roughly the time when he made this theoretical error). Moreover, Stalin was an intelligent man. Why then should he come to so stupid a conclusion that with the successful establishment of socialism, the struggle of the enemies within the country would intensify? This could only be based on the implicit assumption that the country was full of potential spies and saboteurs – This assumption was not true, or at least it was an exaggeration out of all proportion to reality. But to a man of Stalin's psychotically suspicious temperament, it seemed quite natural to come to such a conclusion. Moreover, once he came to this conclusion, it provided a very convenient way to eliminate his enemies – he just had to frame them for espionage and sabotage.

Of course, the trouble did not originate entirely in Stalin's unfortunate character. Because of the primitive conditions of Russian society, the country was full of people who had tendencies to be suspicious, so that they readily agreed to the implicit basis of Stalin's incorrect theoretical analysis. Moreover, because of the heritage of Czarist terror and absolutism, it was not hard for Stalin to obtain absolute authority, so that nobody could dare to oppose the insane aspects of some of his ideas. Moreover, as to the "cult of the personality", this was made easier by the fact that the Russian peasants used to worship the Czar as their "Little Father". What could be more natural than to substitute J.V.Stalin for the Czar? Instead of the ikon on the wall, there appears a picture of Stalin. Thus, one sees how a primitive mentality of the people paves the way for the "excesses" of a leader who loses his balance.

There is no doubt, however, that with a leader like Lenin, the worst aspects of the tragedy could have been avoided. Therefore, I do not agree with the idea that perhaps the Russian revolution had its failure built into it. No doubt, the structure of Russian society (plus the situation of attack from without and poorly developed industry within) contained all these dangers implicit as potentialities that were very likely to be realized, except in the event of special or unusually favorable conditions. Such special conditions could have been, for example, the existence of a group of leaders (such as Lenin) who understood the real situation very well, and who could see just what the dangers were, so as to take specific steps to guard against them. In other words we must not accept the fatalistic view that a certain economic, political and social situation necessitates the failure future of socialism, with an absolute and inescapable necessity. It will be, in general, an absolute necessity mainly to the extent that there are no people who have looked at the problems honestly, analyzed them, and tried to figure out what to do. For in this case people will be at the mercy of circumstances. But to the extent that people in general, and their leaders in particular, are able to understand the real problems and are free to come to the necessary conclusions concerning these problems, then there is much hope that such errors can be avoided. In Russia, there was (and is) neither sufficient understanding nor sufficient freedom to come to conclusions that might embarrass the positions of the bureaucracy, which now controls the country. In this regard, the problem is not very different from that in capitalist countries, where, for example, "social scientists" are not free to come to conclusions that would weaken the positions of the capitalist bureaucracy.

This conclusion points up the crucial role of the development of theory, especially with regard to the possibility of socialism in the Afro-Asian nations, as well as in Europe and America. We have seen what some of the dangers are. Up till now, theory has been too mechanical in its stress on economic factors, and its denial of autonomous causal action to what is, in Marxist terms called the "superstructure" of ideas, traditions, culture, myths, superstitions, religion, science, et al as well as in the emotional make-up of the people. But we have seen how such factors almost wrecked socialism in Russia, and how they could produce still bigger effects, for example in the Near East. We need to work out a theory that takes such effects into account. For example, if the question of socialism in a primitive country comes up, we must consider a number of important points: (1) What measures will be needed to insure efficient work (2) What measures are needed to overcome irrational hatreds, not only in the poor and uneducated people, where the presence of such hatreds is obvious, but also in educated and advanced "progressives" who like to imagine that they have left all their irrational prejudices and feelings behind them. Very often, progressives like to say that all such problems are secondary to economics, in the sense that (e.g., in the case of anti-negro prejudice) they have a basically economic foundation. Thus, when socialism is established the reasons for such a prejudice will cease to exist, and eventually the prejudice will disappear. This is no doubt true; and yet it leaves out another essential side of the truth. For especially in primitive countries, it may take several generations to establish the material and ideological foundations of socialism among the people as a whole. What will happen during those generations? If this process is not properly planned, the whole thing may be wrecked before it can get to first base. Thus, when leaders full of prejudices and irrational hatreds come to power, they are unable to see reality as it is, and continue their previous habits of blaming the results of their own weaknesses and deficiencies on "sabotage" due to evil schemers. Moreover, the cruelty and sadism that they previously learned can now be indulged in freely, under the guise of the severe discipline that is needed if the people are to be made to work hard enough to build Socialism.

More generally, one must keep in mind the following important problem: If one is to adopt a transitional organization of society on the road to socialism, which will remain transitional necessarily for a number of generations, one must consider the effect of this organization on the emotional and mental make-up of the people. Thus, when one considers rigid discipline enforced by harsh and severe penalties, the following problems will arise: Will this discipline not make many people too timid, too sheep-like? Will it not arouse fierce resentments in others, which, in a time of crisis may endanger the society? Will not these fierce resentments be hidden by many people, to come out only when they rise in the bureaucracy, and have an opportunity to take such resentments out on their subordinates? Here we come to another problem which has been totally neglected. What is the effect on the leaders of their wielding absolute (or almost absolute) disciplinary powers? Will they not be corrupted by it? Will it not bring out many of the bad qualities that are latent even in the best of people? In other words, the planners must take into account the effects of their measures on themselves (and more generally on the people who will wield the power with the aid of which the plans are carried out).

I do not wish to imply here that it is absolutely impossible to construct socialism in a primitive country. I only wish to point out that there are some really difficult problems involved in trying to do this and that they must be faced honestly and with care. The facile answers that Communists have given to this problem are evidently not adequate, as experience has shown.

Then there is the question [missing word(s)] capitalist countries. How can Socialism arise there? Evidently the Marxian idea that the proletariat would take over is inadequate. A whole century has passed since then and precisely in the countries whose industrial development is greatest the workers are least interested in changing the structure of society. Why is this? Marxists have done almost no theoretical work on this question because they tended to think only in terms of century-old slogans which have lost most of their relevancy. A quick glance at the situation shows a few of the new factors that have come to be important.

Thus in America, Britain and a few other countries, the workers are, on the whole not too badly off, certainly much better off than they are as yet in any socialist country. It is true that certain special sectors of American + British [missing word] are in depressed conditions, but experience shows that even in a Socialist country, large sectors of the people may remain poor (and get still poorer) as a result of mistakes made by bureaucrats with almost absolute power.Thus, the average worker is quite right when (consciously or unconsciously) he concludes that it is by no means certain that he would benefit from a transition to socialism. Naturally in poorer countries (e.g., France and Italy) the situation is different; but this only serves to emphasize my point – that Marx's analysis is correct only for special conditions of extreme poverty, which no longer exist in certain very important sectors of the world.

Does this mean that there is no hope for Socialism in the more advanced countries, except perhaps as a result of eventual military or other pressures from the less advanced nations, which will have changed or will change to socialism and which will eventually get ahead of the nations that are at present more advanced? If this were true, then there would be nothing to do in the advanced countries but to mark time for 25 or 50 years until Russia, China, and other countries that may later go socialist will have got ahead of the USA in their industrial and technical development. But I believe that such an analysis is false. For there are other reasons for socialism, besides the extreme poverty of the proletariat and peasantry in large sections of the world. For example, there is the fact that people are actually completely interdependent, because of the structure of modern industry. They may as well recognize this fact, and plan the society as a whole in accordance, rather than allow this interdependence to grow in its own partially chaotic way, as it now does. Enormous benefits would come from such planning. For example, one could have not only a higher standard of living, but a qualitatively different and better sort of life. Cities would be planned for convenience and healthy living, with recreational facilities, education, rational solution of traffic problems, etc. Scientific research would be pushed to the point where most diseases would be conquered, life lengthened etc. People could live in mutual trust, happiness and satisfaction coming from working together. Routine mind-destroying work could be reduced to a minimum; and each person would have a chance to make his contribution to society in a way that permitted fulfilment of his own personality. Education and training would prepare people for the new possibilities created by the new way of life. One could go on like this for hours. But the essential point is that even if present boom conditions continue, capitalism cannot do these things. At best, it may give people more of the necessities of life, along with certain luxuries. Some progress in science is possible. But to coordinate all these developments in such a way as to bring happiness, satisfaction and the free development of our essential nature as human beings, it is clear that the framework of capitalism is not adequate. Some kind of socialist framework is needed. The precise nature of this framework has not yet been worked out, but it is clear that it must be democratic for the very essential reason that the mode of being of the new society that we are talking about is such that it cannot exist in an undemocratic form. Of course, there are many practical problems for the transitional stage, just as in the case of primitive societies. But these practical problems are different from those in the primitive societies and must be worked out accordingly. Thus, the better living conditions of the masses makes it harder to interest them in socialism. On the other side, the higher level of industry, technique and education means that once socialism is established, a transitional period as long as several generations may not be needed.

You can see that an enormous number of new theoretical problems needs to be worked out. In this process, it is necessary to take into account all the existing experiences, failures as well as successes. As you said in your letter, very often, experience gained on the basis of a wrong theory can lead us a 'long way' toward an essentially correct theory. Nevertheless, this will happen only with those people who are not afraid to face the fact that the old theory is wrong. At least this much honesty is needed, and I am afraid that the CP is, on the whole, not yet in a condition to have it! Thus far, they seem to be more interested in re-establishing "rigid discipline" and in protecting the positions of party bureaucrats (Both in Russia and elsewhere) than in frankly appraising the sources of past errors. Some errors have been admitted, it is true, but in no case has criticism been allowed to go deeply enough to reach the essential sources of the trouble. For example, in the case of Stalin's "errors", the CP seems to have stopped rather superficially at the "cult of the individual" and "the mistake made by Stalin of supposing that with the establishment of Socialism in Russia, the internal struggle against the bourgeoisie would intensify." No real effort has been made to go into the sources of these errors, for the simple reason that to do so, one would have to subject to a merciless examination the whole theory of the structure of the party, including the question of whether the present organization is proper for the new tasks that confront it today. Such an examination would put in danger the positions of a whole host of bureaucrats, who naturally would not like to see this happen, and who use their very great powers to prevent it from happening. Likewise, one would have to criticize frankly the primitive mentality of a large number of people, and would have to admit openly that in the upper party bureaucracy, it is often the law of the jungle that prevails. One would have to go into the question of why people were so ready to engage in "the cult of the individual", and of why so many Communist leaders became cruel and stupid as soon as they had a great deal of power at their disposal. Unless one examined these and a great many similar questions, one would not be able to get to the heart of the problems that are involved in developing new theories. And as long as one is not allowed to be free to consider all the relevant facts, it will not be possible to make much progress in this direction.

It is clear from the above that what is needed in the left-wing movement today is a certain measure of disengagement from Russia. Russia has made an enormous number of errors. We must be free to examine and admit them, not only in order to make correct theories to guide us in the future, but also to convince people of our honesty. Any group that does not do this is very likely to lose most of its influence with the people in the country in which they live. Thus, there are strong tactical reasons for such a disengagement, as well as reasons of principle. Moreover, one can hardly predict with great confidence that the Russian gov't is unlikely in the next few years to make some more serious mistakes. It is not wise to continue a policy of reticence about such things, because it is like trying to walk with your feet tied up. Of course, this does not mean that an anti-Soviet policy is called for. A real left-wing movement must be pro-Soviet potentially (i.e., to the extent that the Soviet Union develops toward true Socialism) but when bad things are done there, it should be quite ready to call a spade a spade. In this way, the members of the movement will leave themselves free to learn from all past mistakes, whoever may have committed them; and it will earn the confidence of the people as a whole in the honesty of its intentions. Moreover, it will in the long run be doing the Soviet Union a service, by not hesitating to offer its free and frank criticism (naturally, it must be ready to accept reciprocal criticism which experience has shown, the Soviet Union has never hesitated to offer).

This raises the question of the probable future of the C.P.'s throughout the world. The Afro-Asian parties will probably grow, although perhaps not as rapidly as one might think on first examination. The Western parties are certainly in a declining stage now. If they continue their present policies of complete submission to a few bureaucrats who get their direction from Russia (as from some new version of the Comintern) they will probably wither away. These tactics simply won't work any longer. The next few years will tell the story. In any case, what I feel is that regardless which way this particular point goes, it is necessary to develop a new genuinely Socialist movement in the West, which will be in its essence independent of Russia (although potentially friendly). This movement must develop a new ideology, which takes Marx, Engels and Lenin into account, which digests all the developments since Lenin, which criticizes, both positively and negatively, the Socialist movements and governments which have been set up thus far, and which tries from all this to draw conclusions concerning the setting up of Socialism in the West. This movement must also analyze and absorb into itself the best elements of the culture in the Western nations; and it must somehow appeal not only to labor and to farmers, but also to the large lower middle class-element, which is almost in a majority in America, and quite large in Britain. A tough job, you may say – but it seems to be the only way to do it.

Well, I guess that is about enough for the present. I hope that this will not anger (or bore) you. I assure you that I am doing my best to try to come to a rational view of the whole problem.

Give my regards to all

Love – Dave

P.S. Saral sends her regards. She plans to write you later when she has time.

Letter 55. Folder C49, not dated.

[Probably about November, 1956 (returning late to work. Hungarian uprising was from 23 October until 10 November 1956, Israel invaded Egypt on 29 October)].

Dear Melba

Many thanks for your very welcome letter and for your help, advice and comfort.

Well, we are now in Israel again. Whether it was a wise decision, we can't know. Many factors entered into it. First the English university people hinted to me that if I broke my contract with Haifa, I might find it difficult to get a job there. Later, they changed their mind, but when I heard of it we were all ready to go. Then, I feel that it is bad to break from a distance in this way.[2] If the situation gets very bad there will still be a chance to leave. Moreover, all of Saral's people are here, and she didn't want to simply lose all possibility of seeing them, perhaps for a very long time. Then I didn't want to sit by myself in a very unstable European situation while Saral went back to visit her folks nor did I want to suggest to her that she should not see them. So all in all it was a tough decision and who knows these days what decision is right.

The people here are quite calm, and have behaved very well. Services are returning to normal. The Israeli army has captured fantastic quantities of equipment including enough gasolene for the country for a year, 1 million blankets, besides tanks, guns, etc. People here are convinced that the Egyptians were getting ready for a very big offensive.

There is no doubt that the Israeli's were wrong to attack, even if one restricts oneself to the purely practical question as to whether they are now worse off or better off for having attacked. Yet one must consider the provocations which put the people behind this course of action and the fear which came from the frequent threats of the Egyptians to annihilate Israel with the aid of large quantities of arms supplied by the Russians. It is my opinion that Russian arms in the Middle East did not contribute to a peaceful resolution of the problems here. The arms (exchanged for the Egyptian cotton crop) did not raise the standard of living for the recipients, but they did contribute to making a solution of the Israeli-Arab relations more difficult. It

---

[2]Classes are going on as normal: but I am already two weeks late.

has been suggested by some quite leftist people in Europe that if the same resources ($ 600,000,000) had been poured into Poland and Hungary, instead of being wasted in arms, then perhaps the peoples of these countries would not have been so inclined to revolt. In any case it seems inconsistent for a country like Russia which has always claimed to want only peace to be engaged in the supplying of arms in large quantities to other nations.

With regard to the Hungarians, there is no doubt that what you say is correct. Fascist elements did take advantage of the anger of the Hungarian people and incited what was originally only a protest against terrible conditions and repression into a wild and savage revolt. Once the situation passed beyond a certain point, the Russians probably had no choice but to repress it, although they did this with what I think is unnecessary brutality. However, this is not the main point, as I see it. The main point is that under the Stalinist regime, the Russian gov't went to fantastic extremes of cruelty in the way that it treated people. I have personally talked with people who visited Russia, Hungary, Poland, etc. People who are as leftist as one could hope for. People have been tortured, beaten, kept in isolated confinement, have had their fingernails torn out, etc. And all of this has happened on a very big scale. Moreover, millions of people have been arrested without reason and sent to slave labour camps in Siberia. I have it on the word of a man (who is still a convinced Communist), who spent 10 years in one of these camps that the use of slave labor was a regular factor entering into the five year plans. The cost of this period in twisted lives, insanity, destruction of independently thinking people in the Communist movement, etc. is beyond accounting.

Of course, one will say now that it is all over, that K. and B. [Khrushchev and Bulganin] have reformed all of this. And I have no doubt that they have made sincere efforts to do this, efforts motivated in part by the necessity for correcting the Stalinist errors before their consequences became irrevocable. Nevertheless, it is not so easy to correct these errors. 90% of Soviet gov't officials, authors, artists, etc. were Stalinist bureaucrats. One cannot replace these people in less than a generation. Meanwhile, they still fight to maintain their old policy and their old positions, taking advantage of every weakness and difficulty in the policy of K and B et al. I know that a very bitter struggle is going on in the leadership of the Soviet Union. I know also that a similar struggle is going on in the French C.P., where the Stalinist leadership has effectively forbidden discussion of such questions in the French C.P. It is a fact that nothing can be discussed unless it has been printed in Humanité or in some other official Communist source. Since neither the Khrushchev report nor the Gomulka report have thus far been published in this way, anyone who refers to them is accused of using bourgeois lies (which become proletarian truths on the day that they receive official approval). As a result, there is an effective censorship on discussion, and no one can raise such questions publicly except at the risk of being expelled. Meanwhile, the French C.P. is rapidly losing much of its influence, and has become completely isolated. The country is in danger of Fascism, and yet because of its refusal to denounce Stalinism (along with its refusal to face the Hungarian problem realistically) it cannot effectively unite anti-Fascist forces, and could not even protest against the war in Suez. Humanité insists day after day that the revolt in Hungary was mainly a Fascist coup. This position is incredible; where were the Hungarian people! Why did they not rally to the C.P.? Why do they continue to strike and to make the formation of a new gov't impossible?

The basic problem seems to be that these bureaucrats don't want to give up. They still have great power; and the execution, imprisonment, and intimidation of so many independently thinking Communists has cut down the forces of those who oppose Stalinism. Then there has been enormous corruption; even more in Russia than in France. Those who go along get easy honors, high salaries, etc., while those who oppose are ruthlessly swept aside, imprisoned, tortured, etc. One may ask the question "How can people who are implicated in so many crimes bring about Socialism and Communism?"

No doubt many people thought that what happens to individuals and even to whole peoples is not important, as long as the ultimate goal is attained. But can this be true? Here, they lose sight of an important dialectical principle. For the people who carry out a certain line of action are themselves eventually transformed by their own actions. Those who have lied, framed people, tortured and imprisoned them, etc., all in the name of a better society where people could live together in a comradely way, must eventually lose their devotion to the goal and be motivated finally by other purposes. And here, one of the basic things that went wrong was the use of the systematic lie. As soon as a man opposed the direction of the C.P. he became a traitor, guilty of the most heinous crimes. Confessions were manufactured and extorted on a large scale. The truth had nothing to do with the case: what was published was only what would be convenient for the interests of the gov't. This was a direct perversion of the principle that dialectical materialism should be scientific and objective. Perhaps some people said that false confessions served the interests of a "larger truth'. Similarly, Humanité still publishes lies about Hungary, quite cynically, since the truth is evident. It is clear also that the Russian gov't publishes whatever it thinks is convenient about world affairs. Perhaps they have already ceased to lie consciously, and they may be only deceiving themselves. In fact, there is an article about the experience of the Polish gov't stating that the Russian delegation to Poland (including Khrushchev, Molotov, etc.) had a complete ignorance of the real situation in Poland, and that after they were told the real facts, they became much more reasonable. Thus, the Russian gov't became isolated from the real world, and lived in a dream world, built up by a tissue of lies sent in by cynical agents, who told the Russian gov't what it wanted to hear (while people who told unpleasant truths were dealt with by the police).

The use of conscious and systematic falsification was I think the biggest crime, and the most unforgivable, committed by the Stalinists. And they are still doing their best to prevent the truth from being known, even at this late date.

Now a very important question is whether the Russian gov't can make the difficult switch from Stalinist methods to something approximating Socialist methods. I very intensely hope that they can do it. The future of Russia, and very probably the future of civilisation depends on this. Yet, it is by no means clear that the change will work out in this direction.

Let us first consider the positive achievements of the Soviet gov't. They are very big. A very backward people has been raised from primitive conditions to something approaching a European standard of living. In time, no doubt, they will reach and surpass the British standards (perhaps 10 to 15 years). Their science, and education have reached a high level (although the level of culture in literature and art has not risen, mainly because of intimidation and repression). A war against the Nazis has been won during this time, amidst tremendous destruction and suffering.

But now comes the difficult problem. As Kruschev himself indicated, the Russian society has outgrown the possibility of functioning in the Stalinist framework. It is urgent that this framework shall change. And many steps toward a change did take place. But under the stress of recent events, there has been a tendency for Russia to return to the older methods, and very probably a strengthening of the large group of Stalinist bureaucrats who probably still constitute the major part of the Russian gov't. Russian diplomacy has begun to revert to bluster and threats. It was particularly bad when they threatened England and France with rocket bombardment, because whatever the Russians may have had in mind, the average person in Europe regarded it as a threat of atomic bombardment. Thus, the work of ten years of the peace movement was wiped out, since for many people it was by no means clear that the Soviet gov't would hesitate to use atomic bombs if they thought that they could win in this way. Indeed, for the moment, the U.S. gov't seems to have had more of a general orientation away from war than does the Russian. In any case, I myself often wonder whether people who were ready to imprison and enslave millions of people on falsified charges and extorted confessions would have strong qualms about using atomic weapons. And despite the recent very wholesome change of Soviet orientation shown by the twentieth Congress, it seems likely that many of the people responsible for all these evils are still in high positions inside the Soviet Union.

Besides the struggle between the new trends and the old Stalinist methods, many people believe that the Red Army is playing an ever bigger role in determining policy. This seems likely in the light of the fact that events in Eastern Europe constitute such a big threat to the Soviet military position. The Red Army leaders are very likely pushing for a "strong" policy toward the peoples of Eastern Europe, a policy that is not entirely consistent with the process of "destalinization" that is needed if Russia is to be able to progress to a higher industrial, scientific and general social level.

So all in all, my opinion is that while there have been many hopeful signs inside Russia, the situation is still unresolved and very complex. I hope that they will move toward real socialism, eventually to pass beyond all of this cruelty, stupidity and repression that has been so common in their gov't. But I am not so sure that they will succeed. There is a struggle inside the gov't, and from here, it is difficult to gauge the strengths of the various conflicting tendencies. The fate of civilisation depends on a good outcome to this struggle because if Russia goes back to Stalinism, I don't see how a third world war can be avoided. (Of course, it cannot go back exactly to Stalinist methods, but the general trend toward repression, brutality and stupidity could win out if the external situation got tough enough).

The Hungarian situation is particularly tragic. It is clear that by now the entire Hungarian nation to a man must hate the Russians like poison, and must identify Communism with a grim hard life and cruel repression. If the Hungarian people were given their choice, they would very likely elect a Fascist gov't. Yet, this would make the situation worse still. Thus, there is no visible way out, since no gov't supported by Russian arms will be supported by the people and any gov't supported by the people would be impossible for other reasons. Perhaps time will eventually reveal a way out.

Now for some general ideas to which I have come. It seems clear that it makes no sense to impose socialism by force on people who don't understand what it is and have not yet learnt to want it for themselves. This is what happened in Poland, Hungary, etc.; and all that was achieved was to get a lot of boot-licking opportunists to join the Communist party. They made such a mess of the countries economy that life became hard and grim for the people. This made repression necessary. They had to utilise a secret police with arbitrary powers. Such men tend to become very cruel and brutal. Besides they exterminated all those who made criticisms and in this way terrorised the people; thus further re-enforcing the tendency for gov't officials to become inefficient, opportunistic, cynical bureaucrats, who got the economy in such a mess.

It seems clear that a crucial factor now is the development of a real understanding among the people as to what socialism is and why it is necessary. Here, we must be careful to avoid the notion that its main purpose is to raise the standard of living. It is true that Socialism did raise the Russian standard of living during the past few years. But capitalism did the same thing much better not only in America, but also in West Germany (which experienced a much bigger rise in the standard of living than East Germany, where everyone who visited found life very grim). Of course, socialism will raise the standard of living. But it will do much more. It will enable people to live together in an organized, planned, friendly way, feeling secure, working together for the common good, each person free to contribute in his own special way. Instead of economic chaos, we can have a planning for the needs of society – good homes for all, intelligent location of industry, transport, etc., good schooling, giant research programs to attack disease and lengthen life. Instead of a society where each man feels alone and lost even though he is always dependent on others, we shall have one where each man feels that he has his place, that he is wanted, treasured, and loved. And he in turn will learn from childhood that the highest satisfaction in life is to help others and to contribute toward the common good to the utmost of his ability.

It also seems clear that none of this is likely to come about in a society where people fear, mistrust and hate each other, as is true all over the world, in America as in Russia. The use of cruelty, repression, and the extortion of false confessions is especially likely to interfere with the achievement of this goal, as is also the adoption of an attitude permitting the ruthless sacrifice of individuals and whole peoples to some distant goal.

It all adds up to the conclusion that some very different ideas are needed. What do you think?

Please give our best regards to everyone
Love
Dave and Saral

Chapter 20
Letters to Miriam Yevick, October-December, 1951

Letter 56. Folder C116, dated Oct 20 [1951].

Avenida Angelica 160
São Paulo, Brasil

Universidade de São Paulo
Faculdade de Filosofia, Ciências e Letras
Caixa Postal: 8105
São Paulo

Dear Miriam

I have been in São Paulo for about a week so far, and on the whole, I like it a lot better than the U.S. Of course, the American state of war hysteria is not present here. The people seem to be less tense, and more friendly than in the U.S. São Paulo however is a very busy and energetic city, perhaps the most "American" in South America. A great deal of construction is going on, and the traffic is terrible, especially because the drivers go at terrific speeds, and refuse to slow down for anything. Instead, they continually blow their horns. It is dangerous even to cross the street on foot, and as for driving a car, this would take some practice, and much aggressiveness.

On the whole, I do not feel as if I am in a fundamentally different society. The city rather resembles San Francisco in some ways. The language problem is of course difficult, but I am studying the language intensively, and hope to be able to get around in a month or so. There are many wealthy people in São Paulo, also many poor people. It is not so bad as George said it would be however. I think that I would rather live here than in the U.S. although conditions are by no means ideal. It looks as if there will be many interesting people to meet, once I know the language a little better. The department is full of people who are very far to the left.

I am now living and eating in a pension, for about $120 a month. The food is extremely good, much better than I had in the States. The room is however small and too noisy. I may look for a better room soon.

C. Talbot (ed.), *David Bohm: Causality and Chance, Letters to Three Women*,
DOI 10.1007/978-3-319-55492-1_20

The University is rather small (about 3000 people) and split up into many parts. The Faculdade de Filosofia, Ciências e Letras has about 1000 students. Eventually, however (in about 2 years) they will complete a new campus (Cidade Universidade) outside the city limits. This will be very beautiful. But meanwhile, we are in an old building near the middle of the city. I have met a few of the mathematics people, but don't know what is going on there yet. A big reorganization of all courses is now being proposed, and I am using my influence to reduce the excessive amount of abstract math. now taught to <u>all</u> students (Even physics students take differential geometry, projective geometry, topology, etc., in their undergraduate work). However, most of the faculty agrees that a change is needed. We are now trying to plan new courses in physics. The students however are very undisciplined, not being used to doing exercises, etc., and somehow this must be corrected.

I haven't done much work in the past few weeks, but have gotten a few more ideas on the relation of statistics to causality. These ideas apply both to the plasma theory and to the new interpretation of the quantum theory. In fact, I am beginning to see a very close relation between plasma theory and quantum theory.

Please write me how you are feeling, what you are doing, etc. I am sorry for you, in that you have to stay in Washington. Now that I am out of the country, I can see how bad it really was.

Dave

Letter 57. Folder C116, dated: Nov 20 [1951].

Avenida Angelica 160
São Paulo, Brasil

Universidade de São Paulo
Faculdade de Filosofia, Ciências e Letras
Caixa Postal: 8105
São Paulo

Dear Miriam in far away (and long ago) America

I just received your 2nd letter, and have not yet received your first. I hope you received the letter that I sent you about a month ago. I suggest that you write to my home address, or better yet, to the Faculdade de Filosofia, Ciências e Letras, as there are 5 independent physics departments in this university. (Each Faculdade has its own), and heaven knows in which one your 1st letter is now sitting.

Right now, I am recovering from a second attack of diarrhoea plus a bad cold and a high fever, so I don't have much strength. I'll just begin to answer your letter today, and finish later.

Your letter is like a fresh wind from home. Its good to hear of all these ideas again, people warned about being too bourgeois, etc. So far, I haven't had much of a chance to get out and meet the right sort of people, what with my two illnesses, and a trip to a scientific conference at Belo Horizonte (300 miles away). However, I hope to begin soon. Meanwhile life is a little dull; after one gets used to the new environment.

I am studying the language and can read moderately well. In addition to reading the bad news in the papers, I am reading "Diale'tica e Natureza", borrowed from the collections of Schönberg. However, my conversations are still at an elementary level, because people do not pronounce their words "clara e distintamente", as I so often tell them to try to do.

I am not very proud of the way I acted toward you when I was in the States, and I am very sorry. However, I was all mixed up in my emotions. First, as you know, I like George a lot too, and don't like to play such a dirty trick on him as to suggest that you leave him, since he seems to be in love with you. I think life would have been much simpler and better if you and I had met before you married George. Then again, just as you have a close tie with George, I found that my ties with Hanna were closer than I had thought. I think that there is more to her than you or George think. As far as becoming bourgeois fast is concerned, that is a moot question. I have the feeling, and in a way the hope, that enough troubles are in store for me to prevent me from becoming bourgeois. I was very lucky [words missing] in having just enough trouble to help me develop but not enough to overwhelm me.

I guess you are right that I am rather self-centred. Perhaps I will be able to change this too.

I was really shocked to hear about the death of Murray Slotnick. I can hardly believe it. Somehow, the thing doesn't seem to make sense, particularly the fact that they refused offers of help originally. Life hangs on some very fine threads sometimes. He was really a basically good guy.

Your story about the Prince of Egmont and the Prince of Orange is interesting. But really there would be little use in my staying in the States now, without a job, and facing a good chance of being sent to jail for one or more years. Of course, being in jail, I would have a certain nuisance value, because people could always refer to me as the physicist who was in prison. But I am just not that unselfish. Perhaps I can return to the States later. In fact, I really would like to come back some time.

Right now, I am in no condition to understand the mathematical points that you raise, but I'll try later when I feel better. I think that Khinchin makes some interesting suggestions about statistical mechanics but doesn't follow them through. I'll try to tell you about them later also.

I hope that I shall never as you say in your letter, be "lost and gone forever". Life in America seems to be so distant, except for a few very good friends like you. I miss you, the long and peaceful talks we had. You once told me that you had a feeling of peace with me. Well, I had the same sort of feeling with you, especially when we used to go for drives in the [illegible] through the countryside. I have a desire for inner peace, but also for excitement and struggle with nature, and gaiety in relations with people. But none of these comes to pass, except occasionally to a small extent. Many of these things are lacking here in South America, partly because I haven't taken enough initiative in meeting people, and partly because of language difficulties and partly because the people who I am now in contact with are not really suitable.

It is very hard to define the differences and similarities as compared with life in the USA. Sâo Paulo is an industrial city, resembling a cross between San Francisco and Chicago, except that is isn't as clean as San Francisco. There are beautiful apartment houses in the midst of extreme poverty. In my trip to Belo Horizonte, I saw much greater poverty, for people lived in shacks without glass windows and often with thatched roofs. They looked quite undernourished and spiritless. Generally speaking, South Americans (even the middle and upper classes) don't have the push and the "oomph" of North Americans (not even in Sâo Paulo, where people do work pretty hard). Life is definitely more primitive, except for a favored few of whom, alas, I am one. I can't talk to people much yet, especially because the poor people's pronunciation and grammar are so terrible that I despair of ever being able to follow them. Also they don't know the words that I learn in books.

I am now continuing this letter several days later. The illness became a lot worse leading to high fever, etc., so I called a doctor. He prescribed for me a new drug, chloromycetin, which is literally worth its weight in gold (in cost that is). It is an antibiotic like penicillin, but can cure all kinds of things like typhus, typhoid, atypical pneumonia, etc. Apparently, it is good for the bacillus "salmonella" which the doctor supposes is responsible for my trouble. I now feel much better, although I am still weak. It will really be necessary for me to be very careful about my food. When I feel a little stomach upset or tiredness, the meaning may be very different from in the States, because it doesn't go away when I rest, but slowly and inexorably gets worse, leading to fever, nausea, and diarrhoea. I seem to have no resistance against these damned bugs.

I received your 3rd letter this morning, but thus far not the first letter. Your analysis of the problem of chaos was very good. However, I am not yet in a condition to discuss science and math, so I think it is better that I send this letter now and answer your scientific questions later.

I shall also try to answer your questions about the value of scientific research later. Right now I shall mention only one small point. I knew a fellow once who caught atypical pneumonia. He was in the hospital for a few days, and seemed to be recovering, but suddenly got worse and died in two hours. He left a young wife and child. This was before the discovery of chloromycetin. 20 or 30 capsules of chloromycetin would have saved his life. It is on these apparently small and irrelevant material threads that the highest goals of humanity often hang, because life is after all a property of matter, that arises out of matter, and whose laws of development are based on the laws governing matter. Unless the material basis of life is understood and controlled, life cannot attain its full potentialities,and since matter is so interconnected no study of its properties in any one aspect will be fruitful in the long run unless the properties are also studied in apparently unrelated fields. It is my opinion that humanity has no choice but to continue its efforts in the direction of understanding and control of matter not only for reasons of its own health, etc., but also because it is in a sense, the <u>mode of existence</u> of humanity to do so. In other words if people do not try to satisfy their curiosity and increase their control of nature, they stifle an important part of themselves and cannot become complete human beings. This doesn't mean that life consists only of research, but I do believe that the lack of interest of most people in nature is a sign of a kind of illness, or atrophy of certain parts of their mental and emotional structure, and that they are actually taught to lose interest. Knowledge by itself doesn't lead to necessarily good results, but without understanding and knowledge even the best of intentions must lead to disaster in the long run. And I repeat, it isn't enough to to understand human beings at the level of human beings alone, because human beings are made of matter, must have matter to live, and must have their whole ideological and "spiritual" life strongly conditioned by the existing material basis. If humanity does not control this basis, it is blown like a leaf by every chance wind. In order to really become human, people must control and understand inert matter, not least to the extent of being able to provide against famine, disease and various types of natural disaster. This knowledge is the pre-condition for the existence of a real humanity, but it requires the concerted and intelligent efforts of people to bring about the social changes that are made possible by our new understanding and control of matter. However, without knowledge, then nothing would be open to humanity but famine, war, slavery, and culture on a very low level. Tremendous possibilities have been opened up by scientific progress, and to lose sight of them is to fall victim to the propaganda that the reactionaries put out – namely, that a change in the material basis does not really make possible a change in society.

I hope you will excuse the general confusion in the last paragraph as I am still not able to organize my thinking with energy (because of an inadequate material basis, namely the bugs in my blood)

Now for a few remarks on Brazil. The people here have a character of greater gentleness than in the U.S. Some of this is docility, because of the tremendous poverty suffered by most people. So many people do not even conceive of the bad deal that they are getting. However, on the whole, the degree of frustration does not seem to be as great as in the US. You don't see the worried tense anxious baleful looks on people that you see in the U.S. Somehow, though, as I said before, I feel that they lack a certain zest, a "push" that you find in the U.S. (Perhaps I am romanticizing the US in my memory).

The University is highly disorganized, but there are a few good people in the physics dept. There is going to be trouble in the field of atomic energy, because the military people have just concluded an agreement with the USA whereby the USA gets Brazilian thorium in return for a small pile, to be built by the US in Brazil. Most Brazilian physicists oppose this pile; (a) because it will siphon off too many physicists for almost no purpose, scientifically speaking (b) because it will contribute to military control and secrecy (c) because the country is not ready for it; since more elementary forms of research need to be developed first. However a few "big-time operators" like Lattes are apparently backing the military.

Incidentally I met Feynman at Belo Horizonte, and convinced him that there might be something to my interpretation of quantum theory. I also got a letter from Pauli, indicating that he is coming around a little (but just a very little). I have some new ideas on the subject, relating to plasma theory, but will try to tell you more later.

Love
Dave

Letter 58. Folder C116, dated: Nov 23 [1951].

Avenida Angelica 160
São Paulo, Brasil

Universidade de São Paulo
Faculdade de Filosofia, Ciências e Letras
Caixa Postal: 8105
São Paulo

Dear Miriam

I am feeling lonely and tired tonight, but I will have to substitute for a visit by writing you. There is still a trace of this damned bacterial infection, which leaves me dopey and spiritless. I really must begin to widen my circle of friends, as I know that there must be hundreds of good people around here that I would like to know, many of whom speak English.

With regard to your question about the value of knowledge about the material world, I mistook your meaning in my last letter, partly because I was confused on account of illness, and partly because I have been having arguments with an American Research worker named Smith, who often asks whether mankind would not be happier in the primitive state. Smith is a good part, but not very sophisticated in things political. Nevertheless, he is honest, and can see the terrible results of American policy.

I am still not ready to answer your question because I must think some more. In my dispirited present state, I have a tendency to give in easily. Also, when I read the news, I no longer feel anger, fear or horror. The magnitude of the catastrophe and its apparently inexorable development just leave me almost numb. What can be done by anybody in the USA? Today without any proof, and even in the face of warnings from Gen. Ridgeway, there seems to have arisen in the US an angry readiness to drop atomic bombs, just because some Colonel said that American troops have been murdered by the Chinese. The newspaper stories give one the impression of an atmosphere charged with hatred, which is ready to precipitate out on the slightest disturbance. This hatred has nothing to do with anything specific, it seems to be the result of the cumulative frustrations of American life, for which people need a scapegoat on which they can vent their feelings. Another point that I want to raise now just for the sake of communicating it to you is that in South America, much of the motive for "revolutionizing" physics seems to have left me. In fact, to some extent, I am now sure that my work on the new interpretation of the quantum theory and on the collective treatment of electrons had the emotional significance for me as a form of protest against the existing irrational order. I wanted to show that the concepts implicitly behind all American life, which had seeped even into scientific research, were in actual disagreement with the material structure of the world, and were therefore likely to end up sterile and fruitless in the long run. Here in South America, the irrationality is not so maddeningly obvious, and one easily falls into a habit of forgetting about it. To some extent, this is good for my scientific development at present, because I need to develop these ideas in peace for a while. But there is a lack of "ideological" stimulation for me here. Also, an important part of my way of thinking is through discussions with people in which I try to make the ideas clear, and this is difficult with people whose English is poor, as the flow of ideas more or less loses some of its "momentum". As for the book on dialectics, this also could not be done under present conditions, because of lack of adequate interest and stimulation, as well as because of difficulties in communication. That will have to wait for the future I am afraid.

It may have been, as you suggest, a bad idea to leave the US, but perhaps it would have been equally bad to stay, considering my probable fate. In return for your question to me, I ask you "What would have been gained if I had stayed, in comparison to what would have been lost?"

Next day: I feel a great deal better this morning but it will probably be a day or two before my strength returns to normal, as I still need to sleep about half the time. I can now see that some of my periods of exhaustion plus intestinal disturbances in the States may have been due to bacterial infections that weren't serious enough to lay me out. The unfamiliar bacteria down here however seem to be more dangerous to me and I'll have to watch out for them. You just can't do good scientific work when half of the life is sapped out of you.

I was interested in what you said on strong laws, and would like to hear more. Khinchin says something similar, when he refers to events which have a measure zero, or a very small measure. I have a specific question however: Give me a specific simple case in mechanics where after an infinite time, certain classes of events happen only a finite number of times.

I have some ideas on the relation between plasma theory and quantum theory which would be hard to communicate by mail, because they are not as yet very coherent. As for your summary on chaos and order, I would add only the following point: Because of the existence of an infinite number of levels, the deterministic laws of order at each level probably follow only as a result of conditions of chaos existing at lower levels. If the lower-level conditions of chaos could be altered, then the very framework of description of the higher level laws would also have to be altered. Thus, we are led to a more dynamic concept of the laws of nature; for because of their infinite complexity, richness, and depth, the applicability even of certain very general forms of laws at a particular level may depend on conditions at other levels, which are in principle subject to our prediction and control. This experience should ultimately be repeated at any given level, however deep, as our knowledge is extended.

I still haven't heard from the French physicists who were working on the relation between quantum theory and Einstein's field theory. However, Tiomno and I have worked out a preliminary formal [word missing] that may be on the lines on which they are working. Also, I sent a brief article to Massey with the suggestion that he publish it in Nature, and telling him that I hope to visit England this June.

As for statistical mechanics, I don't think there is a really good book on the subject. I was thinking of writing one here, but wonder whether I will stay long enough. I think it is a good idea for you to read up as much as you can on statistical mechanics and kinetic theory. As you suggested, a study of dependent probabilities is a step in the analysis of chaos. In fact, a principle problem is to devise a way of describing this dependence quantitatively in a system having many degrees of freedom.

Please continue to work on these problems and to write to me.

Love
Dave

Letter 59. Folder C116, dated: Nov 28 [1951].

Number on photocopy: 4

Avenida Angelica 160
São Paulo, Brasil

Universidade de São Paulo
Faculdade de Filosofia, Ciências e Letras
Caixa Postal: 8105
São Paulo

Dear Miriam

Well, I have recovered more or less completely from my illness, so I am ready to cover some of the things that I didn't answer in my other letters. I shall be taking an apartment with my American friend soon (Smith), but I shall continue to eat at Avenida Angelica 160. Either this address or the one printed on the letterhead is a good one.

My illness left me somewhat shaken in my convictions about the value of my work. Also, I am somewhat lonely, and a serious problem will arise in that in March, Tiomno, the best of the people here in theoretical work, is leaving for Rio, so that he can join his wife, who is teaching physics at Rio. This will leave me even more on my own than I already am.

Nevertheless, I still think that scientific work by me might be of some value, especially considering how little I can do in the USA in other fields (because I am a marked man). People's ideas are in the long run influenced by their concepts of the fundamental properties of matter. And the idea that the properties of matter are understandable in a rational way, as well as of much interest and intellectual beauty, will help move people's ideas away from the confusion and mysticism in which they are now mixed up.

Thus, a characteristic attitude of people toward life is a cynical one, "relativistic", in the sense that morals and responsibilities are said to be determined only by the prevailing society. If one happens to be in a Nazi society, then one naturally adopts Nazi morals, etc. In America, one adopts the prevailing "American Way of Life". This is the counter-part of positivism, for it says that there is no objective material basis for morals, but that all is determined by a commonly agreed upon convention, which introduces "order" into the system of behaviour. The materialistic point of view is that morals grow out of the material structure of human beings and out of the material relations between human beings in society. Although there is much room for variation in such a development, there are certain limits (at any stage of development) which must be satisfied, if development is to proceed in a constructive way, and if decay of the individual and of society is to be avoided.

Another important concept that must be gotten across is that of the infinite number of levels, that must be used in describing the behaviour of matter. Such a point of view automatically prevents us from closing our concepts, at any particular level. The social counterpart is obvious. Thus, people have adopted the point of view implicitly that human development will take place within a certain closed framework, i.e. in which each person is primarily interested in the welfare of himself and those close to him, and in which, because of altruism, or intelligent self interest, he may be willing to cooperate with others. This naturally leads to an extremely limited view of the possibilities of socialism, a view which makes it hardly worth aiming for, since it just would make everyone a minor sort of capitalist. On the other hand, we know that there are other levels of personality, hardly tapped in our society, which could make a close relation to society and the service of society one of the most powerful satisfactions to the individual, and which could release new powers of thought and action, now largely suppressed.

Now you may ask, "what has science to do with these general ideas". The answer is "a great deal". Science now has a great deal of prestige, which it lends to ideas of positivism, "relativism", etc., all of them leading to discouragement and to the general idea that there is no over-all goal, but just a bunch of isolated individual goals. Moreover, just because the philosophical attitude of science is so lukewarm and negative, its influence is much less than it might be, because such doctrines are not attractive to most people. Also the lack of clarity of such doctrines reduce their effectiveness. A clear, sharp, optimistic, materialistic point of view, promising infinite possibilities of development, and the possibility of growth of a common human purpose having objective existence within the human race, as a developing thing, would have far more appeal, particularly if it could be backed by some striking success in dealing with nature. It would be a powerful talking point to be able to say "the old tired, discouraged, defeatist, negativist (so-called "positivist") point of view failed, obviously because it closed our eyes to all the possibilities in nature at the atomic level. How much more do we expect it to do the same at the human level? At both levels, progress is possible only if we adopt an optimistic, dialectic forward-looking point of view – dialectical because it must take into account the infinite possibilities for development that flow out of the opposition of all the opposing elements in the universe.

You may say that whether or not we have war is the most important question of all. In this, I agree with you. Yet you must admit that just because of my peculiar position, I can have singularly little influence on this question. On the other hand, I may be in a position to have much influence on these scientific questions. Whether or not there is a war, these philosophical questions will be important in the long run, because I do not believe that civilization will actually disappear, however great the destruction. And if one is considering a scale of time that goes beyond one's probable life-time, one may as well make it a long scale, since to the individual concerned, time will then make no difference.

I hope that this gives a good idea of my point of view. I am very anxious to hear what your opinion is, as I find that this question affects me much more than I thought. If I can't justify my existence here, then I just won't be able to stand it, and will have either to go elsewhere, or face whatever I have to face in the USA. So please try to give me some good advice.

As for the book, I am constitutionally unable to write it by myself. I really need criticism, stimulation, and encouragement.

Now for some of my ideas on probability. I believe that Khinchin's most important idea is contained in the region around pages pp 57-61. The basic idea is that the use of an ensemble arises because as the point moves in the space of φ and ψ, then the integral of the motion

$$I = \varphi\psi' - \psi\varphi'$$

can have an incommensurable set of values when the system takes on integrally spaced values of φ and ψ corresponding to the same state of the system.

A rather similar idea which is I think closer to the actual mechanism of the origin of chaos is as follows. Consider a phase point $(p; q)$, which we take for simplicity to be two dimensional. If the system is defined within a region $\Delta p \Delta q$, then Liouville's theorem states that this domain $\Delta p \Delta q$ is a constant of the motion, even though the shape of the domain may change in a practically arbitrary way. At first sight, it would then seem that no need for probability could ever arise, since the phase point is, after all, always in a small but definite domain having a volume $\Delta p \Delta q$. However, under conditions of chaotic development (which would actually tend to occur mostly with complicated systems having many degrees of freedom, but which we may discuss 2 dimensionally for the sake of argument), the shape of the region, $\Delta p \Delta q$, will become very complicated. Such conditions will arise when:

(a) A slight change of boundary conditions will ultimately produce a large displacement of the phase point.

(b) There is no appreciable correlation between the phase point, $p(p_0, q_0)$ and $q(p_0, q_0)$ reached by the system initially at the point (p_0, q_0) and the phase point $p(p_0 + \delta p_0, q_0 + \delta q_0), q(p_0 + \delta p_0, q_0 + \delta q_0)$ reached by a system initially at the neighboring $p_0 + \delta p_0, q_0 + \delta q_0$. Under these conditions, a region $\Delta p \Delta q$, initially simple and rectangular like this:

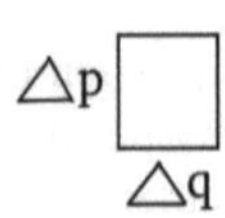

would be drawn out into a fine and tangled thread having, however, the same volume

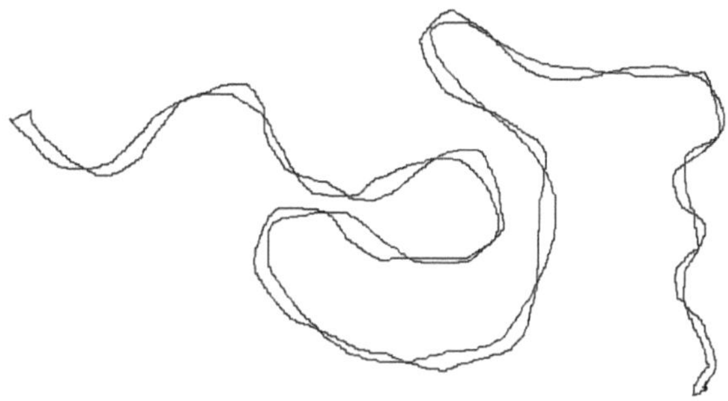

Ultimately the thread should grow so mixed up that it would be likely to cross any moderate sized region such as $\Delta p \Delta q$, many times. Thus, we should say that even though a phase point started at a fairly well defined place, and even though the region of phase space in which it can be remains equal to $\Delta p \Delta q$ in volume, we cannot for practical purposes predict the actual location of the phase point, but can instead set up a measure of the probability that it enters a given region. In order that probability calculations be useful, it is necessary that correlations between different possible positions shall tend to vanish as time goes on, or that (for example)

$$\overline{\overline{p(q_0 + \delta q_0, p_0 + \delta p_0) p(q_0, p_0)}} \longrightarrow 0 \text{ as } t \longrightarrow \infty$$

where the average is taken over δq_0, and δp_0.

If one could carry out the program I sketched above for an actual system, one would have a fairly good quantitative description of chaos. However , in statistical mechanics, such a description would still be too detailed, because most of the regions of phase space lead to large scale averages (such as pressure, temp., etc) which are not significantly different. As far as statistical mechanics alone is concerned, one would ask the question "Suppose a system started out with a fairly well defined point, in a small region, $\delta p \delta q$. As the volume element became tangled in phase space, we would want to know whether these tangles would correspond to appreciable possible changes in macroscopic quantities, and we know from experience that for an overwhelming measure of possible positions, no appreciable fluctuation in macroscopic quantities would occur. Whether such a treatment would be adequate in the non-equilibrium case is however not clear, but from my present intuition on the subject, I believe that it would often be adequate, although for some purposes one would be interested in following the more detailed implications of the motion of the phase point, and would not be satisfied with a description of the changes in macroscopic quantities.

Please let me know what you think of these ideas, and let me hear from you soon.

Dave

Letter 60. Folder C116, dated: Nov 30, 51.

Number on photocopy: 5

Avenida Angelica 160
São Paulo, Brasil

Universidade de São Paulo
Faculdade de Filosofia, Ciências e Letras
Caixa Postal: 8105
São Paulo

Dear Miriam

I am just writing you a small addition to the answer to your question about the value of scientific work. You may have noticed that the pope recently announced that scientific calculations verify the existence of God. I quote a headline in one of the more sensational papers of Sâo Paulo "The existence of God proved scientifically." And how was this miracle accomplished and by whom? Apparently by none other than G.Gamow, who works for the George Washington University, which is, I believe a Catholic Institution. On the basis of almost pure speculation, Gamow has spun a pretty theory, of the following kind. There is much evidence that something unusual happened several billion years ago. (a) Radioactive evidence in the rocks. (b) The stars are each receding from another at a rate which if extrapolated backwards suggests that they all came from the same general region several billion years ago. These are well verified facts, and with them, one can have no cause for dissatisfaction. But Gamow now proceeds to extrapolate the general theory of relativity, and deduces that the universe began as a ball no bigger than your hand at a certain time about 2 or 3 billion years ago. The mathematics indicates that there is a certain time, $t = 0$ before which there was no time. Gamow (and others) therefore deduce a "beginning of time" and an "act of creation", which was the event that started the universe expanding, and caused "time to begin". Before this time, there was presumably no time, no space, or no differentiated matter of any kind, and the intervention of an external being was clearly needed to get things going. The pope is very happy with this for two reasons: (a) The paradox of the beginning of time is beyond human understanding, thus creating a new mystery paralleling the trinity in its fathomless depths (b) There is clear evidence that matter did not exist for ever, but was effectively created by God in the act in which He caused time to begin. The pope says these things himself plainly and openly.

In the midst of all this beatific satisfaction with the new reconciliation of science and religion, a number of weak points in Gamow's work appear to have been forgotten. (a) The present laws of nature, especially the general theory of relativity, were deduced from facts obtained with a generally low density of matter. Let us admit that all the evidence suggests that the density was much higher several billion years ago than now. But if the density was as high as postulated by Gamow and the Pope, it is very doubtful that it is correct to extrapolate the present theory of relativity into these domains. It is very likely that if the correct laws were known, they would not lead to the paradox of the beginning of time, but would just describe a situation in which the universe underwent a cataclysmic change from one general regime of organization to another.

(b) As to be expected from reactionaries, Gamow and his colleagues neglect the possibility that at such high densities of matter, the properties appropriate to as yet unexplained levels may come into play, thus totally altering the situation, and making meaningless the extrapolation of the laws appropriate to the levels <u>now</u> known to us. Gamow like so many other physicists, closes his conceptual structure, then extrapolates his concepts into new domains, and then miraculously comes out with unfathomable mysteries, which are almost certainly a consequence of his neglect of significant factors of which we are as yet unaware.

In any case, it can be safely said that the facts known to science do not warrant the conclusion that

(a) Scientists have proved mathematically that there must have been a beginning of time

(b) That God must therefore be responsible for the initial act of creation

These conclusions are a pure web of fantasy, but it is spun so skilfully by men like Gamow that even to many capable physicists, conclusion (a) appears to be sound. And from (a) to (b) is only a small and easily taken step.

Now comes my point. As you know, the Pope is a clever and skilful tactician, fighting a desperate losing battle to save the Catholic Church. Apparently the Pope does not share your doubts about the importance of scientific research. He knows that science has much prestige, and that it is ideologically the principal enemy of religion (which latter is of course ideologically one of the main props of the existing order). He therefore welcomes scientists who effectively turn traitor to science, and distort scientific facts to reach conclusions that are convenient to the Catholic Church. And the fact that papers in such a backward place as São Paulo give this statement the first headlines shows that newspapermen also understand the importance of scientific concepts, as they bear on the general human point of view. Enough said.

Dave

Letter 61. Folder C116, dated: Dec 3, 1951.

Number on photocopy: 6

Universidade de São Paulo
Faculdade de Filosofia, Ciências e Letras
Caixa Postal: 8105
São Paulo

Dear Miriam

Thank you very much for your letter. I am waiting for your answer to my argument about the possible value of scientific work at this time, also for the "more and more interesting news in a few days" that you talk about. My health has improved considerably, but I am still always on the verge of becoming sick. I have a continual tendency toward diarrhoea but not enough to make me sick. In many ways, however, this was almost my normal state in Princeton. I am convinced that this is due to bacteria, because I can clean it up with chloromycetin (at a cost of $ 100 a gram) but it always comes back in a few days. I am sorry to hear that you too are not feeling well, and hope that you will get over these difficulties soon.

I am beginning to meet more people, and am beginning to like it better here. Most of the people I meet seem to be socialists of a sort, but strongly (more or less) anti-communist. There are many refugees from Austria, Germany, Hungary, etc. There are many communists here, even in the University; but I have not yet had occasion to meet any. Life in Brazil is best characterized by the phrase "mais ou menos", which literally taken means "more or less" (i.e. approximately), but which may be applied to the way everything is done; that is, without accuracy, force, or conviction. It is a national habit, reflected in the gov't, the way the railroads, planes, etc. run, and everything else. The atmosphere is therefore not too stimulating intellectually, although São Paulo is definitely better than the average American city, since even though it is not positively stimulating, it is at least not stultifying. One thing I miss very much is conversation with people who can share my ideas and who speak fluent English. I depend much more than I thought on such conversation, in order to bring out dormant ideas into a definite form, and to prevent them from continuing to "sleep" forever, as they tend to do without that stimulation. I find that the "flow" of ideas is important in this process. Thus, a smooth flow causes each idea to lead to another, and I come out with conclusions which I would never have expected at the beginning. This fluent talk is almost as important as the opposition and stimulation of a person who is strongly interested in the same problem, and who understands the background well. Last night, I discussed philosophy with a man who spoke German but understood English. So I talked English and he talked German. Even though it was a bit hard, it was a real pleasure to discuss with someone without having to go slow, to hold back one's speech, in order to permit the listener to follow. And I learned quite a bit, sharpening up my ideas, etc., even though the man's understanding of the problems was only moderately good.

As for Portuguese, I am learning it. I can read quite easily, but cannot speak very much. I shall have to make an effort to practice speaking. In the house where I live, English is the first language, German the second, and Portuguese a poor third.

I am moving into an apartment with an American friend, Smith, who is also working at the University. He is somewhat young and naive, but I rather like him. And an apartment will be much more pleasant than a room in a pension. However I shall continue to eat my meals at the pension (except for breakfast) because the food is very good. I suggest that for the time, you send letters to the address printed on this letterhead. It is a perfectly good address, except that no mail is delivered there on Saturdays.

As for your discussions about making fundamental changes in your place of life, I will now repeat the advice you once gave me and urge caution. This is a complicated problem and difficult to discuss, particularly by mail. Perhaps it is even unwise to "commit your conscience to what you may write down in a letter". After all, your conscience should be a private thing, and not shared with too many people.

I was very disturbed to hear that Hanna is having a hard time. As I told you, I still feel attracted to her in many ways, and I feel worried about her, because I don't think she can cope with the problems that with which she is facing what must seem to her an almost hopeless and meaningless struggle, it is almost as if I am feeling the same things myself. I feel sure that it would be unwise for Hanna and me to marry, if only because of the terrible irritation that develops in me when she begins to express political and other opinions without any basis, repeatedly and without any effort to obtain more information. But I still cannot avoid feeling a desire to help her solve some of her problems, if I can. Also, as I said before, I think that if she could solve these problems, she would have considerable talents in certain directions.

I would now like to begin to write down a few of my ideas on probability (in lieu of the much more pleasant process of telling them to you directly). I may repeat what I have said before to a large extent, but I want to try to present a complete picture.

1.The notion of "probability" and a "statistical ensemble", while useful in many problems, suffers from several disadvantages. First, it is an abstraction from reality to define an individual as a member of an ensemble. As you pointed out in your letter, the ensemble can be used only because the causally determined properties of the individual lead to a chaotic behavior that can be numerically described to some extent by a probability function. 2. The concept of an ensemble applies only to a static situation, in which a certain range of conditions do not change, as an experiment is repeated time after time. In non-static (or non-equilibrium) systems, the concept of the ensemble will generally fail to be useful, because to use the static statistical terminology, the "probability distribution" is changing in accordance with the causal laws and with the precise non-equilibrium conditions with which we started. 3. The concept of probability has an intrinsic vagueness that makes it unsatisfactory when applied to <u>real things</u>. Of course,the abstract definition is precise enough, but the relation to the real world is extremely vague. Let us define probability of a given event as the limit of a fraction of the number of times it occurs in a series of tries as the number of tries approaches infinity (provided that the limit exists). But let us [text missing] large. Suppose that we obtained a number of "successes" very different from the probable number calculated according to certain hypotheses. What can we then conclude? We can never conclude from this that the hypothesis is <u>certainly</u> wrong, since implicit in the use of a probability function is the assumption that no matter how improbable a given series of events is, there is no contradiction with anything if it actually occurs. All we can conclude then is that if a "highly improbable" series of events (as calculated from a certain hypothesis) occurs, this hypothesis is "probably" wrong. But what can we mean by this? We are led around and around in a conceptual circle, and the real meaning of none of the concepts can be pinned down (that is, the meaning as applied to the <u>real events</u>, and not as applied within the abstract definition made by the mathematicians).

Now, from the causal point of view, all becomes simple and clear. We are led to use probability (**a**) When we do not define a system completely (e.g. when we measure the pressure of a gas, we leave 10^{23} variables, the molecular positions and velocities undefined).

(**b**) When the undefined variables are large in number, and when the causal laws are such as to bring about a chaotic behaviour of these undefined (or hidden) variables.

If, for example, we start with a gas in a small container, and allow it to expand into another evacuated container, the usual theory says that the system will "very probably" expand to fill both containers approximately equally. However, from the causal point of view, we say that we have not defined the system very well. Most of the molecular configurations consistent with our definition of the pressure will lead to an expansion in which the gas will fill both containers approximately equally. This follows from the intrinsic tendency for the degree of chaos to increase, as a result of the form taken by the causal laws. In some few configurations, however, it is possible for the degree of chaos to decrease for a while. The "probability" assigned to such an event may be 1 in 10^{100}. If this event happens, then we say that is no contradiction with the causal laws. If such an event happened again and again, we would begin to suspect that our assumptions concerning the causal laws needed to be modified. We could then make investigations of the individual atoms to see whether the causal laws did in fact need to be modified. Thus, we are not led into a vicious circle which the purely statistical point of view implies. For if we stick to a purely statistical description (as in the usual interpretation of quantum mechanics), then no matter how often an improbable event happens, it is not really a contradiction of anything. In practice, the physicist is led in such a case to consider a modification of the causal laws by modifying the wave equation, which in turn modifies the predicted "probability distribution". But there is a conceptual problem here that can be solved in principle only by an appeal to the possibility of testing the causal laws in each individual case.

I don't know whether I made my ideas clear above, but to sum up, I believe that the application of probability to real material systems suffers from an intrinsic lack of conceptual clarity as to the meaning of the terms, which can be removed only by appealing to the concept of the complex and chaotic individual system. In many applications, this lack of clarity is of no practical significance, but nevertheless, the lack of conceptual clarity hides a real problem which becomes important in those cases not treated effectively by present methods, i.e. the cases of "improbable" events, or of systems far from statistical equilibrium.

Another case arises in statistical mechanics in connection with the "H theorem" (see Tolman for a discussion of this). The H theorem states that a certain quantity, H, (equal to the entropy in the equilibrium state) tends very probably to increase, if it should happen to have a value differing from its maximum possible value. But here we get into trouble. For the laws of motion of the particles making up the system are reversible in time. Thus, we conclude that the quantity H most probably decreases as we go back into the past also. This means that the "probable" value of H looks like this as a function of the time

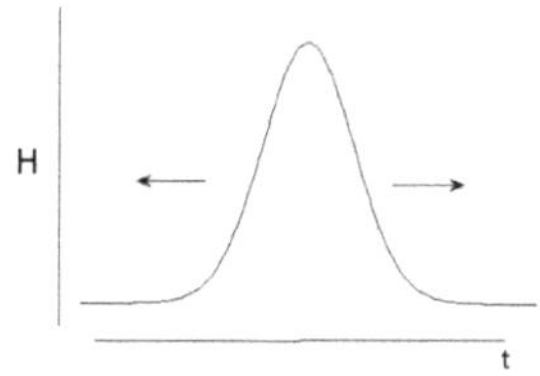

Physically, this means that if *H* differs from its equilibrium value, it is most probable:
(a) That it will fall back to its equilibrium value as time increases
(b) That it differs from its equilibrium value because of a highly "improbable" fluctuation, which occurred in the past.

In cases which occur in practice, this description is fantastically far from the truth. A system differs from a more chaotic equilibrium configuration, not because of an "improbable" fluctuation, but because it has been treated in a definite way in accordance with definite causal laws. Thus, if we consider our example of a gas allowed to expand into an evacuated chamber, the initial presence of the gas in only one of the chambers is not a result of a fluctuation of very low "probability", but it is a result of the essentially non-equilibrium character of the initial state of the gas. Now in the gas, there are as I have said a (very large) number of coordinates not defined by specifying the macroscopic parameters. It would be wrong to define these by the implicit assumption that every non-equilibrium system is the result of a fluctuation from equilibrium. Instead, the facts are that every non-equilibrium system is a product of an increase of chaos in a system that was initially even further from equilibrium. The distribution of molecular parameters therefore depends (in general) in a fundamental way on the specific history of the system, its rate of approach to equilibrium, etc. The purely statistical point of view becomes inadequate here, and leads to a fantastic description in which, for example, the earth and everything on it (oceans, waterfalls, people, etc.) are a highly "improbable" fluctuation away from statistical equilibrium. Clearly, the abstraction of a statistical ensemble fails when applied to systems far from statistical equilibrium. Here we need to use causal laws, combined with the concept of the development of chaos.

I hope I haven't bored you by repeating these things too much, but I am also trying to clarify them for myself.

Let me hear from you soon
Love
Dave

Letter 62. Folder C116, Dec 5 [1951].

Number on photocopy: 7

Universidade de São Paulo
Faculdade de Filosofia, Ciências e Letras

Dear Miriam

Something alarming happened today. While I was in the physics dept, a man came from the American consulate, and asked me to go down there. I went with him, and there they asked me to register. After I registered, they refused to give back my passport, saying that I could stay in Brazil, but that I could get the passport back only if I returned to the United States. They refused to give any more information.

So now I am up in the air. The best interpretation is that they just want to keep me put. The worst is that something is really cooking. I would appreciate it if you would watch the papers carefully to see what is happening in regard to the "Weinberg case", etc. Of course this means that the trip to England this summer is off.

This event leaves me feeling a little numb. The complete uncertainty about the future is especially bad. Perhaps I shall have to accustom myself to the idea of staying here for a long time. Incidentally, when we were driving to the consulate, I talked a bit to the man, and he seemed to know a number of things about me, which suggested that they are keeping some kind of check on me. How far this extends I don't know, whether as to mails, travelling, etc. Incidentally, they may get around to checking your mail too. At least this is a possibility.

Meanwhile I have to keep my balance, by trying to meet more people, and by working with more spirit than I have had before. I am particularly interested in pushing my new interpretation of the quantum theory, as well as the theory of probability, etc. I have written to many people to find their reactions but have received no answers yet. Incidentally have you heard what Mr Sheikin [name not clear] and that Indian fellow think about my theory. Neither of them has written to me at all, and you promised they would. So many people are skeptical that it is always refreshing to hear from people who are willing to consider a new point of view. And unfortunately, these days you can find few Americans who are willing to do this. It is particularly necessary at this time for me to hear the opinions of other people, as it will help keep me from becoming stale and from feeling isolated. So would you please let these people know that I am still interested in hearing from them. I have answered Pauli and am awaiting an answer from him, and am still awaiting an answer even from de Broglie (Thus far have I sunk!) Incidentally, I sent a manuscript on quantum theory almost a month ago to Massey in England, with the suggestion that he have it published in Nature. I still haven't heard from him yet. I have a suspicion that he is losing his nerve, and is both afraid to send it in and afraid to tell me that he is not sending it in. But in a few weeks I shall know (That is, if I am not kidnapped and spirited back to the good old USA as that fellow in Holland was). Seriously though, I doubt that any such thing will happen, but believe that they have now learned more genteel ways of doing the same thing.

Now that I am in danger of being forced to leave Brazil, I find that it is not such a bad country. There is much misery and poverty here, but from the little I know, I doubt that the poor people are as unhappy as they are in the U.S. Life is not so tense and full of fear. Everything is disorganized and not much gets done, but Brazil is the first place I have felt safe and relaxed for more than 10 years. Even if the feeling of security was illusory, it reflects better relations between people. I am just beginning to learn the language, and I can see that there are many nice people here to meet(except that it could be dangerous for me at this time to meet some of them in public). The people I know now are also very nice (as human beings) but there is a feeling that there must be a barrier that prevents the free sharing of everything. But I know that there are people here with whom this barrier is unnecessary.

I can feel my character changing again. My new-found sense of security is leaving me, as I face the long arm of the U.S. carrying the American State of terror and irrational hopelessness down into Brazil and through the whole world. I can now remember vividly the state of mind I associate with the U.S., particularly with N.Y. I wonder, will there ever be peace for us? I wish I had made better use of this brief interval of peace in Brazil. When will we see each other again and under what conditions? Is there any hope that this world-wide catastrophe can be averted? Sometimes I feel as if my life must be near its end, because I cannot see the possibility of maintaining life in the times that seem to be coming.

To come back to a more mundane subject, I must let you know that we have a salary now. I shall now get 9,000 cruzeiros a month, 12 months a year (Roughly $ 1000 at official exchange rate, $ 800 in purchasing power).

(Next day)

On thinking things over last night, things do not seem quite so alarming. The consulate claimed that they had sent me a letter on Oct 12, asking me to report to them to register. I never received this letter, but if they sent it (as they probably did) then this means that they had in mind to take away my passport as soon as I got to Brazil, even while I was in the States. Therefore they probably do not plan any immediate action, and nothing has probably changed since I left the States. However, they are watching me. Smith told me that he saw a car remaining in the neighborhood of the house and circling around for an hour or more last night. They probably wanted to see my reaction to having my passport taken away. I may assume that they will continue to watch me, at least to some extent, but this is of no importance, since I am not going to do anything. Nevertheless it is unnerving (a bit). Incidentally, what did you say in that 1st letter. I hope it didn't have as many references to touchy subjects as your later letters had.

This thing is not getting me down very much, (as long as it doesn't get worse). I got your latest letter this morning, but will answer it later. I got another offer with a letter containing an offer of a job at the University of Southern California (subject to approval of the university authorities of course). Your discussion of relative effectiveness in different places is interesting. I think that I will still have about 6 months at least before troubles get so hot that the present unstable situation changes. Meanwhile, it is necessary to think, and to meet as many people as possible, to see how life goes here. If it weren't for the continual threat of Congress, I might even be able to get a job in some place like Southern California, but the idea of re-entering the American Nightmare is not attractive to me. Nevertheless, as bad as it is, the hatred generated in me by this terrible way of life stimulates me to work, almost as my only means of fighting it. The Brazilian way, which is "mais ou mores" (neither here nor there) sort of leaves one half-awakened. As I have been told, for example, practically everyone here is Red in speech, but it doesn't really mean much in action.

Perhaps as much as I dislike what is going on in America, I still have the American tendency to do things in a definite way (at least in physics and politics, I hope).

[ends. . .]

Letter 63. Folder C116, Dec 24 [1951].

Number on photocopy: 9

Universidade de São Paulo
Faculdade de Filosofia, Ciências e Letras
Caixa Postal: 8105
São Paulo

Dear Miriam

Thanks very much for your last letter – it was much delayed, mailed on Dec 6, and received on Dec 21. Most of my letters come thru in from 5 to 7 days. I shall watch to see if your letters are hereafter systematically delayed.

I have moved into an apartment with Smith. The address is Rua Brasilio Machado, 380, Apto 703, but for the time, I suggest that you send your letters to the address printed on this letter. The apartment is very nice, although a bit expensive. Time will tell whether it's worth it. Anyway, I may as well spend all this money now while I still can.

Your inferences about perfumed paper in the last letter mystify me. Unfortunately, they are completely unjustified. As you know, I have some problems in meeting new people, and there are additional problems here. First Sâo Paulo is a remarkably "bourgeois" city, more so than any city in the U.S. In this, it is totally unlike, for example, Rio, or most of the rest of Brazil. Most of the educated (and therefore wealthier) Brazilian girls are raised in a sort of caricature of 1850 British middle-class morality, with the double standard admitted, etc. Their role is to be breeders of children + keepers of the home. There are actually many cases of such girls who were disowned because they were discovered having intercourse with a man <u>before</u> they married him (even though they were in love). This leaves the field mainly to foreign girls. Also, they are much more likely to speak English, which is a tremendous advantage. I think George has a rather elementary idea of relationships of this kind, so he would say that language is not even needed. But this is not my idea. There are some likely looking girls around here in the Physics Dep't + Math, but now is vacation and they are seldom around. Besides this, once I get depressed, it is very hard to take the initiative needed to get out of this state. I have been $\frac{1}{4}$ sick most of the time. It is probably the water, which contains bacteria. It is not as bad here as in Rio, where the faucet water is visibly muddy, but even the Brazilians get sick about once a month. I am getting used to it, and slowly coming back to normal strength. The heat is also a nuisance. It's summer here, and we're on the Tropic of Capricorn, but this lasts only until March. Then the weather will really be beautiful, I am told the 3000 ft altitude helps a lot.

Your discussion about the value of science was most interesting. One point struck me especially. You say that perhaps my work in science is not so important after all – it will be done eventually by others anyway. This may well be true, but the same statement applies, with ten times as much force to my possible effects in politics. After all, so many people have died in concentration camps, after untold suffering, etc. One more will really not make much difference, really. Even if I should take your advice and plan to spend the next few years in jail, I would produce a much bigger effect if I first became a well-known scientist. Then it could be said "Well-known scientist being persecuted". But so many people are going to be persecuted that one more who is hardly known is also not so very important.

I hope you will excuse me for joking about this point, but you must admit, your argument works both ways. Actually, the most important argument that you make is that I may become too depressed to work effectively. But I think that within a month or two, I will meet some more interesting and stimulating people.

As for the I.F.Stone article, I am inclined to feel that it is best not to do it for the present. I think that you must have more patience in this scientific work. Without some concrete discovery, you can waste years trying to get people to accept a new philosophical point of view, but with one it will be much easier. Incidentally, I have been in communication with Régnier + Schatzman. They tell me about all sorts of wonderful discoveries using these new ideas, but as yet no details. I have sent them letters recently urgently asking for details.

You must remember that not every fight succeeds. Every day on the Korean front, many people die and suffer, probably because of mistakes made by people in power. Tremendous resources are thrown away, human and otherwise with little concern. And perhaps there is no way to avoid this. Yet, as long as I have the opportunity, I choose to fight in my own way. I am convinced (intellectually at least) that this is the best way, and a similar emotional conviction is slowly growing. If it doesn't work, then it will be just one more failure in the list of 100,000,000 or so that have occurred in the past 20 years. But there are already indications of success, such as the achievements of the French physicists (if true, as I hope), and I have some ideas for further progress here that I am working on. Also, I finally got an admission from Pauli that my point of view is logically consistent (Please tell this to George). Pauli still rejects the idea on philosophical grounds.

Please tell me more about Greenberg. There may be possibilities for jobs here, unfortunately not at a very high level of pay, but enough to live on. As for the math. dep't. I am not yet familiar with what the situation is there, but in a month or two, I may know better. I am sure I could get George a job here at roughly the level of ass't prof, but I doubt very much whether he would want it, considering the situation existing between the three of us.

Please let me hear from you soon

Love
Dave

Chapter 21
Letters to Miriam Yevick, 1952, Part 1

Letter 64. Folder C117, dated: Jan 5.

[January 5, 1952].

Number on photocopy: 10

Universidade de São Paulo
Faculdade de Filosofia, Ciências e Letras
Caixa Postal: 8105
São Paulo

Dear Miriam

I was very glad to get your last letter. It really looks as if there is nothing to worry about in this passport business. I think that I shall even begin to try to see what I can do about getting to England this summer. As for the "right to keep your passport" I already know about it, but I doubt that it's worth much in Brazil. The police here work in close cooperation with the US gov't. While there isn't much repression in Brazil, this is only because the police are pretty inefficient. But if someone were to get them interested in any particular individual, they can, as experience of various people here shows, be rather unpleasant. On the whole, I figured that it is not wise to call attention to myself by such a defiant stand that would involve much publicity.

I am very glad that you told me about "strong" laws in probability as I got an idea out of it (even though I do not yet fully understand your problem). Life would be easier for me if you would explain yourself in words, and not in the customary math. notation, which is very unfamiliar to me. Anyway here is my idea:

Consider a complex mechanical system represented by a point in phase space $(x_1, x_2, \ldots, x_n; p_1, \ldots, p_n) = \underline{P}$. Now, I wish to discuss the development of a Maxwell-Boltzmann distribution of velocities in the long run, without reference to such vague and subjective concepts as "a-priori probability", etc. I merely wish to use the causal laws which determine the motion, to predict the development of

C. Talbot (ed.), *David Bohm: Causality and Chance, Letters to Three Women*,
DOI 10.1007/978-3-319-55492-1_21

this distribution. To achieve this result, I first define the actual density of particles in phase space (this is not a probability), which is

$$N(x_1, x_2, \ldots, x_n; v_1, v_2, \ldots, v_n) = N(\underline{P})$$

Now we know on general grounds that as a result of collisions, etc., most of all of the possible initial conditions (represented by corresponding points, $\underline{P}_0$, in phase space) will lead to a chaotic distribution of particles in the long run. Moreover, as we we also know, the condition for development of chaotic instability i.e. a small change of initial conditions produces a big change in the final position (in phase space). Now we also know from previous work that the final distribution for N is nearly Maxwellian (a Maxwellian distribution corresponds to a Gaussian error curve in velocity space $N(x, v) \sim e^{-\frac{mv^2}{2kT}}$, where T is the temperature). Let us now ask the following question:

What range of initial conditions, $\underline{P}_0$, will lead to a deviation of N from the Maxwellian which is larger than ϵ?

Now, if the motion is unstable, this means that if we consider a given small deviation ϵ in the end result, the possible initial range of positions (signified by $d\underline{P}_0$) which can lead to this deviation must become smaller and smaller as $t \to \infty$. This is because for a given $d\underline{P}_0$, the change in final position becomes larger and larger as $t \to \infty$. Thus we conclude that as $t \to \infty$, the possible range of initial conditions that lead to a deviation from the Boltzmann distribution approaches a set of measure zero. Physically speaking this means that it is very difficult to produce a set of initial conditions which lead to appreciable long-run deviations form the equilibrium Boltzmann distribution. Thus, we can treat the entire problem without recourse to such subjective procedures as defining "a-priori" probabilities, ensembles, etc.

The above is only a sketch of how the problem can be treated. I propose to begin with a simpler specific problem where the motion of each particle can be solved explicitly, i.e., the perfect gas of non-interacting particles contained in a "rough-walled" container. A "rough-walled" container is one whose walls contain many irregularities on the molecular scale; thus
(this is a good approximation to an actual wall). When a molecule hits this wall, it is deflected by an amount that depends very sensitively on exactly where it strikes. The position reached by the molecule after it leaves the wall will thus undergo very big changes if the initial position of the molecule is only slightly altered; (see diagram below)

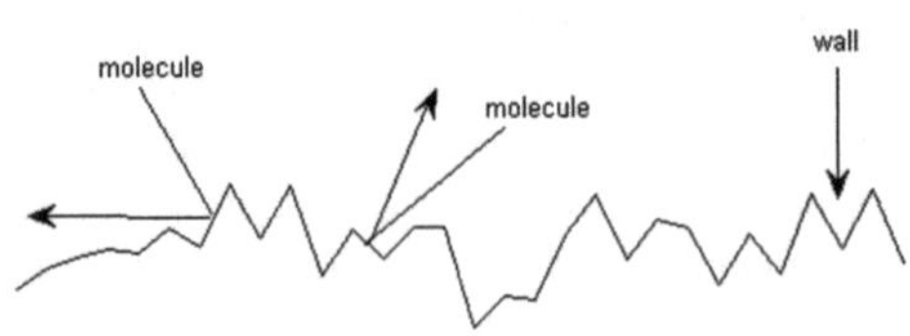

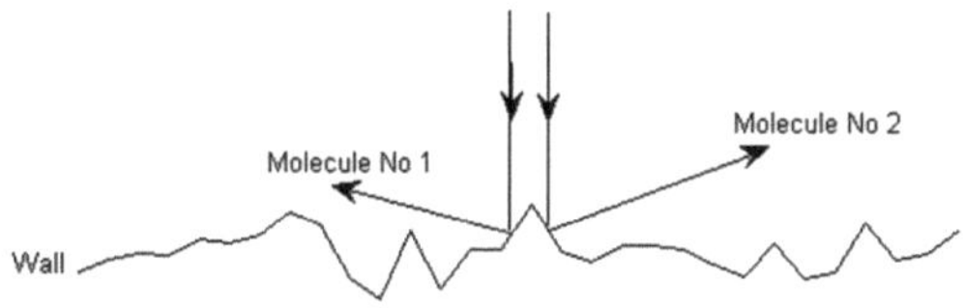

In this way, we obtain a simple problem in which the essential feature of instability appears.

I now propose to deal only with the spatial distribution (in order to illustrate the method) and to assume for the sake of simplicity that all molecules have the same speed. It is clear that as a result of the extreme instability in particle orbits after the particles bounce of the walls, a chaotic distribution in space occurs in the long run. Moreover, if there are many molecules, there is, as we know, only a small probability that the density will deviate much from its mean value,which we denote by N_0. We now ask the question: what range of initial conditions are consistent with a deviation of density greater than ϵ in any specified region of space? This problem can be solved exactly, and it should be possible to show that as $t \to \infty$, the region in question should have very small measure. Thus, in some ways, the problem resembles what you are working on. From my point of view it has the advantage of being very specific, and capable of exact treatment, as well as leading to a clear physical picture of the principles involved. The essential difference is that the mechanism leading to chaos is treated explicitly, i.e. the reflections off the rough wall and the subsequent motion, which are responsible both for the extreme instability and the "scrambling up" of particle orbits. In the math treatments that you talk about, the chaos is only postulated in the form of an assumed "random" walk, but the problem of the origin of the "randomness" is not considered. Of course, there are certain advantages in abstracting the problem in the way you mathematicians do, but in this case I have hopes that the treatment of something approaching a real physical problem may suggest new ideas that will even be useful in the more abstract problems thus far treated in mathematics. This is because, from a materialist point of view, one expects the actual inter-connections existing in the material world to be rationally understandable, so that if one tries to solve an actual problem, one is led on the track of inter-connections whose trail is covered up if we abstract the problem too much. The essential point is to know when to consider the material problem more closely and when to take advantage of the simplicity made possible by further abstraction. I believe that mathematicians have lost this skill as a result of lazily trying to produce abstractions that are subject to a simple solution, instead of breaking their heads on real problems that might lead to new concepts and new modes of inter-connection of these concepts.

I shall be interested in hearing more about the "strong laws" of probability theory. Is there a text simple enough for a poor physicist to read? My main trouble is that I detest notation.[1] I have struggled for 20 years to learn what notation I now know, and

[1] My principal trouble in writing math has been the irksome job of finding a good notation. It seems a waste of time to have to find a notation to [words missing] to me.

I don't like to have to double my fund of notation just to learn about one subject. My experience has been that I can go over a book like Khinchin and get the ideas without following the development of theorems, notation, etc. in much detail. I like to see a book centered around the concepts and their inter-connection, while the notation should be restricted as much as possible to demonstrating the precise formulation of the theory. If possible, the notation in a given section should be introduced after the author has explained what the section is all about.

Well, I guess that's all about this problem for the present. I am going to set a student working on it too. We are jointly going to give a course in stat mechanics and try to iron out some of the confusion that has arisen in this subject, because of excessive abstraction, which gives a fictitious aspect of generality to its results. It is now necessary to come back and solve a few real specific problems, however simple they may be to show that the reserves of statistical mechanics are far from finished!

Incidentally, I would appreciate it if you would try to write a little more clearly, as I have much trouble in reading your writing. You said that my plasma papers had been commented on favorably in the latest "[unreadable]" reviews. Which reviews are these, pray tell? I tried first to assume that the letters are in English, then I tried Greek, but this didn't help either.

I have answered de Broglie's criticisms (published in Comptes Rendus) of my article by sending a letter to the Phys. Rev., which should come out in a few months. De Broglie apparently didn't read my article, but simply re-iterated Pauli's criticisms, which led him to abandon the theory, but did not point out my conclusion that these objections are not valid. He is going to look a little silly. That's what he gets for rushing into print 5 months before my article came out. (My article comes out in Jan 1 issue).

It is hard to predict the reception to my article, but I am hoping that in the long run it will have a big effect. My experience with Feynman indicates that nobody ever reads an article unless it says something they already expect to hear. When I met Feynman, he thought that the idea was crazy but after enough talk, I convinced him that it is logically consistent. It is really necessary for me to give talks, in Europe if possible, and perhaps even in U.S., if Europe is not possible; or else nobody will take the trouble to read it. I am hoping that I can stir Pauli into writing a letter against it; this will stir up interest, and now that Pauli has admitted its logical consistency, he can do no real harm, since the main point has been conceded by him. In this connection I am glad that de Broglie gives me an excuse to write a letter to the Phys. Rev. I also got a chance to say in this letter that Pauli had (in a private communication) admitted the consistency of my point of view. What I am afraid of is that the big-shots will treat the article with a conspiracy of silence; perhaps implying privately to the smaller shots that while there is nothing demonstrably illogical about the article, it really is just a philosophical point, of no practical interest. Or they may say (with Oppenheimer) that "everybody is of the opinion that, if anything, nuclear phenomena will display even less causality and definiteness than atomic phenomena", forgetting that it is a misapprehension based on the usual interpretation, to say that atomic phenomena are not causally understandable.

I am feeling much better these days, as I conquer diarrhoea with various drugs such as sulfasusadine, sulfatalidine, etc. I spend an average of $10 a week on sulfa. If I get really sick, I am reserving a new "super-drug" chloromycetine, which, according to the label, can cure almost anything. As it comes from the U.S., it is expensive. In fact, by weight, it is 3 times as expensive as gold. To cure an attack of diarrhoea costs $ 40 with chloromycetine, so I generally use sulfa instead. I used chloromycetine once when I was really sick with about 4° of fever, and it cleared up pretty fast, whereas for a week, the sulfa drugs had not been able to do any good. But usually the sulfa drugs are good enough.

I have hope of meeting more people soon. I already know a few girls, but they don't appeal to me very much. However, I don't propose to stay by myself much longer. Smith and I have a beautiful apartment overlooking the city, and while it will take us some time to finish furnishing it, I shall soon have a large surplus of money. I'll save up enough for the trip to England in 3 months, and then have a bit left over. (The plane fare is about $1500 a round trip).

The weather is surprisingly cool for mid-summer in the tropics. We are now in the rainy season, but we can expect beautiful weather starting March, crisp and clear with strong sunshine.

Well it's getting late, so I'll close

Love
Dave

Letter 65. Folder C117, dated: Jan 7, 1952.

Number on photocopy: 11

Universidade de São Paulo
Faculdade de Filosofia, Ciências e Letras
Caixa Postal: 8105
São Paulo

Dear Miriam

I should like to make a few comments on what you say about my book. On the whole, most of your criticisms seem to stem from you being a mathematician. With regard to the argument on p63 on showing that $\Delta x \Delta k \cong 1$, I must say that the objective at this early stage is <u>not</u> to give an absolute proof of this relationship, but merely to give some arguments stating what actually happens, in order to bring out a more vivid concept of the the meaning of this relation. An exact proof of a more general relation appears in Chapt 10, Sec 9 (pp 205-207). However, as far as a physicist is concerned, the exact proof is less important than a statement of the correct result, plus a general argument which enables him to see how the relation comes about. It is best if this argument is rigorously logical, but a qualitative "plausibility" argument is more valuable at an early stage in the text than a precise argument full of a forest of symbols, which does not permit the student to grasp the essential meaning of the result. In the case of $\Delta x \Delta k \geq 1$, the real meaning, <u>as far as quantum theory is concerned,</u>

is that a highly localized packet in x space is made up through the interference of a range of waves having different values of k. This concept is crucial for a grasp of the meaning of the quantum theory, in any terms that are other than purely formal. The student will later see that this plausibility argument is buttressed by a more precise argument. I admit that the language did not indicate precisely my intention in giving this argument; this was a mistake on my part. However, there are two points I wish to adduce in my favor:
(1) Physicists learn to understand the real meaning of such arguments with experience, so that to a physicist, this argument very probably does not give a serious misconception of the true state of affairs. Since the purpose of the text is communication; i.e. to set in motion the proper train of thought in a typical reader's mind, small technical errors are not serious, if the typical reader is used to errors of this kind (and I assure you that they need arguments of this kind by the hundreds, as they must, if they are ever to learn physics). (2) I have tried to reduce these minor technical errors to a minimum, by having the manuscript read by as wide a group of people as possible. But this particular point escaped all readers so far. If I write it again, I would say as follows: "It can be proved generally that $\Delta x \Delta k \geq 1$ as will be done in a later chapter. We now give a discussion showing the plausibility of this result, and showing in a physical (an intuitively conceptual) way how the result comes about. Such an intuitive concept proves to be very important for grasping the significance of the uncertainty principle and other aspects of quantum theory, which will be treated later." I think that if you ever try to write a book of this length on a subject that is intrinsically muddled up, you will find you cannot obtain perfection the first time. I hope you appreciate that the arguments that I use here had to be developed from an almost non-existent basis; the literature is very formal. Probably you would find nothing wrong with a book like Dirac or Kramers; the math. is quite clear, but the origin of the math. is lost. You must also remember that much of the mystification in physics today results from the idea that nature can be understood only mathematically. But there is no concept of how the math happens to be connected with reality. Schrödinger, Dirac, and other great geniuses, who are in tune with the fundamental harmonies of the universe, pulled these equations from heaven, but it is meaningless to ask for a concept of what is behind these equations. After all, you can predict the numbers that come out of experiments, and what more could you possibly want? My book was a conscious effort to counteract this point of view within the framework of the usual interpretation of the quantum theory. If it did not succeed entirely; this is at least partly because (as was not known to me when I started the book) the usual interpretation is intrinsically inconsistent with my objectives. Nevertheless I think I did a pretty good job of it, and can without too much conceit quote Einstein's comment "It is the best that could be done with the usual interpretation." Moreover, I don't think it's right for you to estimate the book before you read enough of it to see what it is driving at. Moreover, if you will read Dancoff's review of my book in "Reviews of Scientific Instruments" (Sept or October, I believe) you will see statements in there highly praising the formal elegance of the treatment. So you see, different standards must be applied in physics books + in math. books. Also, you should remember that formal elegance was my smallest interest in writing the book; in fact, in some ways,

it was what I was fighting (in its false aspect as something that dominates the work; its true role is as a tool in exposition and application of the basic ideas but only where precision is needed).

Some of your other criticisms are valid. In some places the book is a bit wordy. The error on p 50 I noticed before; it is the printer's fault. With so many formulas it is difficult to get them all. You must remember that much of the work on proof reading occurred while I was awaiting an indictment, so that it was hard to concern myself with getting all these formulas correct. A list of errors would be helpful.

I don't intend to write the book over again, until the new interpretation is pretty well developed (as will certainly happen in time). Then it will be rather a different book. Some of the formal arguments can be retained, but much of the physics must be altered.

With regard to the Pope's speech, his discussion of instability of matter is dialectical enough, but unfortunately, not materialistic. The pope likes the concept of matter as something that can spring out of nothing (i.e. be "created"), and which changes in a way that is beyond human understanding. The concept of matter elaborated in his speech is almost that of a "phantom", so that matter begins to take on the shadowy character of ideas. In other words, he de-materializes the concept of matter. Now, to retain the concept of matter, we must above all retain the idea that in some aspects at least, matter is indestructible and uncreatable. How then do we explain the prevalence of change and the transiency of material things? This is done by the notion of endless transformation. The "things" at each level, are made up of smaller "elements" at a more fundamental level, and it is the motion of these more fundamental elements (not usually directly visible to us, except with the aid of elaborate scientific research) which causes the appearance and disappearance of the "things" existing at a higher level. These more fundamental "elements" however, cannot be permanent, but must be made up of still more fundamental "elements" and so on ad infinitum. Thus, we can see that every "thing" that exists may at some time come into existence and later go out of existence, but there is always a deeper level, in terms of which this change can be viewed rationally as a transformation of a more elementary form of matter, which is not itself basically altered in this particular transformation. Nevertheless, no single "thing" is uncreatable or indestructible. Only matter as a whole in its infinity of properties and potentialities is eternal.

Experience in a wide variety of fields in physics suggests that the stability of each "thing" is due to a conditional and temporary balance of opposing effects or forces, each of which by itself would destroy or alter the "thing" but together maintain it in existence, as long as the balance persists. In general, because of the connections of the "thing" with the rest of the universe, the condition of balance cannot be maintained indefinitely, so that the "thing" eventually dies out of existence, to be replaced by new kinds of "things" that are stable in the new "environment".

An example is the plasma. Here, the opposing effects are the forces of interaction which tend to create rigid order among the myriads of electrons, and the thermal motions, which tend to destroy this order. The balance between the two leads to semi-stable oscillations, but ultimately the effects of chaos over-balance those of order, and the oscillations die out. New oscillations can be brought into existence by external disturbances, which further alter the balance between chaos and order.

Right now, I am working on a causal theory, in which space is assumed to be filled with a form of matter much more fine-grained than the "elementary particles" and also much more dense. Matter as we know it is only a small disturbance in this uniform background, stabilized by the balance of forces leading to order and by motions of the fine-grained particles leading to chaos. Under proper conditions, an "electron-positron pair" can lose its balance. It will then disappear and be replaced by a wave of light, which carries the energy; or the reverse process can occur and the energy of this wave can be "condensed" by the formation of an electron-positron pair. Thus, we can hope to understand in a rational way the "creation" and "annihilation" of matter as we know it now. But note that the rationality is retained by the postulation of permanent existence of matter at a deeper level. (The pope would prefer a theory in which matter sprang, like a shadow, out of nothing). Of course, this matter at a deeper level can in turn be "annihilated" in still other processes not yet known, but this can be understood rationally in terms of a still deeper level; etc., etc.

There is already much evidence for the existence of a space-filling "substratum" of matter, or "ether" as it has been called. George has asked me about this in a recent letter, and I shall soon send him a letter, explaining some of the evidence. Ask him to show it to you.

One more point: An essential aspect of dialectical materialism is that "a thing is not equal to itself." This statement seems nonsensical at first sight, but a little analysis shows that it is a surprisingly deep insight into the real nature of matter. Consider the statement "A stone is a stone". Is this a tautology? From a grammatical point of view – yes. But actually, the word stone changes its meaning when it crosses the word "is". The first word refers to an actual object, a "stone", in its infinity of properties, which are all inter-connected and united in such a way that they can be understood only by conceiving of a material object existing in a certain region of space. The second word "stone" refers to our concept of stone, formed from our general knowledge of stones, etc. This concept is finite. Usually it includes only (in addition to the filling of space) massiveness, hardness, roughness, a certain feeling, some color, etc. [following phrase crossed out: for a scientist it might include the . . .] all united, but only a finite number of aspects. For a physicist, you might add the concept of the stone as made up of moving molecules, held together by forces. This is already a deeper concept, since if the stone is melted, it is changed into "non-stone" (i.e. a liquid) (or it may often be dissolved in water). From the atomic point of view, there is a transformation involving properties of matter at a deeper level. But in turn these atoms can be destroyed, and to understand this, we require a still deeper level, etc. ad infinitum. Thus, no matter how detailed the concept meant by the word "stone" following the word "is", it cannot include the qualitative and quantitative infinity of properties and interconnections referred to by the first word "stone". This is also the dialectical materialist solution of the problem of the particular and the general, since the first word "stone" refers to a specific material object, and the second word "stone" refers to a concept of a general category, which includes in principle all that is now known about stones, but which may also be only an approximation to reality, since it neglects an infinity of levels. Thus, there is room for each object to be individual and different in an infinity of aspects, and still belong to a certain general category of objects.

With regard to my need for speech, you are right in that from an emotional point of view, I need a sympathetic person to listen to me, to understand to some extent what I am driving at, and to criticize weaknesses in my arguments without attacking the basis; that is, to criticize from the point of view that the basic goal is sound, but that the method may in various aspects be inadequate. I particularly need someone who can detect logical weaknesses without implying that they invalidate the whole structure. As you know, my way of thinking is not step by step, but rather through the inter-connection of various aspects of the whole. Thus, I may be led to guess the end result by a general "intuition" that it is the only theory that seems to be consistent with the situation as a whole. Then I like to demonstrate directly (without a long series of analytical steps) that this end result is right. From this, you can usually see that various other results follow, because they are needed for logical consistency of the whole. But you can't explain it this way in a paper these days. So I am then forced to write out the analytical steps. This bores me greatly, since I never write down any formulae before I am already sure of the result on qualitative grounds. However I guess a suitable number of intermediate steps, but since I already know the answer, these may be wrong in detail (signs, etc; things left out). However the general line of reasoning is almost always correct, in the sense that if anyone follows the same steps independently and correctly, he will be led to prove the given result. This used to infuriate the students who helped correct the manuscript, because after finding 14 errors, they would invariable see that the final result was correct. Thus, if someone would point to a detailed error and then were to want me to give the whole thing up because of this, he would very probably be making a mistake since the analytical process was usually not important in obtaining the result!

I see by your [word crossed out: last] letter that you are discouraged and depressed. I hope very much that this is only a temporary thing, due to not feeling well, or something like that. I know the feeling very well, and how easy it is to get into a state like that. But it really is giving in to the other side, in a way.

I have one more bone to pick with you. I think you have a tendency shared by many intellectuals, to "bend over backward" in your attitude to theory. That is, you tend to deprecate the role of understanding and theory, and, the way I see it, to romanticize the achievements of those who are visibly doing something in the role of direct action. We all agree of course that theory unconnected with action is sterile, and that action not guided by theory is usually destructive and dangerous. A theory such as the dialectical mat. point of view, drawn partly from inferences obtained in a study of nature, was admittedly of tremendous influence in guiding great social transformations, even though at first, the number of people who possessed this theory was minute. Since that time, much has changed, both in the social situation, and in our knowledge of nature. New problems have arisen such as Fascism, not foreseen in the past. The dialectical method requires continual growth and enrichment of theory, with new synthesis at ever higher levels. Such syntheses have not been forthcoming.

One of the symptoms in USSR is, to my way of thinking, a rigidity of thought that prevents the growth of new concepts, in some ways. I ask myself the question "Why in 25 years didn't someone in USSR find a materialist interpretation of quantum theory?" It wasn't really very hard. De Broglie + others suggested the 1st step in 1926, and the 2nd step (theory of measurements) is also not really difficult. But bad as conditions are in U.S. etc, the only people who have thus far had the idea are myself in U.S. and Vigier in France. Yet, in U.S.S.R, there has been much criticism of quantum theory on ideological grounds, but it produced no results, because it may have scared people away from these problems, rather than stimulate them.

The point I am coming to is this: It must be admitted that the direction of the left-wing movement has been on the whole very unskillful, and many mistakes have been made, both internally and externally, in socialistic countries. These mistakes, I think, are not random in cause, but spring partly from an inadequate theoretical background, a background going back to concepts elaborated largely in the 19th century. New theory, and new concepts of humanity are needed. As in the 19th century, these concepts must come in part from a better understanding of nature in its interconnectedness, and partly from direct experience with people, society, etc. Even if only a few people hold them at first, they will exert an influence that is much greater than you imagine. Moreover, such concepts play a dual role, since they will not only provide a better basis for action, but they will also help win over many people who are now skeptical. You must admit that the world is now in a vicious circle, and that thinking on all sides has rigidified to a serious degree. Science is important here not only because objectively it helps provide much material for these new concepts, but also because of its enormous prestige, which can be used to influence many people, if the proper basis is laid. One cannot guarantee results, of course, but neither can anyone who dies in a concentration camp be sure that his death will help matters. At best, an individual can play only a very small role in such things, but in science is a place where the role may in some ways be bigger than one thinks.

Love
Dave

Letter 66. Folder C117, dated: Jan 23, 1952.

Number on photocopy: 12

Universidade de São Paulo
Faculdade de Filosofia, Ciências e Letras
Caixa Postal: 8105
São Paulo

Dear Miriam

I am flattered at your estimate of my potential effects in your last letter. If I were really that powerful, then it would be necessary to consider carefully again whether I should remain way off here in Brazil, when I ought to be in the thick of the fray, etc. However, I'm afraid that you overestimate the effects of carefully reasoned editorials and underestimate the importance of the basic points of view that help lead people into all sorts of difficulties. A very simple interpretation of history and politics, once suggested to me by Lomanitz, is that everyone acts according to his own interests. But this is too simple, for what are my interests? How am I to know what my interests really are? In fact, if I did know, I might become a different person and thus obtain different interests. So what does this statement mean, if anything? Now this question is important, because the American people (and to a lesser extent, the people of Europe, etc.) are acting in accordance with what they now believe to be their own interests. You cannot convince them otherwise by rational arguments. So are we to conclude that the failure of America and of "Western civilization" generally is due to cowardice, laziness and selfishness of the individuals who make it up?[2] This idea has some truth in it, but it cannot be entirely true, for at certain times, Americans have acted quite otherwise. Moreover, Americans are basically no different from other people in their fundamental constitutions. Another idea, hinted by you in your last letter is that the American left (like all of America) is very shallow, so that they do not know their own interests. This point is also very important, but one must remember that a shallow person cannot resist the forces of society that twist him and turn him into something very different from what he might have been. Even if initially he has courage and good will, a shallow person cannot appreciate how his character can gradually decay as a result of the adoption of a point of view whose rottenness he cannot see. Moreover, he does not see that to some extent his shallowness of understanding is a "protective device" endemic in the society, which serves to prevent him from facing these questions. In other words, a great deal of "shallowness" is just simply a dishonest way of refusing to face the facts, with the dishonesty hidden one level below full consciousness.[3] The whole thing is from one point of view a form of cowardice, a refusal to think of things that might offend people, make you uncomfortable, and get you in trouble. But the shallowness in turn re-inforces this whole tendency in any one individual and facilitates the corruption of other individuals.

Yet from another point of view, we can see that the interests of the individual are being thwarted in this way. No one can realize his potential, everyone has a sense of loneliness, insecurity, and futility. All human relations are poisoned. The need for psychological treatments multiplies, the rate of insanity rises. But the very process of thwarting their underlying interests makes most people cynical and thus turns them further toward the same path, where they contribute further to the whole process in their relations with other people, etc.

So you see, the whole problem is complicated. It doesn't do much good to reason with people who are caught in such a mess. There are two ways out. One is to show them a way of action that has some hope of success. This does not seem to exist now. (It actually does exist but the average American leftist will tend to refuse to see it, in order to protect himself from immediate danger). The other way is to try to help lead people to a new concept of themselves and of their relations to each

[2]So that it is in the interests of Americans to act in a cowardly, lazy and selfish way.

[3]And for this reason is very difficult to change by logical arguments alone.

other and to nature. It seems clear that if people really knew what they were and what they could be, they would not long be satisfied with this half life that they now have. Now, in America, one of the principal difficulties is a concept rooted in "the American way of life" that "things are what they are", and that there is therefore no use in trying to change them fundamentally. Thus, there is a certain limited concept of human nature and human society, corresponding roughly to the existing forms (or some idealization of them [4]). It is true that American technique is revolutionary, but it is no accident that people have adopted the narrow positivist point of view that technique is just manipulation of certain "operations", and that there is no relation between what we learn about nature in these manipulations and what should be our assumptions concerning other questions. On the other hand, the materialist point of view stresses that our "operations" deal with a material unity of infinite richness of interconnection, so that there are a multitude of relations between technique and the form of society (and not only the relatively obvious one that a certain level of technical development calls for a certain form of society). Now if you tell people that our techniques are being thwarted by the present form, and that we may therefore perish, the average person would probably feel "So what! Human nature requires the present form, and if we perish, then this just shows that it was a mistake for us to aim at such a high level of technique. Therefore, next time we ought to kill the scientists before they let such things loose again, and go back to the good old ways." (He probably wouldn't formulate it that definitely, but that is how he would feel). Therefore, it isn't enough to say that our present technical level requires socialism. This must be supplemented by a different idea of "human nature" and of the meaning of human relations. A crucial point in getting across this new idea is to break the bond of thinking that "things are what they are." As I showed in my last letter it is essential to think that things are not only "what they are known to be", but also a whole list of different things connected with the infinite number of levels not known to us. These other things may be thought of roughly as "what is coming into being" since it is in the future form of the thing that the underlying factors will ultimately manifest themselves.

Moreover, the stability of all existing forms of matter is due to the temporary and conditional balance of opposing forces or processes that tend to change it. Thus, in order to change something in a <u>fundamental</u> way (and not just to change the arrangement or form of already existing things) it is necessary to get at the underlying and opposing factors which cause things to be "as they are", and to aid the forces producing the kind of change that is desired. In society, these forces are quite manifold and complex, but one of them is just the concept that people have of themselves and of society. Even under the best possible conditions, this concept cannot be perfectly accurate, but at all times, it is a guide which leads people to act in a certain way, to attempt certain things, etc. The experience gained in this way make possible a further growth of the concepts which leads to new experience, etc., ad infinitum. But people's concepts have grown up (and must continue to do so) not only in interaction

[4] [words missing] even socialists do not usually see beyond spreading "benefits" of the "American way of life".

with society, but also in interaction with nature, in which (as one must not forget, even if you live in New York City) society is embedded, and out of which society is formed and nourished. And by "nature" I mean not only the trees and flowers, but also the earth and the stars, the sea, the rocks, etc., which man has had to understand and master (to some extent) in order to form a society. It was in dealing with nature that man was forced to produce his first objective and clearly thought out concepts, and in this way developed a form of thinking that we hope can now be applied to human beings and to society. But the experience gained in this pursuit up to now has largely been seen in terms of the distorted idea that "things are what they are, and nothing more". Mechanical materialism is a form of this idea. Lately, with so much violent change in all fields, people have been forced to admit the existence of changes. But such changes are viewed again in a distorted way, for people (using the positivist point of view that only what is obvious to us can exist) now regard these changes as irrational, not caused by anything, and therefore a form of magic. This is so not only in the quantum theory, but also in politics, where people tend to think of what is going on in Asia and Europe in terms of arbitrary decisions of people who want power. People therefore admit the fact of change, but now these changes present the aspect of shadows, or even nightmares. Thus, people see in change only a re-inforcement of the idea that "things are what they are", but in order to hold on to this idea, they are forced to imagine that both external and human nature possess a terrifying and utterly irrational instability, because they are trying to explain what happens only in terms of what they can see now. In order to avoid this terror, it is necessary to understand the levels underlying the visible level, to know the factors responsible for these changes, and thus to bring them under rational control. Otherwise, one may be led to imagine, as many erstwhile liberals, etc. do, that there are mysterious and unfathomable depths in human nature that prevent people from being able to live together in a society. [5] But in order even to formulate the problems correctly, we need to have some systematic experience with matter, showing in a specific way the importance of the "levels" concept, tied together with the notion that a thing is "not only what it is, but also, in a sense, what it is becoming, or what it would be capable of becoming, under different conditions". This experience can now be gained in a very striking and dramatic way, in connection with the understanding of the existence of the most stable things known to us; viz, the "elementary particles". Surely, if anything is "what it is", it ought to be an elementary particle. But I am sure I can show that this depends on at least one deeper level, and very plausibly on an infinite number of them. Moreover, if we ever get to understand these deeper levels, we can "create" forms of matter that have never existed. But the transformations at this more elementary level may make available energies which cause atomic energy to look like the flare of a match. Moreover, I can see possibilities for matter to move faster than light, and for the transcending of all sorts of limitations. But the general idea I am aiming at is this: As in the structure of "elementary" forms of matter human beings contain an infinite

[5] This was Freud's final conclusion, viz, that the "id" was basically anti-social. Freud started toward the concept of deeper levels, but made the mistake of prematurely rigidifying his concept of the "id" (as a thing that "is what it is") before he proved [words missing].

number of at present unknown (or poorly known) levels of complexity of behavior. This fact has two important implications: (1) The most obvious, that by scientific study, we may ultimately learn to control some of the factors at any particular level, and thus to produce startling changes in human nature (including even ourselves) (2) Before this can be done, the different levels will manifest themselves in that people cannot correctly be regarded as "being only what they are", but that they can also undergo fundamental transformations of character with changing conditions. Our present practical experience is enough to show both the possibility of a more social form of behavior, and its necessity for the health of the person as a whole. Of course, changing social conditions will ultimately change humanity. But more important right now is that part of the conditions determining "human nature" are the concepts that people have of "human nature". Thus, it is possible for a new society to grow in the midst of an old one, because with the growth of new concepts, people can change their "natures". In fact, it is only to the extent that this happens in a certain fraction of the people that a change will ever be possible. Up to now, it has been possible to some extent for the new concepts to grow in a haphazard way, as a result of experiences of people working together, etc. This has always been slow and unreliable, since it often gives people such a shallow understanding that they are easily confused, especially by pressures as subtle and all-pervading as those in America today. It is necessary that people be "armed" with a point of view that is so solidly grounded that they can deal with such confusion. An essential part in the development of such a point of view is to turn our view of nature "right-side up" so that science becomes an aid to the forces favoring the growth of a new form of "human nature", and not as it is today, in part a source of confusion and despair. Although I disagree with George's idea of "a scientific society" I think that science will play an integral role in social changes; and not only through its effects on technical development but also through its support for new ways of thinking that will help get mankind out of its present vicious circle. However the scientific developments alone are not enough. After they have been carried out will come the time for the development of a certain general point of view such as I have very crudely and dimly outlined here. During this time, we all hope that war can be avoided, but even if it is not, the development of such a point of view would play an important role, no matter how it turned out. One must not underestimate the importance of concepts leading to hope in a period of chaos, misery, and hopelessness that would result from this war, whatever its outcome.

Well, so much for this discussion. I am making slow progress on the scientific problem. Every day I see some new piece fit in, but there is still much to puzzle me. My way of working is to get an idea and then to let it stew, trying to connect it up with other things. This is the way that I came to the causal interpretation of quantum theory.

As for the book, you must try to imagine the situation when I wrote it. You suggest that I may have had some dishonesty, perhaps some desire to please the "big shots" in writing it, and that this led me to back up the usual interpretation of the quantum theory. You must remember several things however: (1) When I wrote this book, there did not exist anywhere a clear statement of the basis of the theory. There existed some books which made ridiculous abstract mathematical postulates that no one could possibly understand, and there were other discussions, such as those of Bohr, which aimed at discussing the physics, but in an incredibly vague way. A student at Princeton once told me that Bohr's statements not only cancelled out with regard to their meaning in the first order, but also with regard to connotation in the second order. It was therefore necessary to go to the third order to find what Bohr meant. When I first started to study this subject 15 years ago, it fascinated me and puzzled me. I had no reason to suspect that the "big shots" had muddled up the subject, since after all, had they not been astonishingly successful in predicting experiment after experiment? Above all, I never got over being puzzled by the theory. When I started the book, I was in no position to see through the matter, because I still hadn't made complete sense of it. All I knew was that there was one school, which utterly repelled me, in which one was supposed to introduce abstract mathematical postulates, and be satisfied if the calculations agreed with experiment. Against this, Bohr's school seemed to be a big improvement, because at least he tried to explain the physical meaning of the theory. Moreover, there was an element of dialectics in Bohr's point of view which attracted me. It seemed progressive because it broke the old mechanist materialist determinism, which left no room for growth and development of something new. After I had written the book, I finally began to grasp the full meaning of the theory, and could see that it leads inevitably to a form of (dialectical) idealism. But this was not so clear when I started, because of the general confusion in the literature. If you tried to read other books, you wouldn't be able to say that you see through this stuff, just because the other books leave things just vague enough so that you don't know quite what you are seeing through. In writing this book, I hope that I have not only clarified the issues for myself, but perhaps for other people too. I suspect that a clear presentation of Bohr's point of view (the first clear one, if I may boast a little) will do more to favor the causal interpretation than to favor Bohr's interpretation. Now with my new point of view, I can see an infinitely better way to get out of the trap of mechanistic determinism; namely through the concept of an unlimited number of causal levels. I would call Bohr's point of view "static dialectics". This is because it is a form of "slinging the lingo" in which the dialectically opposing concepts are made just vague enough so that the contradictions between them are avoided. Thus, one is not faced with the necessity of seeking new concepts that synthesise the opposites, and the dynamic aspects of dialectics (i.e. the contradictions leading to something new at another level) are lost. Finally, I should say that I wrote the book in a spirit of struggle against the obscurantist notion that nature can from now on be understood only in terms of abstract mathematical postulates. The struggle was well worth while, since it led me to a new point of view. I doubt that I could have reached this point of view without firmly understanding the existing quantum theory first, nor could I have carried the new point of view to its logical conclusions either, and I find it is only by studying something sympathetically that I can understand it (even if it is wrong, one must begin by giving it the benefit of the doubt, and criticize it later). I have also drawn one other lesson from this experience. One must not over-estimate the importance of success, not even of a long string of successes. From 1925 to 1935, all of nature seemed to be giving its blessing to the usual interpretation of the quantum theory, with the verification of one prediction after another. Yet what was being proved was the correctness of the wave equation, and not the correctness of the full interpretation. But I had a tendency to be overwhelmed when faced with people who showed success in practice, so that I accepted conclusions that went beyond their practice. This is a common American habit, like listening to Henry Ford as an oracle in all fields because he was so consistently successful in making money. This is a danger that all intellectuals must avoid. The importance of practice is continually being dinned into them, with a sort of sneering implication that after all, intellectuals are only theorists. Thus the practical and successful businessman, politician, scientist, labor leader, or leftist, may draw general conclusions very confidently because he has succeeded in practice, and the fledgling intellectual timidly loses his nerve and accepts the most arrant nonsense, since, after all, isn't this a practical man speaking from successful experience. One must be ready to evaluate the experience of practice independently. Remember, it is the objective practical experience that is all important and not necessarily the opinion of the man who had this experience (although this too must be listened to, but always with a critical ear). One must not fall into the trap of supposing that a man who has a lot of successes to his credit is infallible in all his conclusions.

As for the rest of your letter, it is too late to answer it now. Don't get discouraged. I am sure that things will come out in the long run. Thanks very much for the clippings. I appreciate them.

Love
Dave

Chapter 22
Letters to Miriam Yevick, 1952, Part 2

Letter 67. Folder C118, dated: Jan 28, 52.

Number on photocopy: 13

Universidade de São Paulo
Faculdade de Filosofia, Ciências e Letras
Caixa Postal: 8105
São Paulo

Dear Miriam

Thanks a lot for your letter. Your objections to my book don't annoy me so please continue with them and keep on sending the clippings. Smith appreciates them, as do Tiomno and the others here.

I have begun to make contacts for the purpose of recovering my passport, but this will be slow business. I am setting Nov. 15 as the date for leaving for England, as this is the beginning of our summer vacation. Also this should leave enough time to get these things done. I wish that I can see you in England, as it would give us an opportunity to face the problem of our relationship in a definite way. I believe this stay in Brazil has made me more able to face these problems (for complicated reasons perhaps). I have had some time to think and can see things more clearly, and also I feel more secure internally. There is nothing like having money and a position of assured respect[1] to make you feel secure. I must guard against going too far, but one of our principal weaknesses is this sense of insecurity, born of being in the middle class. Genuine aristocrats often have an unbreakable self-assurance that others would do well to obtain if possible. I can begin to see how everybody's character would change if they lived in an environment where everybody was sure of a livelihood and the respect of others, as long as he followed the basic rules needed

[1] University people get much more respect than in U.S.

C. Talbot (ed.), *David Bohm: Causality and Chance, Letters to Three Women*,
DOI 10.1007/978-3-319-55492-1_22

for the functioning of the society. Be that as it may, I am very anxious to see you again regardless of what we may decide to do about our relationship in the future.

Now I shall begin to answer your long letter, but probably will not finish tonight. First, about the book. I already explained a lot in my last two letters. I appreciate the fact that you like abstract math. thinking, but would like to warn you against certain overwhelmingly serious dangers which attend its application in physics. First, it grants an illusory sense that as you say, "nothing has been put over on you." Actually the whole of the usual interpretation of the quantum theory, for example, has been "put over" on physicists with the aid of abstract math. that appears in Von Neumann's little book that everyone agrees [2] is "the last word on the subject". You must remember that to get this sense of infallible proof, you must begin by making abstract postulates. Thus, for example, Von Neumann makes the postulate that every Hermitian operator in the quantum theory corresponds to an observable physical quantity. With this and a few other similar postulates, he proves a great many things in an impressively ironclad way, including for example, that no causal interpretation of quantum theory can possibly lead to all of the results given by the usual probability interpretation. But has anyone ever proved that these postulates are correct? No! Moreover, it is impossible to check or prove these postulates. Their only justification is that von Neumann found it easy with their aid to set up an abstract basis for the quantum theory. These postulates could easily be totally wrong. For example, thus far, in experiment, all observable quantities correspond to a very small number of Hermitian operators, and there is no proof, and indeed there is no way of ever proving, that an arbitrary Hermitian operator is observable. Moreover, from the point of view of my interpretation, most of von Neumann's assumptions are just plain not true.

The second big danger is that because the formulation is so abstract, no one (and not even von Neumann) knows just what is being assumed. Thus, in his proof that there can be no causal interpretation of quantum theory, von Neumann implicitly assumed that there can be no "hidden variables" in the measuring apparatus. Nobody could have seen from von Neumann's math. that this was being assumed. I was able to see it because I had a counter-example which I knew violated von Neumann's theorem, so that I knew, by God, that von Neumann had "put something over" on everybody, including himself. With the aid of the counter-example it was not hard to find the "hidden assumption" with which von Neumann forbade the existence of "hidden variables". So you see that these abstract postulates are really dangerous, and the more clever the man who uses them, the more dangerous they become. In my opinion, quantum theory might be 20 years further ahead, if von Neumann had refrained from "cleaning up quantum mechanics". There is much room in science for hard clear mathematical thinking, based on clearly defined and well established postulates, such as the equations of motion in classical physics. For with the aid of carefully reasoned analysis plus imagination, new solutions can be found, often having unsuspected properties, and thus a genuinely creative element is introduced into physics. Moreover the finding of precise solutions is also a great service, since it

[2] It infuriates me that even leftists like Furry show an absurd deference to von Neumann in the [text missing – CT].

facilitates comparison with experiment. But heaven preserve us from the "bishops" who wish to impair the imagination and limit our mode of thinking without our realizing that they are limiting it, by putting the basic postulates of science into incomprehensible and in principle unverifiable forms. Let these "bishops" confine their attention to math. where people have already given up the idea that their work should have any relation to reality.

It is hardly necessary for me to add that there is nothing I love less to read than von Neumann's papers. They strike me as an ugly mess of symbolisms, with the idea completely hidden so that it is practically incomprehensible. I sort of chafe at the mouth when I am expected to believe that mechanical manipulation of symbols according to a prescribed set of rules constitutes a contribution to our understanding of a subject. So I think that in my statement, $\Delta x \Delta k \geq 1$, much less has been "put over" on you than has probably been done by von Neumann, in some of his infallible proofs.[3]

Let me finish up with the book before proceeding to other questions. Let me say that most of your criticisms have a large element of correctness. I have to explain again that when I started the book, my objective was to show that quantum theory can be understood qualitatively and imaginatively, and not only in terms of abstract math. postulates. It did not occur to me then that the whole idea might be wrong, so I tried to do the best I could with the existing point of view, feeling that certain points were perhaps still not clear, but might be cleared up in the future (as is necessary in every piece of scientific research or writing that has ever been done). Your criticism of the "collapse" of the wave function, asking what happens to an individual case, was actually the starting point that led me to seek a causal interpretation, for I decided that the usual interpretation said that nothing described this individual case until an observer had looked at the apparatus and become aware of the actual result. This meant that the usual interpretation is basically idealistic. But to come to this conclusion, I had first to develop a theory of the process of "collapse". This is not treated elsewhere (not even by von Neumann, who merely decrees in a very obscure, roundabout, and incomprehensible language, that it must occur). Before having this theory I always felt that the problem could somehow be treated, and felt that I would try to do it some day. But most other physicists do not deem this question important. Being good little positivists, they say that after all, only the results of measurements need to be treated, and what happens to the apparatus when nobody looks at it is a "meaningless" question. But by developing the theory of measurements (which the Institute physicists claimed in a talk I gave had all been done long ago by von Neumann) I was able to remove the vagueness and nail down the real meaning of the usual interpretation of quantum theory. If you look at any other book (including v. Neumann) you will discover that this process of "collapse" is carefully avoided, because it is not understood. Instead, everybody is satisfied to say that "The mean value of an observable, A, is $\int \psi^* A \psi d\vec{x}$. What more should a theory deal with?"

Your point about the possible existence of things that did not average out in the classical limit is also important, but it can likewise only be understood in terms of

[3]The abstract mathematician "puts everything over" in his first step, when he makes his assumptions.

a concrete model, like my interpretation, because the usual interpretation requires you to assume that this possibility cannot occur. This assumption is contained in the statement that the wave function $\psi(\vec{x})$, contains all possible physical information, and that $|\psi(\vec{x})|^2$ yields the probability of a given position, while $|\phi(k)|^2$ yields the probability of a given momentum. If you made $\psi(\vec{x})$ into a very narrow wave packet, $\phi(k)$ would become a very broad packet (according to $\Delta x \Delta k > 1$) and the momentum, $p = \hbar k$, would become very indefinite. Thus, all significant physical properties cannot become defined simultaneously. But in my interpretation it is only a conditional fact that the probability, $P(\vec{x})$, is equal to $|\psi(\vec{x})|^2$ and $P(\vec{x})$ could differ from $|\psi(\vec{x})|^2$ if there ever arose a situation in which Schrodinger's equation had to be modified in a fundamental way. Similarly $P(\vec{k})$ could differ from $|\phi(k)|^2$.
Thus, the relation $\Delta x \Delta k > 1$ would cease to imply that in an individual case all significant variables cannot be defined simultaneously.
Now, let us come to the question of the appearance of chaos in the problem that I mentioned of molecules bouncing off an irregular wall. I have some ideas for a solution, but can only indicate them here:
(1) Begin by defining the density of particles at the point $\vec{x}$

$$\rho(\vec{x}) = \sum_i \delta(\vec{x} - \vec{x}_i)$$

where $\delta(\vec{x} - \vec{x}_i)$ is Dirac's delta function. $\vec{x}_i$ is the coordinate of the ith particle. Let us Fourier analyze this (in a large box of volume V, which may be taken as unity).

$$\rho_k = \int \rho(\vec{x}) e^{-i\vec{k}\cdot\vec{x}} d\vec{x} = \sum_i e^{-ik\cdot x_i}$$

We see that setting $\vec{k} = 0$, we get

$$\rho_0 = \sum_i (1) = N = \text{total number of particles}$$

The remaining ρ_k describe fluctuations about the mean density, N, of particles. It is particularly convenient to describe these fluctuations by means of their Fourier components.

Now, between collisions with the wall, each particle moves in a straight line with constant velocity vector $\vec{v}_i$. If $x_i{}^{(n)}$ is the position of the particle in its nth collision with the wall, and $\vec{v}_i^{(n)}$ is its velocity after this collision, then between the nth and the $(n+1)th$ collision, the position is $x_i = \vec{x}_i^{(n)} + \vec{v}_i^{(n}(t - t_i{}^{(n)})$, where $t_i{}^{(n)}$ is the time of the nth collision of the ith particle with the wall. We get

$$\rho_k = \sum_i e^{-i\vec{k}\cdot\vec{x}_i^{n_i}} e^{-i\vec{k}\cdot\vec{v}_i^{n_i}(t - t_i{}^{n_i})} \qquad (1)$$

where ${x_i}^{n_i}$ denotes the fact that the ith particle at the time t has made n_i collisions (Different particles, having different paths, can have made different numbers of collisions).

Now, the first question is, how much does the time average of $\rho(\vec{x})$ differ from zero as $t \to \infty$? Since the average density ought to be N, this is equivalent to asking the mean value of all the ρ_k as $t \to \infty$, i.e. $\overline{\rho_k} = \frac{1}{T}\int_0^T \rho_k(t)dt$ must be shown to approach zero.

If we put in the value of ρ_k appearing in Eq. (1), we see that because of the oscillatory character of the ρ_k (in time), the integral over time will be limited in value. Thus, between collisions with the walls, the changes in ρ_k are bounded. In fact, the integral of ρ_k between collisions of the nth particle with the wall is

$$\int_{t_n}^{t_{n+1}} \rho_k \, dt = \sum \frac{e^{-ik\cdot x_i^n}}{k \cdot v_i^n} \left(e^{-ik\cdot v_i^n (t_i^{n+1} - t_i^n)} - 1 \right)$$

To find the total integral, we must sum this up over an infinite series of collisions. Now, one may make the plausible conjecture that except for a set of initial conditions (given by x_i^0) of measure zero, all paths correspond to a value for this sum, which approaches infinity less rapidly than T. Intuitively speaking, this is because the points at which the particles strike the walls are in fact distributed at random, so that we will be adding random phase factors, and one would guess that this integral would at most approach $\sqrt{T}$, and in fact it is probably smaller still. If, for example, it approaches infinity as $\sqrt{T}$, then we get

$$\frac{1}{T}\int_0^T \rho_k \, dt \longrightarrow \frac{\sqrt{T}}{T} = \frac{1}{\sqrt{T}} \to 0$$

Now, to prove this conjecture, we must show that the paths leading to a value of the integral that approach proportionality to T are a set of measure zero. To do this, let us assume that we have found such a path. Such a path can exist only if the phase factors at the end of each collision are commensurable, so that they can add up in a systematic way. We must then show that because of the instability of motion, the slightest change of initial conditions will in enough time destroy the special commensurability of the phase factors, and thus prevent the integral from being proportional to T as $T \to \infty$. The problem is thus reduced to a problem in analysis (and perhaps theory of numbers) which I feel sure is not difficult to solve. I would appreciate it if you would give me your opinion on how you would solve this problem (or better yet, of course, if you would solve it).

There are a number of additional problems that one can now raise. For example, what is the “probability” that the ρ_k will at a given time differ from zero by an amount between ρ_1 and $\rho_1 + d\rho_1$, $\rho_2 + d\rho_2$,... $\rho_k + d\rho_k$ ... This is equivalent to asking for the probability of a given set of fluctuations from the mean density N_0. To answer this question, we must find the fraction of the time spent by the system between these limits as $T \to \infty$. If we do this, a reasonable conjecture is that this fraction is the same for all trajectories, except a set of measure zero. (If this conjecture were not true, the average wouldn't be very useful, since it would depend on the detailed microscopic values of the boundary conditions; the observed independence in such systems is an "a posteriori" demonstration of the plausibility of the conjecture). Our problem now is to prove the conjecture. To do this, we need again only demonstrate that the instability of motion is such that if there is a trajectory for which the mean time spent in $d\rho_1 d\rho_2 \ldots d\rho_k \ldots$ differs from the phase average, then the slightest change in initial conditions will destroy this difference, given enough time.

The third point is to show generally that time averages of arbitrary functions of the ρ_k are equal to the phase averages. (This has been shown thus far only for linear functions of the ρ_k, since here it is easily shown that the phase averages are zero and therefore equal to the time averages).

If the program indicated above could be realized, then we would have (for this case) justified all the results of statistical mechanics in a dynamic way, without making any assumptions, other than that the laws of mechanics hold. The next step would be to generalize to wider classes of systems.

As for the question of jobs here, I think it would make us all very unhappy if George were to come down here. If you should come down here, I feel that a job could be arranged, but no matter how it worked out, I cannot see the possibility of a happy solution if George also came down. If you remained with George, the situation would be terribly complicated, and if you left him, George would be stranded in South America. I am sure that if George is tired of the U.S. he can get a job in Europe, where he could meet more people, and generally enjoy himself more.

Thus far, I have offered a job to a man in Syracuse, highly recommended by Peter Bergmann, and I shall probably offer one to a former student of mine named Weinstein. It is not that I wish to be unfriendly to George by not offering him a job, but that I wish to avoid a hopelessly mixed up situation here.

As for the other questions you raise about "strong laws" in math, I haven't yet had time to look up the books. Generally speaking, new books are not easily available here, so I'd appreciate it if you sent me some from time to time.

As for a letter to I.F.Stone, I'll see what I can do about it in the near future.

Love
Dave

P.S. We are running into some interesting connections between causal interpretation of qu. mech. and general relativity, by a method very different from Vigier's. However, I have not yet been able to understand Vigier. I'm afraid that your analysis of French physics holds for him too.

Letter 68. Folder C118, dated: Feb 15 [1952].

Number on photocopy: 14

Universidade de São Paulo
Faculdade de Filosofia, Ciências e Letras
Caixa Postal: 8105

São Paulo

Dear Miriam,

I have been working very hard last week, writing papers on plasma theory, plus numerous letters. The weather is cooling down already, and I think it will soon be a little too cool for my tastes. As soon as the rainy season starts (about March 1) I hope to make some trips into the mountains and forests around here, so that should make life more interesting. I think that I am getting over some of my extreme restlessness that [unreadable] made work difficult; but it would be nice if I could meet some nice people now, with whom I could establish a real contact on many levels. I am always very glad to receive your letters, and in fact, I usually devour them in about five minutes, so please write often, as it helps relieve the lack of contact with people & ideas, that is still my problem here.

As for my touchiness about criticism, there is some truth in what you say. But it is exaggerated now, because I am consciously trying to build up self-confidence by refusing to accept criticism too easily. I know how to accept criticism, when something important is at stake. For example, in writing the book, I sought and acted on the criticisms of many people. But I am now trying to counteract the false modesty one learns in our society, which makes one feel too insignificant to attack a real problem. Perhaps I am overdoing it. In any case, don't stop criticizing the book as (a) you have already made a number of good criticisms, (b) You give me practice in rushing to my defence, and unashamedly saying that the book is good, etc., etc. Incidentally, what are my other two faults?

Alas, I fear that there may be more than "three" blemishes on my character. Sometimes I feel as if the space between my intestines and my lungs is one big emptiness,and that instead of sincere feelings and desires originating there, it is all action by habit and hypocrisy. So there is another fault! I know that if it were not for certain intellectually based ideas & beliefs in the future possibilities of humanity, I would have caved in long ago. These beliefs are like a strong but very thin steel framework that supports the empty region. Now I need to gain some real live desires in this region, and not just the desires I know are correct. In times as confused as the present, I do not see how anyone in the middle class (or anyone with middle class aspirations) can fail to have his fundamental drives, desires, etc. corrupted, no matter how good such things may be under ordinary conditions. Thus, desire for accomplishment, desire for love, desire for security, desire for children, desire for the society of cultured people, desire for scientific knowledge & research, all of these things which could in themselves be good, now constitute tentacles attaching the individual to a decaying society, and causing him to tend to decay with the society. No matter how good a person's intentions, he cannot resist this decay unless he first understands it and then, as a result of this understanding, is determined to resist it, by resisting the process outside of himself in society, as well as in himself. But once the decay gets going, then the person welcomes confusion, obscurantism, and anything that will prevent him from facing reality. These people who can be argued with for $5\frac{1}{2}$ hrs without much effect are generally pretty far gone (but not always, since the willingness to argue $5\frac{1}{2}$ h shows that at least the person is still interested). But long arguments are not really what will move people. (I only wish I know what it is that will really do the job).

In one of your letters, you mentioned the need of hard grinding work for many years on this problem. While this may be true, I have certain doubts about this in America. Or at least, before we start grinding away too far, we ought to get some idea of what is the trouble with all the grinding that so ineffectively took place in the past; and to try, if possible, to avoid such mistakes. My way of working (which I am finding effective in science) is to try to avoid exhausting yourself by headlong attacks on insurmountable rock walls. Instead I like the method of ceaselessly feeling out various aspects of the problem, seeing how things fit together, looking for cracks and weak points. When you find a crack, explore all the surrounding region,and if it looks as if there is a possibility there, you hit that crack with all you've got. Sometimes you are gratified by seeing a very substantial split in a problem that previously seemed impenetrable. But sometimes this means delaying decisive action for a long time. I have found it very profitable to delay an "all out" attack on the plasma problem for about a year and a half, while I slowly absorbed various aspects, got various ideas, some right and some wrong. Now it is all coming out beautifully. If I had tried to do this a year ago, I would not have known the weak points, I would have got all mixed up in a hopeless maze of calculations, and I would just have become discouraged.

Whether political problems will succumb to similar kinds of attack is not clear. But one thing I know is that whenever I am thrown into something, willy nilly, I always make a botch of it. I cannot do anything without time to think and absorb the various aspects of the problem. So if I am going to do something in any particular line, that is the way I have to work, or otherwise, it will be a waste of effort. But right now, I am slowly maturing certain concepts & ways of thinking drawn from science, which will, I think, be useful some day in helping find the "weak point" in the apparently monolithic wall of American society. I am convinced that the peculiar failure of American society arises just as much from the pragmatic (positivist and empiricist) ideology that is absorbed unconsciously by everyone, as from various more easily visible factors, such as the large middle class, etc. This pragmatic attitude was of some help in building up the industry + technique of the society, but is of tremendous hindrance when people need to learn to make long range plans, contemplate fundamental changes in society, etc. (It is also, as experience shows, a great hindrance in theoretical science). I have at present only the vaguest idea as to how the concepts gained in science may affect this general ideology. I admit that I am quite discouraged about the possibilities very often. But I have started a number of things with even vaguer ideas, and have been led to successful conclusions. In all these cases, however, I had a feeling, based mostly on the way that things seemed to fit together qualitatively, which suggested that this might be a fruitful track. And I feel the same about this problem of general concepts and ways of thinking. I need a year or two more of primarily scientific work, and then I'll have enough ideas to begin something else, perhaps in the form of a book, such as we once discussed. Whether the time will actually be available to me, no one knows, but since any other actions that I could take would hardly influence the course of events, this is of no significance in determining my course of action. My feeling is that this is a direction that needs to be explored, and probably will not be explored by anyone else in the world for many years, unless I do it now. And I think that a different point of view of humanity toward the world and toward itself is very necessary, especially in the kind of crisis that we are now facing. We tend to accept current ideas at face value (even when we think we are critical), and we often don't realize how these ideas are limiting our consciousness, blinding us to the real possibilities, and, for example, leading us to accept decay and loss of belief in humanity. Even people who are uncultured absorb these points of view in their daily life. In other countries, the problem was somewhat less important, because working people had a real solidarity, that tended to lead to the necessary form of action, to some extent without conscious direction. But in America, middle class aspirations have cut deeply into everyone's point of view, and the trap of individual separateness & loneliness will have to be overcome with the aid of a real conscious understanding of the possibilities of humanity long before much pragmatic experience is available to the American people, which demonstrates the truth of this understanding. And in this task, the concept of the infinite possibilities of all matter, conceived of as real completely interconnected, and capable of endless development, will be important in shaping people's attitude toward humanity. At first, only the more educated and intellectually curious will be influenced, but these will be very important, as from them will come most of the leaders of society; and also, through them, concepts will diffuse through all levels of society. These ideas alone are not enough, but they make certain tasks a lot easier.

In plasma theory, I just made a step which will be useful at 3 levels (a) The plasma; (b) The problem of constitution of "elementary" particles; (c) The level of general concepts that I was just talking about. I have found that a strongly interacting aggregate of electrons not only shows collective behavior in the form of acting like a "field" that covers all space continuously, but also that new kinds of individual particle coordinates are established. These coordinates, which are complicated and highly inter-connected functions of the electronic coordinates, act more like individual particles than do the electrons themselves (i.e. they act more nearly in the role of individual coordinates, which increase linearly with the time, as would the coordinates of a free particle in space). I call them the "collective individual particle coordinates", abbreviated "cipc". If anyone should try to measure the coordinates of an electron in such an aggregate by means of processes involving distances greater than the mean inter-particle spacing, he would really be observing a "cipc" and not an electron. Only by going to distances less than the inter-particle spacing could he obtain the electronic coordinates. Thus, we have here the 1st case in physical science of where the "individual" is collectively conditioned, both with regard to character, and with regard to the number in existence. If, as seems likely, all space is filled with a dense aggregate of interacting "substratum particles", then the above result suggests that the particles that we now see may be collective structures, built from the underlying particles, and that the substratum particles can be observed only by going to very small distances indeed.

Of course, you may say that as long as we can still analyze the cipc in terms of "true" sub-stratum particles, we have not obtained a genuine collective conditioning of the individual, but merely have a new way of classifying certain functions of the true individual particle coordinates. But if you postulate an infinite number of levels, then it becomes likely that at each level, the "true" individual particles will ultimately be discovered to be collectively conditioned. Since there is no end to the series, no matter at what stage the theory is, we shall have to use the idea that the most fundamental "individuals" that we now know are collectively conditioned. We then come to a deduction having general importance. The universe cannot be analyzed into a series of components, each of which are the constituents of the next higher level, and each of which determine the higher levels in a purely analytic way. For the higher levels will also always help determine the character of things that may exist at the lower levels. Thus, every level is in a sense, just as real as every other, since the "whole picture" cannot be deduced by starting at the "lowest level" and working upward. The significance of this concept when applied to society hardly needs to be explained.

I find your explanation of strong laws in probability theory quite clear. As you say, the problem for me is to show how they can be deduced from the causal laws without any assumptions of randomness. You ask how we can do this when we drop the postulate of randomness. The answer is already supplied in my previous letter. Briefly, it is that each one of the "unusual" results, (of measure zero) corresponds to a very special set of boundary conditions, in which the positions + momenta of all the particles must be fixed exactly at prescribed sets of values. The instability of motion is such that the slightest deviation will ultimately lead to chaos. Now, it is a very weak assumption to suppose that no physical system can have a certain set of parameters with exactly prescribed numerical values. All experience proves this. Moreover, it is quite reasonable to assume that all physical systems contain at least a trace of the chaos that prevails at the molecular level. This trace is sometimes called "error" or "irregularity". The theorem that I wish to prove is that the slightest trace of chaos leads ultimately to complete chaos.

Well, I guess that's all I have to say for the present. I am thinking of "strong laws" and "chaos" in the background of my mind, but after I finish some work on plasma theory, I hope to work systematically on this problem.

Love
Dave

Letter 69. Folder C118, dated: Feb 16 [1952].

Number on photocopy: 15

Universidade de São Paulo
Faculdade de Filosofia, Ciências e Letras
Caixa Postal: 8105
São Paulo

Dear Miriam

I have just heard from a friend that J.O. may be called before a committee soon (of course, he has ways of getting out of it perhaps, but I don't know what will happen). I would appreciate it if you would watch the papers for this, for discussions of immunity statutes, etc., as I am naturally concerned.

As for me, I am a bit depressed now, because war seems very likely indeed. Brazil is too close economically to the U.S. to make me feel comfortable. I cannot get over this feeling of foreboding + depression that weighs down on me. What use is this work of mine really? It will be 20 years before it can make any difference, if our society comes crashing down in a war. This is a bad time to be lonely. Letters from a few friends would help, but there has been a slightly ominous absence of letters for over 2 weeks. Perhaps you might try [address in Brazil], and see if this speeds up delivery.

I have heard a few details about people's reactions to my paper, and I am interested in hearing more. Von Neumann thinks the idea consistent, and even "very elegant" (The unprincipled bum!). The elder Bohr told Art Wightman it is "very foolish" but the younger Bohr thinks it consistent, although "rather similar to introducing absolute velocity into relativity", unless I can find new applications in the domain of 10^{-13} cm, which "would be wonderful". Rosenfeld told a friend that it is "very ingenious, but basically wrong". As to just what is supposed to be wrong with it, my friend tells me that Rosenfeld is very vague. Schrödinger did not deign to write me himself, but he deigned to let his secretary tell me that His Eminence feels that it is irrelevant that mechanical models can be found for the quantum theory, since these models cannot include the mathematical transformation theory, which everyone knows is the real heart of quantum theory. Of course, his Eminence did not find it necessary to read my papers, where it is explicitly pointed out that my model not only explains the results of this transformation theory, but also points out the limitations of this theory to the special case where the equations are linear. In Portuguese, I would call Schrödinger "um burro", and leave it for you to guess the translation.

I feel particularly slighted because I have not heard from Jehuda about the quantum theory, and his attitude toward it. Is it possible that like so many other people, he takes no interest in a new idea? Or is he just too lazy or busy to get in touch with me? Perhaps he doesn't realize that I am way down here all by myself in Brazil, and need all the support and encouragement I can get, particularly in such depressing & frightening times.

Please let me hear from you soon, and regards to George

Love
Dave

Letter 70. Folder C118, dated: Feb 21 or 25? [1952]

Number on photocopy: 16

Universidade de São Paulo
Faculdade de Filosofia, Ciências e Letras
Caixa Postal: 8105
São Paulo

Dear Miriam,

I was very glad to receive your letter, and will answer you in a brief letter, which I shall probably supplement later with a more extended letter.

I have been feeling very depressed lately, perhaps partly because of the weather which has been extremely damp and rainy, partly because of lack of the right sort of people to talk to, and in an even larger part, because war seems so close. The action of the French Socialists is disgraceful. I can hardly contain my feeling of disgust for people who call themselves socialists, but who form alliances with the Nazis and deliver their country (and the whole world) to a third world war. From now on, I frankly do not see how war can be avoided. It is all a question of when and how it will start. If Taft is elected, there is perhaps a ray of hope that the cautious businessmen will gain the upper hand, and fearing a world catastrophe, may stop this movement toward war. But with Truman or Eisenhower, the game is up. As bad as Taft and Hoover are, the worst they could do is infinitely better than what would result from a war.

I have begun to have some doubts about the value of continuing this scientific work, as the goal is too far off, and it is becoming hard to see how its results can ever reach the people.

I think you do understand my attitude toward money. I don't particularly want money, nor do I even like it. I agree with you that it is very bad to have so much money, when most people are on the verge of starvation. Nor do I want a nice secure "middle-class" life. In fact, I can hardly stand the emptiness of middle class people, even though they are very friendly and otherwise nice. I confess that I did want (and still have a slight desire for) a place where I could work peacefully and fruitfully in a stimulating and friendly atmosphere. In the absence of this, I have wanted to realize some great scientific success, not only to make my own skin more secure (which I confess I would like to do, except at times like now when I don't care much), but also because it would give me more power to do some of the things I would like to do, in the direction of influencing people more. Right now I have (as everybody does) an exasperating sense of ineffectiveness, which embitters life a great deal. As for what I said about self assurance in the last letter, this referred not so much to money, but to the respect that people have here for education, which is absent in the U.S. Actually, I resent the respect that people would have for me, just because I have money. I admit that I fear poverty, working at a dull meaningless job, on the edge of hunger. But what I meant to say is that because of the competitive atmosphere in the U.S., people are conditioned to tear each other down, especially in the academic field, and for this reason, there can be no feeling of assurance (which the aristocrat also used to have, because he knew that he was assured of respect, for reasons of birth). My hope is that there will be a society where everyone will be assured of respect for the simple reason that he is a human being.

As for what American leftists could have done (and could still do) if they weren't so scared, I had in mind that in America, the word "socialism" has an aura of sacrilege about it. The fact is that socialism is the only possible solution to the world's difficulties, and America, a key country, does not know what the word means, and refuses to consider the question for a moment. This stupid attitude stems largely from the pragmatic attitude of the people, who like only to engage in short range modifications based on immediately obvious premises. But there has been no real effort by the American left to bring the concept of socialism before the American people, because it is so unpopular. It was always easier to come out for something like "Progressive Capitalism" the "New Deal" or some other such small change, which was at once safe to advocate, and which promised to bring immediate, if small, improvements. The American policy has been "The means justify the end." In other words, people do things just because they are easy to do, and attach a peculiar virtue to the fact that a given possibility is already in existence. In the last analysis, this attitude stems from cowardice. It is true that most American liberals are so confused that by now they can honestly say that they don't understand the issues, but I ask whether this confusion, arising from a refusal to see the forest for the trees, is not at base a protective device, endemic in the culture. In the days of the new deal, socialism could have been propagated with little real danger, except perhaps, for some disapproval by the more reactionary element, and some decrease in job opportunities. But because this was not done, when it would have been easier, even New Deal liberalism will soon be dangerous now. In other words, things do not remain static, and if the left does not take advantage of favorable currents in its own direction, it will be in a hopelessly weak position when the currents go the other way.

I believe that the main thing now is to raise the people's hopes. Everybody is down-hearted & cynical. One word of hope now is worth a great deal, as it may perhaps be a seed, in the dark times that may come. We can't have hope without first presenting a plausible solution to the world's difficulties, and what can that solution be but socialism? At this point we meet the cowardly socialist and liberal defence by joining in the chorus as to how bad Russia is. There is no doubt that a great deal is wrong with Russian socialism (and many of its faults can be ascribed to the constant need to prepare for war, in the threat of which Western liberals have cooperated). But we must simply point out that whatever is wrong with Russia, we have no other choice, but to go to Socialism. In doing this we hope to avoid the ruthlessness and wastefulness, with which the transformation was realized in Russia. It must be admitted that the Communists, by refusing to admit that Russia can ever be wrong, make the problem unnecessarily difficult, but an American socialist movement must at all costs avoid the trap of becoming anti-Russian. The correct attitude is to admit the good + bad[4] points on Russia, as we see them, and to stress the fact that most of this emphasis on Russia is a blind, an excuse not to remedy conditions here, and also an excuse for cowardly leftists and liberals to ally with their own enemies.

[4]If we do this, we must be prepared for vicious + unprincipled opposition from the Communists, but it is important that our defence be kept on an objective and cool level - i.e., we must not be carried away into the very easy dodge of condemning Russia from a superior moral vantage-point.

Finally, I want to reassure you that I'll never become a real bourgeois. It is not mainly a question of principles as of courage with me, but just that I could not stand it. If I tried it, I would either go to pieces or be forced to give it up. So I guess I'll have no choice but ultimately, as you put it, "to join the fray".

Love
Dave

P.S. Thanks for the clippings.
P.P.S. I saw a statement in Monthly Review "a labor movement without socialism is like an engine without petrol, like a body without a soul". How appropriate today. The crucial thing on the American Labor Movement is to start to transform it into a socialist movement, regardless of how quixotic this may look today. No short range objectives can mean anything any more, and only in this direction is there hope of a solution. I have a feeling (as I said before) that one open Socialist in a union will be like a seed in fertile ground. He may seem to have little influence today, but the time will come when his influence is multiplied by a hundred (provided that he is also respected as a human being and as a worker).

Letter 71. Folder C119, dated: March 1 or 2, Rec March 6 [1952].

Number on photocopy: 22

Universidade de São Paulo
Faculdade de Filosofia, Ciências e Letras
Caixa Postal: 8105
São Paulo

Dear Miriam

I received your letters, and clippings, which were very interesting. Your discussion of normal distributions was very nice. Can you explain to me why the Fourier component of a normal distribution should be easier to work with than the distribution itself?

The news certainly looks alarming. I wouldn't be surprised if they passed this immunity bill against self-incrimination soon, and if they do, there is liable to [be] plenty of trouble. As for the more general problem of war, I cannot estimate it at all, especially because Brazilian papers publish only the official US handouts. What do you think? I had already been looking into the question of Brazilian citizenship. The quickest way is to marry a Brazilian citizen, and then you can get it in 2 years after your date of arrival in Brazil. As a capitalist (who owns $ 5,000 in land) you can get it in 2 years. As a scientist, or specialist, you can get it in 3 years. For anybody else, 5 years are required. However, I am afraid of becoming stale if I stay here too long. Nobody here really goes deeply into anything, and I will have to depend entirely on myself for stimulation. As to exactly how safe it would be here, that too is problematical, since Brazil is tied so closely to American policy. But where can you look for safety these days anyway? Mott seems to be willing to offer me some kind of a job in England. What do you think of the idea? Unless I can find a circle that is more stimulating intellectually here, I shall be inclined to accept it. I find that even my physical health suffers when life is intellectually dull, for the dullness communicates itself to the rhythm of all processes taking place in the body, and pretty soon, I feel tired, dopey and run-down. This feeling is relieved very often when something interesting happens.

I shall comment in more detail on the math question in another letter. For the present, I shall only say the the introduction of $f(X) = \sum_i \delta(X - X_i)$ was suggested by a long line of work in plasma theory which I was too lazy to repeat. However, I feel that your comparing me with Erdos is a sure sign that I am coming up in the world. Just think, some day I too may become a legendary figure! I can hardly wait.

A few more small points. First, Mel Green's objections were, as you say, irrelevant. Secondly, as for paying for the books, it is complicated to send money out of Brazil. However, my father still has a little of my money. I shall ask him to pay Gross $ 200, thus establishing a reserve of $ 100, and you may apply to Gross for payment. Thus, the bourgeois conventions may be satisfied.

I'll write you more later. By the way, what about the Jan 1 issue of Phys. Rev.? Has it come out yet?

Love
Dave

Letter 72. Folder C118, dated: Rec March 13 [1952].

Number on photocopy: 18

Universidade de São Paulo
Faculdade de Filosofia, Ciências e Letras
Caixa Postal: 8105
São Paulo

Dear Miriam

I was glad to receive your letters and I'll try now to answer them in more detail. I am very lonely & discouraged these days, and often wish you were down here. The University is somewhat disappointing, as there are practically no good students and students do practically no work. Moreover, of the three other professors in Physics, 2 are really no good at all. Worse than that, they engage in fantastic plots to prevent the physics dep't from being built up, as they wish to keep things for themselves. If we could get a few more professors, we could obtain a majority, and then keep these two professional destroyers in order. The director of the Faculdade + others are tired of the way these two guys are continually throwing monkey wrenches into everything, but because of "the independence of the chairs", little can be done about it right now. I wonder whether I shouldn't push this idea of Mott's in England more energetically. Besides I need people to talk to. I grow stale without being stirred up once in a while, and I have to talk with people who understand English & Physics well, in order to crystallize my ideas, and to push them forward. Classes will finally begin next week, and maybe I shall feel better when I can have more definite things to do. Also, I shall make a really systematic effort to meet more people. Right now, my impression of Brazilians is that they tend to be shallow, seldom working hard, and do not try to see deeply into any problem. This kind of atmosphere is not good for me.

I am glad that you understand the problem of the rough-walled box & the development of chaos. I am interested in what you say about normal distributions, but have two questions:

(1) In what real problems do you have distributions with Fourier coefficients, with $\alpha \neq i$

$$\varphi(t) = e^{-|t|^{\alpha}} (C_0 + i C_1)$$

(2) What is the meaning of the imaginary coefficient iC_1? I can imagine that you might take the real part, but even here, you would get negative probabilities.

I am glad that you are getting the confidence to beard the mathematical big shots in their dens, and to disregard their use of the word "trivial". The next step is to refrain from worrying too much what they say about your papers, when they make trivial criticisms. As you say, they have a highly exaggerated sense of their own importance. Who will have heard of them in 25 years?

With regard to your self-detected tendency to "shoot off your mouth", you probably exaggerate a bit. However, remember that this is just what all the "big shots" are doing, when they try to show how "big" they are. They are doing it for the same reason that you are, to feel more secure & more important. Everybody does it to some extent and it is even OK to do it consciously in some cases when you have to impress people who refuse to listen to sensible arguments, but it is important not to be carried away to the extent of believing your own propaganda as to how good you are. But you also have to guard against the opposite side of the same difficulty – i.e., to feel an exaggerated sense of humility and disgust at one's own actions. The propensity of some people for self-humiliation is just what feeds the propensity of others for self-glorification, and very often the same person may jump from one role to the other as conditions change.

I shall continue this letter later, as I do not feel too well right now. I seem to have developed a new attack of diarrhoea + headache.

I shall include a few remarks I made yesterday about the Central Limit Theorem, I had intended saying some things indicating how one might try to treat such problems in terms of the concept of chaos, but I'll do that later.

Love
Dave

Letter 73. Folder C118, dated: Rec Mar 31 [1952].

Number on photocopy: 19

Universidade de São Paulo
Faculdade de Filosofia, Ciências e Letras
Caixa Postal: 8105
São Paulo

Dear Miriam,

I am sorry for the delay in answering your letter, but I have been rather busy with classes, etc. I am writing up my notes in Atomic Physics (Senior year) in Portuguese, before I deliver the class. It's as much work as the Qu. Mech. book was, multiplied by 2.5 because of the strange language. But it's a good way to learn to use the language and to obtain at least a scientific (and philosophical) vocabulary – I am still troubled by about one attack of diarrhoea per week, which is usually followed by a cold, resulting from the loss of strength due to diarrhoea so that I lose 2 or 3 days a week in this way.

It is not so easy to find a suitable girl (as you suggested) because (a) not many of them know anything at all (b) Brazilian girls are terribly worried about reputation, etc. [unreadable]. It might interest you to know that Brazilian men almost [unreadable] in the middle class (that is) refuse to marry a non-virgin. As one of them said "you wouldn't buy a used car if you could get a new car, would you?" I told him that he was too mechanical, and that buying a horse will be a better analogy. "You wouldn't take a chance buying a horse that had never been ridden would you?" I asked. (c) There are a few [unreadable] girls in my physics classes, but they are only about 20 years old. There is one who is [unreadable] looking, and who is said to be very far to the left indeed, but I can't help thinking of her as a little child, which she really resembles. Communication is made more difficult by the fact that these little girls are fearful [unreadable] foreign professors. Even Smith finds that he cannot help scaring them, although as he says "I try very hard to treat them nicely". (d) There is as you say, my own lack of initiative.

I was very glad to hear that your articles will be published among the big names & the bright stars of the math. world. This will have 2 big advantages for you (a) You will get more [unreadable] (b) Your self-confidence will go up. But don't overestimate the importance of this step. These great names have hardly ever being heard of outside the limited circle of mathematicians. Nor is this rather abstract work very likely to have extensive repercussions in the future. However, you are certainly justified in feeling that now you are approaching a level of equality with the fabulous big shots. However, there is a real possibility that you can surpass them by a qualitative order of magnitude and this by being ready to look into new directions and to open up new fields. Such a step is made possible less by formal brilliance than by being willing to forego temporarily (and perhaps permanently) the recognition of the "big shots" for doing small things within the existing framework.

You state that you are intrigued by certain "deep" questions arising out of your work. I do not know what mathematicians mean by the word "deep" but I suspect that it is the opposite of "trivial". If so, then a question is "deep" because resolving it will raise your position in the "pecking order". My reaction to this is that in studying even "trivial" aspects of reality, you may be led ultimately to theorems much "deeper" than can be found by trying to raise one's position in the pecking order of mathematicians, which is after all, only a minute fraction of reality. In any case, as I have so often said, the depth of reality is limitless, so if you want to dive into the depths, just follow the trail of a suitable problem based on reality.

At this point, I would like to say a few words about the "infinite number of levels" of reality, which I fear you have treated in a somewhat cavalier fashion. A better word than "infinite" is "limitless" because that is all that we really mean by infinity. With regard to your offer to explain Cantor's theory of transfinite numbers, I am interested but sceptical. The infinities of Cantor consist of collections of separate discrete objects, all alike, or similar, but the levels of reality are all qualitatively different, and each must be treated on its own merits. Moreover, they are all interconnected in such a way that that they form a unity. It is this unity which permits us at any given stage to approximate the limitless number of levels by a limited number; for the as yet undetected deeper levels all work together in such a way that to a certain degree of approximation, their net effect can be described by means of the lowest known level. I do not detect any such possibilities in the theory of transfinite numbers (I may be wrong, but then you must show me). I think that the explicit recognition of a limitless number of levels would be a big step forward in science. Most of the errors of both the positivist and the 19th century "mechanical" materialists spring from an implicit assumption that the laws of nature will some day finally be understood in terms of a limited number of hypotheses. From this comes the nightmare of a mechanically determined universe that follows an inevitable course. To avoid this nightmare, positivists and idealists have given up causality and assumed a "spontaneous" (i.e., uncaused) element in physical processes. This is the real "human" meaning of Bohr's point of view in qu. mechs., which has its counterpoint in Bergson's 'elan-vital', etc. But in practice, because the "spontaneous" element is uncaused, it is beyond scientific knowledge – it can only be "experienced" at each instant of time – and this means that the scientifically understandable aspects of reality are restricted to a limited number of levels. Thus, both points of view tend to cramp the development of science, because they discourage a search for "deeper" causes than those already known. In politics, it is the same. Mechanical determinism leads one to believe that what already exists is permanent, and the "elan-vital" leads us to believe that although changes are always taking place, they are spontaneous, beyond understanding & control, so that it is of no use to interfere with them, a conclusion very satisfactory to the existing powers.[5] The concept of a limitless number of levels suggests, however that the work of science is never finished and leads one at each level to seek the contradictions which can [unreadable] at the next level etc. Thus it provides a motive power for continual development & growth. Moreover, the nightmare of complete determinism is avoided. Although each level is causal, the totality of levels cannot ever be taken into account. Thus, as a matter of principle, we say that complete determinism could not even be conceived of, yet, each level can be determined. Here, we part company with the believers in "spontaneity" for we say that what appears to be spontaneous is caused by factors, in principle, knowable, but now hidden to us. But to be able to say this without implying complete determinism, we must assume an unlimited number of levels. It is the unlimited number of levels which give matter its "non-mechanical" aspects, for if the analysis of physical laws could ever be completed, the theory would either be deterministic + "mechanical", or "indeterministic"

[5] E.g. the Pope's fondness for theories of a mysteriously changing universe.

and "spontaneous". Another interesting point – if there are an infinite number of levels, we can expect that all existing limitations (such as speed of light and uncertainty principle) can be overcome with the aid of more fundamental levels. Thus, by the use of causal laws, humanity can move toward freedom. Whereas, in the ignorance of causal laws, humanity is enslaved either to determinism or to "spontaneity", which, being pure accident, is just as tyrannical.

One other point, a distinction between "determinism" and "causality". Although both words have roughly the same meaning, their implications are different. For causality implies (a) that if you know the causes, you can predict the effects. (b) That if you change the causes, you can change the effects in a predictable way. But determinism implies only predictability. In fact, with complete determinism, it would be impossible for us ever to change anything. Now, if there are a finite number of levels, then complete causality obviously implies complete determinism. But if there are an infinite number, then the two concepts part company. For we can have complete causality at every level, in the sense that we can use this causality to change the world in a predictable way, with the error in the predictions dependent only on our level of knowledge; whereas we can in no sense conceive of the world as completely determined. In this connection, note that the statement that new things can come into existence is consistent with causality, only if what is already in existence has an infinite number of levels. For if we have a finite number of causal levels, then the future is already contained logically in the present, but not if we have an infinite number. The appearance of qualitatively new things with time is possible with an infinite number, because the effects of the limitless number of lower levels can always surge up into a higher level (and vice versa) producing qualitative [missing words] describable as a rearrangement of things already in existence.

Finally I should say that the concept of a limitless number of levels provides a logical base for dialectical materialism. Without this concept, dialectical materialism has certain mystifying aspects, but with it you can see how it is made necessary by the very structure of matter. I do not have time to go into this here, as it is a subject in itself.

Now to answer a few of your questions. I am very interested in your result about the chaotic scattering of particles from a rough wall. Please send me a reference as to where I can find the original papers. About the use of operators in my book, I agree with you that they are probably not necessary, but if I did not put them in, then students would not be able to understand the existing literature, which is couched entirely in the language of operators.[6] Further, we can thank people like Heisenberg, Dirac, and your hero, von Neumann. Indeed, their formulation of q.m. is much more abstract than mine. It goes like this "Corresponding to every observable, q, there exists a linear Hermitian operator, Q, operating on a Hilbert space. This operator is equal to $\frac{\hbar}{2}$ times the classical "Poisson bracket". Then follow rules determining probability, etc." The whole idea is that (a) Only math. eqns. are significant these days (b) This approach is ever so much more rigorous and solidly grounded than the old-fashioned idea of using physical images (dirty word!).

[6]In the causal interpretation operators can be dispensed with entirely.

About Schwarz's inequality, the proof you suggest is neater than the one I used. However math. elegance is not important in physics as you seem to think. If I ever rewrite the book, I shall use the most elegant methods I know, provided they don't interfere with the clarity of the physical "images" (The Portuguese language refers to physical pictures as "images", a considerable improvement over the English word, in my opinion).

Incidentally, about v. Neumann, it is true that I am a little jealous, as you suggest, of his command of formalism, but the case is more complicated than that. I feel that he is a mathematical "virtuoso" and uses techniques just to impress people. This angers me (a) because it confuses the issue (b) because it forces me to waste time translating his techniques into understandable reasoning (c) Because other people are impressed by his techniques, and do not study him carefully enough to realise that a lot of it is just a show, put on to make v. Neumann look big, and everyone else look small.

I think that if we could solve the problem of the development of chaos in a real gas, that would be something enormously bigger than any abstract problem that is likely to flow out of your present line of work. It might lead to a new branch of mathematics, and in any case, would play a part in the necessary philosophical revolution in science, or as Vigier put it in a letter to me, "to turn quantum mechanics right side up" (With apologies to Engels). It strikes me that the 3 principal problems in physics are:

(1) Pursue the materialist interpretation of q. mech.
(2) " " " " of relativity, thru theory of ether
(3) " " " " of statistical processes

All three are intimately interconnected, since the ether is [unreadable] a chaotic system of particles in interaction in such a way as to lead to the laws of relativity and qu. mech at a higher level. In all three, people have been looking at the subjected [sic] in an inverted way – i.e. – by defining everything in terms of what an observer sees or knows, rather than objectively in terms of what really exists. It is a gigantic problem to overturn prejudices in such a big field, but it must be done some day, and as I have so often said, I think that success in this might have much bigger [unreadable] for the future of humanity than would appear at first sight.

Tell George I'll answer his letter as soon [unreadable].

Love
Dave

Chapter 23
Letters to Miriam Yevick, 1952, Part 3

Letter 74. Folder C119, dated: Apr 30, Rec. May 8 [1952].

Number on photocopy: 21

Dearest Miriam

Well, all hell is ready to break loose. This Nazi group has joined up with the local dep't stinkers, and they found out about my past from Europe. I have dismissed my "assistant" and we shall see what his powerful father can do. Meanwhile, they have proposed to another member of the dep't (a friend of mine) that instead of taking Schiller from the US, we accept the 2 germans as assistants. In this way "we could avoid the Dep'ts reputation as "communist" and thus avoid army intervention." The fight is ready to begin – it will probably begin Saturday in the Congregação of the Faculdade, when my request for Schiller's contract comes up. Meanwhile, we are trying to figure out a way of letting the world know what a dirty skunk von Wiseacre [Bohm's spelling of Weizsäcker] is. They are trying to establish a nest of Nazi vermin here. Either they will win or we. And we have powerful backing in the University + elsewhere. It is a fight to the finish, for if they win, physics is finished in São Paulo, perhaps in all Brasil, because there are other powerful forces in Rio with which they will probably join.

Meanwhile, I have made much progress in my theory. I have proved that given an arbitrary probability distribution, $P(X)$, because of statistical effects of collisions, it will ultimately decay into $|\psi(X)|^2$. Feynman was terrifically impressed with it, and now I think he is my friend for life. I'll bet you that Jehuda would be impressed too if he heard of this, but it remains to be seen whether the impression will be as deep as it was on Feynman.

To answer your last letter, I am not as much against beauty in math as you think. In fact, I like it very much. But I object to beauty for its own sake. I think that when a person honestly tries to solve a real problem and is successful, the result is inevitably beautiful, because the inter-connections of both the external world and of human

C. Talbot (ed.), *David Bohm: Causality and Chance, Letters to Three Women*,
DOI 10.1007/978-3-319-55492-1_23

thought processes are beautiful. But when you purposely try to make a "beautiful" paper, you are probably thinking of a meeting of the Math (or Phys) Society, and everybody saying "My, how beautiful! And so deep, too". A poet once analyzed the problem nicely. He said that the essential point is the difference between beauty + loveliness. A beautiful thing aims to be complete in itself, to emphasize its distinction from the rest of the world, including humanity. Insofar as it symbolizes perfection, it symbolizes what cannot belong to one object or to one time, but belongs only to the long run goal of humanity. On the other hand, loveliness is informal, casual, and does not attempt to hide its incompleteness, its connections with the rest of the world. It leaves room for development and change, and does not imply perfection. Thus, all "beauty" has in it an element of evil, and also sadness; for we all know that although beauty implies deathless perfection, all real objects are doomed to decay in time. Thus, a beautiful woman is really in an unhappy position, especially if she emphasizes the perfection of her beauty. But almost any healthy woman who enjoys life and is living a fruitful life cannot help having her own form of loveliness.

In this regard, I always expect that a paper carefully done will be lovely in an unplanned way, as a sunset is lovely, or as an unexpected little spot in the woods may strike you with its loveliness.

Finally, I disagree with George, who seems to think that ugliness in papers is a sign of merit, and I also think that most papers in the Phys. Rev. are inevitably ugly.

I don't think that math is just aimed to physics. It has a real creative role in itself, and other forms than diff. equations are important. But unless there is some objective basis to it, it degenerates into the spinning of would-be-closed systems of thought, aimed mainly at raising the position of the individual in the pecking order.

Now as to the infinity of levels. I feel better that at last you have an inkling of the importance of this concept. Your example of filling space up with denser and denser sets of objects might be useful in some ways, but it is only a pale shadow of the original concept. Hanna appreciated its importance immediately as she saw the possibility of a "common-sense" basis for the most abstract things, which would be tremendously appealing to most people, and also bridges in part the dangerous gulf between scientists and the rest of the people, which may cause people to turn against the strange and incomprehensible mumbo-jumbo that they don't understand. So Hanna ain't so dumb after all!

I had a very encouraging experience in explaining the ∞ of levels (in Portuguese) to my class. They just lapped it up, and saw its significance immediately. The idea here is simple. Originally people supposed atoms indivisible + immutable, though there were 92 kinds of them. Later they were observed to change into each other, and still later to be made of electrons, protons, + neutrons. But now we have 25 kinds of "elementary" particles, which transform into each other. Obviously they too must be made of more fundamental elements. But why should these be indivisible + immutable? Both history + logic suggest otherwise. In other words, it is most probable that there is no end to the divisibility of matter + the number of levels. If mathematicians were wise, they would see that the ∞ of levels is implicit even in the thought process itself. For as Gödel has shown, no self-contained consistent theory is possible which can provide its own logical basis. The only solution is that each level of math can be justified only on a more fundamental level, etc. ad infinitum. So if mathematicians would stop stewing around about the hopeless situation in the foundations of math & get to work, we might see some interesting things.

Now I want to ask you to please write to me more often as I am in a tough situation now, and each letter helps. I may not have time to answer all your letters, but I'll try. Also, try to see what you can do about lining up publicity against Weissacre, but don't do a thing until I say "go".

Regards to George

Love
Dave

P.S. Please – please – please let me hear about your proof of the set of zero measure. It is important in the struggle against V Weissacre + the rest of his rats, as they are trying to attack my scientific prestige in comparison to his. The dep't stinkers go around calling him a "genius" every time he opens his mouth.
P.P.S. If by making you work on chaos, and causing you to face chaos in the future, I keep you away from the Institute, I will do you the favor of saving you from internal chaos, which is the worst possible form of the beast.

Letter 75. Folder C119, dated: Rec May 12 [1952].

Number on photocopy: 22 (same as letter 71)

Dear Miriam

I received your two letters + clippings. The math looks f o r m i d a b l e, but I'll try to get through it. Don't worry too much about losing letters as they do get lost. While waiting at the US Consulate to discuss my passport case, I heard an employee complaining to another that he was continually losing letters in the mail. But please don't include things in letters that you wouldn't want to get lost.

The trouble I described in my last letter is not as bad as I thought. We may soon have it under control. But we need to take the initiative from now on, if we are to maintain any degree of security.

I want to answer only one question of yours now. The rest will come later. The introduction of a rough wall does not mean a "random" wall. I used the word "random" inadvertently. V. Weissacre tried to catch me on this point in a discussion of this problem. My thesis is this: Any "irregular" wall will lead to chaos. We don't need a statistical ensemble of walls, but only one wall. By irregularity, I mean that the microscopic unevennesses in the wall present a wide range of angles in any moderately small region. Thus, a very slight change in initial position of a particle is sure to change the angle of reflection a great deal, and this in turn will lead to a still bigger change of position, etc. It is this instability that leads to chaos, as shown in the figure below.

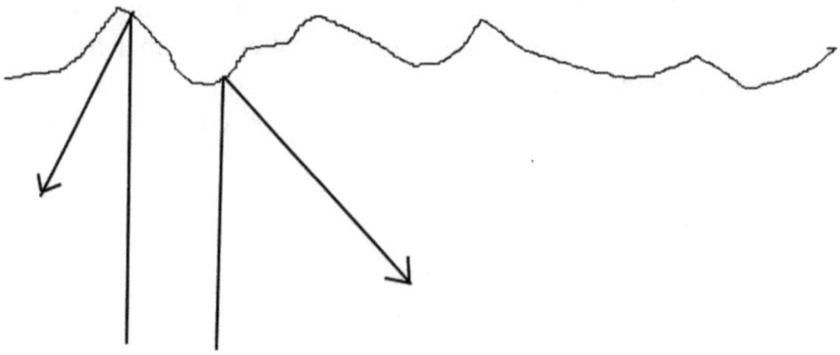

Now what do I mean by "a set of measure zero". I have read some more work on Lesbesgue integration, and I have the impression that it seems to be "much ado about very little". I have never seen any cases where a set of measure zero could not be understood in terms as the limit of ordinary integration over a smaller and smaller set of regions. For physical purposes, it seems best always to imagine that these regions are finite, but as small as we please. Let me give you an example. Suppose that we have a set of unstable trajectories, corresponding to averages different from the usual ones (as shown below):

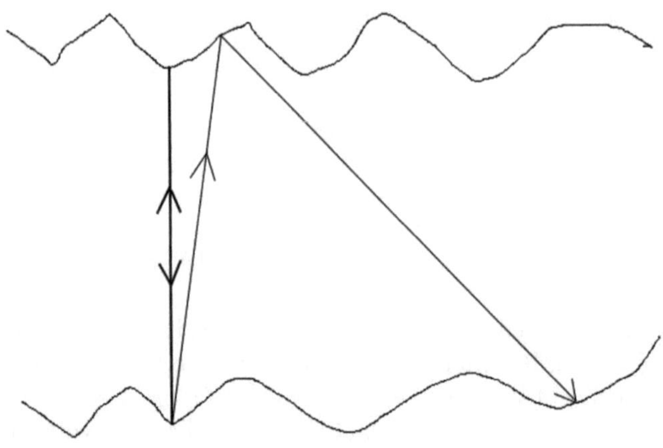

It would be possible for example for a particle to have non-chaotic motion, if it reflected off the points in exactly the manner shown above. But the slightest shift in initial conditions would destroy this state of affairs. Let us now consider a time, T, and a region, ϵ, near the special orbit shown above. In order that a particle remain in the region, ϵ, during the time T, it is easy to see that the domain of initial directions which can lead to this must become smaller and smaller as $T \longrightarrow \infty$. Thus for "$T = \infty$", there is only one direction that works, and we have a "set of measure zero".

Now, if we can show that the combined volume of the space of initial conditions for all points leading to appreciable deviations from the "usual average" approaches zero as $T \longrightarrow \infty$, then our physical problem is finished. For we know that from direct experience that it is impossible to adjust any physical conditions exactly because chaotic fluctuations are unavoidable. Thus, if there is initially the slightest degree of chaos in the system, it will inevitably be multiplied to a maximum.

At this point in the argument, v. Weissacre objected that after all, I still had to use a sort of assumption of chaos at the very beginning. My answer was this: By means of this deduction, we have narrowed down our assumption of chaos (or probability, if you prefer) to a very narrow and well defined region that can be checked experimentally. In other words, we reduce the hypothesis to a point where it is very easy to see just what is being assumed, and how to check it. But the hypothesis of a statistical ensemble in equilibrium is much more difficult to check and to understand, because it is very difficult to know if your hypothesis concerning distributions of particles in ensembles are really right in detail in thermodynamic systems. The agreement with thermodynamics is such an insensitive check that it means almost nothing about the details of the distribution. And it is impracticable to look at every particle, and count the number of times that it is in a certain place. On the other hand, it is quite easy to show by experiment that no system can be maintained exactly on a certain unstable point indefinitely. Moreover, it is always easy in each case to trace the chaotic processes that prevent this from happening. Thus, the theory is put on a more solid basis.

One more point. You complain that I am so far from "understanding the universe" and that if I were closer, you would have more desire to help me. I wish only to point out that if I were really close, I wouldn't need any help at all, and that at this point, everybody would also be more than willing to help. But help is really help when you need it.

Love
Dave

Letter 76. Folder C119, dated: Mailed June 3, Rec June 9 [1952].

Number on photocopy: 23

Universidade de São Paulo
Faculdade de Filosofia, Ciências e Letras
Caixa Postal: 8105
São Paulo

Dearest Miriam

I was very glad to hear from you. Every time I see a letter from you, I say "Good!", but usually it turns out to be only a set of clippings. Then I say "God damn it!" I appreciate the clippings, but much prefer your letters. Also, when the clippings carry bad news, like that of Weinberg's indictment, this is even worse. I think that poor Weinberg is a dead duck, in the atmosphere that exists today. Heaven knows how far they will try to go, and if they will not ultimately (perhaps next year) try to involve the great man himself, whose face once appeared with infinite sadness on the cover of Time magazine. Perhaps he will really have reason to be sad too. As was once said he looked like "Jesus Christ" in the picture. I think a better image would be a linear combination of J.C. and Judas, as of Judas trying to look like J.C. An interesting case of mistaken identification, don't you think?

As for me, I have stopped worrying about such things. I am getting used to the idea that our individual lives count only in so far as they relate to the great transformation of humanity that is now taking place, and that I have already managed to snatch about $\frac{1}{2}$ of it (i.e., my life) from the jaws of chaos, and to do a little work, which might ultimately play a role in this transformation. After all, many people lose their whole lives, either in concentration camps, in sickness, in middle class futility, or in working class squalor and hopelessness. I hope that the rest of my life will be saved too (in the sense of my getting some useful work done), but in a chaotic situation, one has no security about the individual. Only the ultimate general collective outcome is certain.

This brings us to the question of chaos in statistical mechanics. I think I am making progress on the problem. First, there are the following simplifications of the rough-walled box problem. Let us take the case of 2 dimensions, and suppose an ensemble of particles with the same speed. We suppose elastic collisions with the wall, so that their speeds do not change. Now we postulate an irregular wall. Any kind of irregularity will do, but for the sake

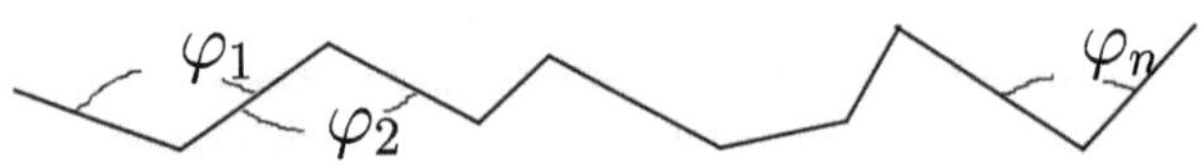

of definiteness, we define the angles φ_n, the changes of direction of the wall, in such a way that the sum of any number $\varphi_1, ..., \varphi_s$ is not equal to π or to an integral multiple of π. We also postulate that for all s, $\varphi_1 ... + \varphi_s$, is not equal to φ_n, any φ_n in the set, or to $\varphi_n +$ a multiple of π. (Remember that the number of angles is finite but large). We also postulate that the angles φ_n are distributed in such a way that the net inclination of any surface to the mean plane of the wall, is never more than some convenient number, say 45^0.

We now define the complex number $z = x + iy$, where x, y are the coordinates of the particle. Every collision with the wall produces a change of angle of motion $2(\theta - \psi)$

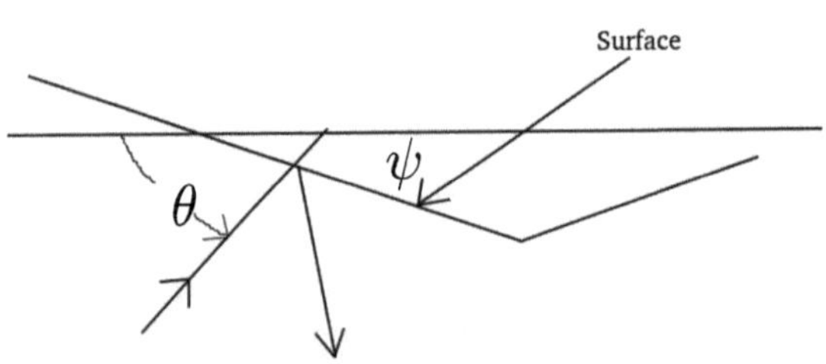

where θ is the angle of the particle relative to the wall, and ψ the angle of the surface relative to the wall.

Now each particle moves in a straight line, with a velocity with components $w_x = vcos\theta$, $w_y = vsin\theta$. Define the complex number, $w' = ve^{i\theta}$. Then $z = z_0 + w't$

After collision, we get $w' = we^{-2i\theta}e^{2i\psi} = ve^{-i\theta}e^{2i\psi}$

It appears to me that the above notation is more convenient than the . . .[ends here]

Letter 77. Folder C119, dated: Mailed June 21, Rec June 27 [1952].

Number on photocopy: 24

Universidade de São Paulo
Faculdade de Filosofia, Ciências e Letras
Caixa Postal: 8105
São Paulo

Dearest Miriam

Glad to get your letter of June 9. I got your other letters OK. Things are still quiet here. I think I shall now start my moves to get back my passport. We'll see what happens.

The letter to you from Chung is disgraceful. He doesn't want to give you credit for anything. I'd say he hates you, and is also a bit jealous of you, and afraid that you might show him up someday. With people like that, it's important not to invite being stepped on by humiliating yourself. Chung deserves to have some major misfortune happen to him, but on second thought, I'd say that the necessity of being Chung is punishment enough.

As for your career, I can see that it's a tough problem. I like children when I see them, but forget about them when I don't see them. You had better delay children for a while, anyway. There is real danger that you may be tied to down in a way that you will regret. Although children are very good, they do not satisfy other needs, as observation of middle class life makes obvious. What happened to the British job you had in mind? In any case, don't get children just because of a panicky fear of middle age. That would be as silly as my marrying Hanna because of a panicky fear of loneliness. Don't forget that whether you want children or not, we are all headed for the same end, and when it comes, having or not having children won't make much difference. The question is whether in the meantime, life would be better with or without children. In your case, this is a difficult question, but you aren't so old yet that you need to worry as much as you are worrying.

I received a letter from Shenstone in which he mentioned incidentally that he hoped I wouldn't do anything irreversible, like going to Israel without a passport. How he got the idea I would do that, I don't know. Perhaps I once mentioned to someone that I would consider a job in Israel. But the situation in Israel is by now so terrible that it would hardly be worth going there – no work could be done there.

As for collective coordinates, I must admit that I don't understand them fully yet either. It is a fertile field for research. I will need some time to understand your idea of a collective probability distribution. There is a great difference between this idea and the collective plasma coordinates; for in the latter case, the collective coordinate is made a good variable, because the potential energy, $E = 2\pi e^2 \sum_k \frac{|\varrho_k|^2}{k^2}$ is a sum of separate terms, one for each ϱ_k. Thus the forces themselves produce a simple collective behaviour, and the stronger the forces, the better the collective approximation becomes. But in your case, you would need some physical cause, such as a force, that was simply expressible in collective coordinates before the collective point of view would be useful.

As for Jehuda, I doubt that he is interested in my problems. Although I would be interested in communicating with him about collective coordinates, I can certainly get along without hearing from him. However, there may be certain aspects of this problem that are now coming up which would be worth discussing with him. But I have heard from a friend of Jehuda and he will try to find out why Jehuda is so negligent about writing.

I now understand the stuff you sent me about random distribution of paths. Hereafter, I suggest that you define your notation in more detail (e.g., the "dyadic" system was very hard for me to figure out, but my mathematician friend here explained it quickly). Also, for example, you should have mentioned on Tchebysheff's inequality that $f(X)$ is a monotonically increasing function. It seems obvious, but how can I tell whether by some "deep" theorem, the mathematicians haven't proved it for arbitrary f?

The theorem that you proved does not refer to my problem, since it presupposes a random distribution of paths and therefore a random distribution of walls. I am interested in showing the development of chaos with a single moderately irregular wall. I can do the problem rigorously now, and I'll send you a proof in a few days. But your letter was very helpful, since it defined certain ideas for me. The idea of "strong" laws is clever, I am reading Feller's book "Prob. Theory + Its Applications" which is about my present speed. He is clear, but shows his positivist bias all over.

Will write soon again

Love
Dave

Letter 78. Folder C119, dated: Rec July 3 [1952].

Number on photocopy: 25

Universidade de São Paulo
Faculdade de Filosofia, Ciências e Letras
Caixa Postal: 8105
São Paulo

Dearest Miriam

I haven't written you (or anybody else) much lately, because I have been in a somewhat depressed state, and don't have much to say. As soon as I finished up my article on the proof that the Probability density approaches $|\psi|^2$ in the Causal Interp.

of the Qu. Theory. I had nothing more to aim for, and as usual, the aimlessness & futility of this life then began to get me down. I don't think I can stand much more of Brazil. The people here are completely without any push or purpose but don't even enjoy life in compensation. While U.S. may be said to be decadent, Brazil is best described as "unawakened". In any case, the confusion and inefficiency here is such that nothing much can ever be done, except by wasting endless amounts of time pushing each detail past the associated bureaucrat who sits on it. Worse still, there is no one for me to talk to here. It is not just [unreadable] girls, but more generally, regardless of the sex, there is no one who has thought about things sufficiently to make his (or her) opinions of any interest to me. In science (as well as in other things) they just learn the formulae which are comfortable to recite, but that is all they will ever do. I suppose I would feel better if I tried to find some friends anyway, but it seems a lot of effort which will lead to nothing. As for physics, I also am losing much of my interest in it, because it now seems clear to me that barring an unlikely stroke of luck, nothing will change much for at least 20 years (The time needed for the present generation of physicists to die off and be replaced by a new generation). It is not the opposition of the "old Guard" like Pauli, etc. which disturbs me so much as the fact that even those who are interested in a new point of view so persistently misunderstand everything, even when it is clearly stated. This question of probability is a case in point. At several places where my papers were discussed, people came to the conclusion that the assumption of the relation, $P = |\psi|^2$ (where P is the probability density) although consistent, was mysterious, and a man named Keller from N.Y.U. even wrote a paper to the Phys. Rev. (which I received) saying that behind my work was an implicit assumption of a "deeper" kind of probability than the usual kind so that I didn't really succeed in getting a new interpretation of qu. mech. All this despite the fact that I pointed out in my papers several times that I thought the ensemble, $P = |\psi|^2$ resulted from collisions, etc., and conjectured that this could be proved. In my latest paper, just finished, I proved this conjecture,[1] but I am sure that the boys will now find something else to get confused about. I have a feeling (perhaps irrational) that most of those who say they favor a causal point of view always seek some point to be confused about whenever they are presented with such a point of view, in order to avoid having to really face the problems involved. I am interestedly waiting to see what else they can get confused about. For example, I was told that Phil Morrison agreed with my philosophy, but thought that Bohr's point of view should be retained, because it is "simpler". This is plain inconsistency, since it is only a positivist philosophy that would justify "simplification" achieved by a loss of causal explanation of the relation between past and future. In any case, it now seems clear to me that any effect I might have on science will be so slow that it can hardly matter. I feel that I am almost completely without allies. Perhaps the

[1] I have been trying for 2 months to get this paper mimeographed, but due to difficulties in getting a typist who can do English, and due to an endless series of bureaucratic confusions in the University, I have just about given up hope. We finally had it stencilled once, but it turned out that the stencils sent us by the University were of such low quality that the papers were illegible. The second batch of stencils were of higher quality, but now it turns out that they won't fit the machine (after the paper had been re-typed).

you know who's will work on these ideas, but I am even somewhat doubtful of this, because their physicists, like ours, take the lead from a few big shots, who might easily decide that they don't like it. So I am open to suggestions as to what can be done. But the international situation now looks so dark that nothing seems to mean anything.

With this discouraging introduction, let me explain my ideas on the problem of chaos. First I want to simplify the problem by making it two dimensional. Secondly, instead of a wall, let us consider a periodically distributed set of scattering centers, located about 10^{-8} cm apart.

As in Chap. 1 in my book, we replace the walls by periodicity conditions. We assume a series of square boxes of unit area, with the initial particle distribution in each box exactly the same as that of the others. This insures a periodic distribution of particles, which in turn insures that the total number of particles in each box will remain constant. This is all that we really need the walls for.

Now chaos is produced by the scattering centers. Since in a real gas, the scattering centers are irregularly distributed, if we can prove the development of chaos for our case here, it will be even stronger for a real gas.

Now we specify the state of the particle by its speed, v, the angle, θ, representing its direction of motion, and by its coordinates, x and y. Let all particles have the same speed v, and let the collisions be elastic, so that v never changes. Then θ, x, y will be adequate coordinates. Moreover, if we can prove the development of chaos on the lines ($y = -1, 0, 1, ...$) on which the scattering centers exist, this will be adequate, since the motions of the particles will (as can easily be proved) then propagate the chaos into all values of y. Since the distribution is periodic in y, the line $y = 0$ will be adequate for our purposes.

Now, let us imagine a particle striking the scattering centers on $y = 0$ at the point, x. We suppose the scattering centers are separated by a distance, a, which is very small (of the order of 10^{-8} cm). Suppose that the scattering is "isotropic". That is, we suppose that the deflection is proportional to the distance from the center until we reach $\Delta\theta = 2\pi$, and then goes back again to zero, as shown below:

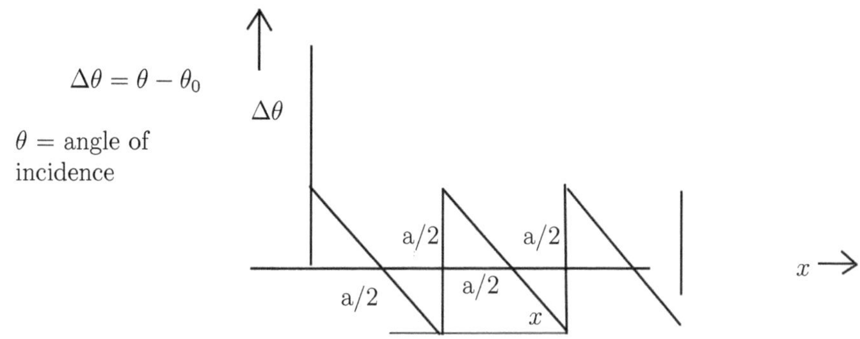

So $\Delta\theta = 2\pi \dfrac{x}{a}$

(The formula adds $2n\pi$ to the angle, instead of having a discontinuous shift, but this makes no difference as physically the two are the same).

Now the essential idea is that the angle of deflection is very sensitive to the initial position, x. In the usual theory of probability, we would suppose that at each point x, there is a certain probability $P(\Delta\theta)$ of a deflection through an angle, θ, independent of x).

In our point of view, each x leads to a definite $\Delta\theta$, but one which varies so rapidly with x that if the position is defined only within a region, $\Delta x \gg a$, the deflection appears to be arbitrary, i.e. determined by "chance", in a gross point of view.

Now our procedure is to suppose an arbitrary initial probability distribution, $P(\theta, x)$, which is large in only a small region $\Delta\theta$, Δx, as shown in the diagram. But although Δx is small,

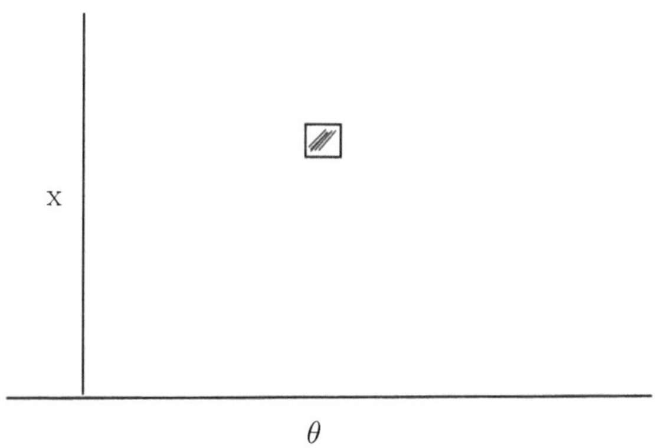

it is much larger than "a". Now when the particles are scattered, they are thrown between 0 and 2π depending on their initial x. The distribution after scattering will be as shown below

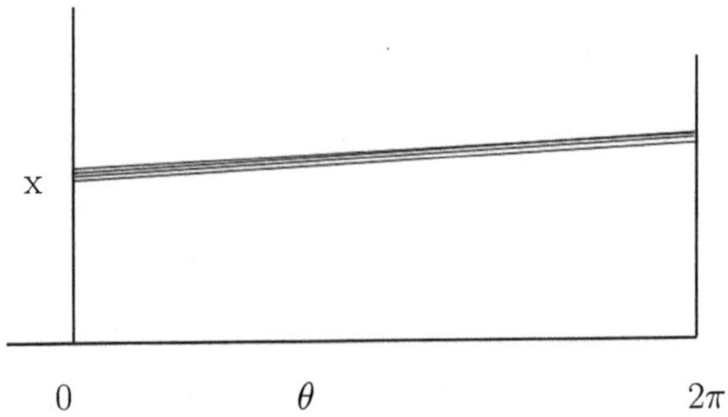

where the initial rectangular volume element has been changed into one consisting of fine threads running between 0 and 2π.

After scattering, the particles will move to the next set of scatterers (a unit distance away). Each particle will move a distance $x = \tan\theta$. The distribution will now look like the figure below, where the fine threads have been stretched over the whole

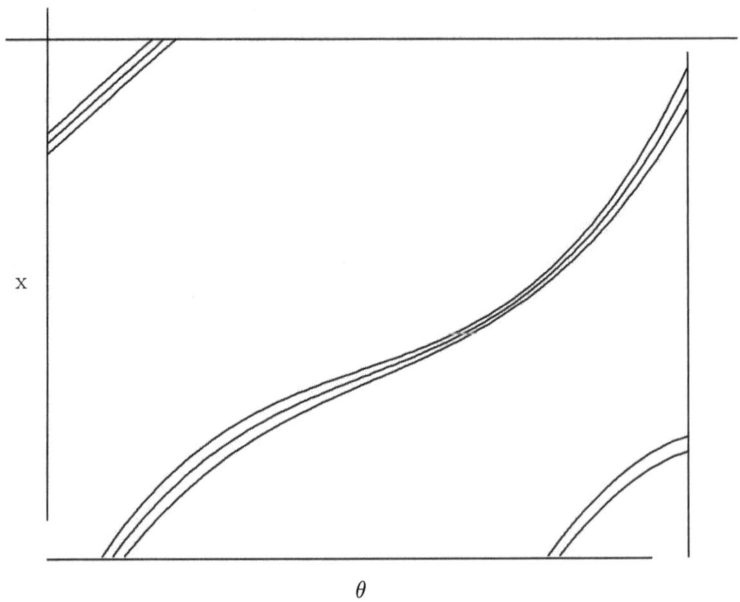

range of values of x. The next step is a new scattering process. Here, each of the fine threads is turned into still finer threads running from 0 to 2π, as shown below

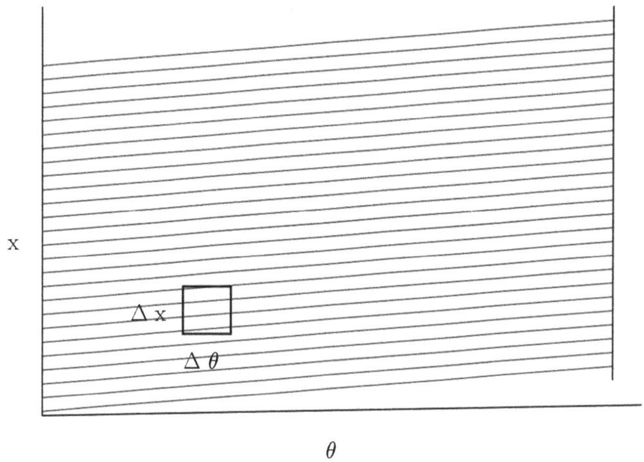

This process repeats itself again + again, and ultimately the whole space is filled with fine threads, which come arbitrarily close to each point. It is now intuitively clear (and can be proved) that a small element $\Delta x \, \Delta\theta$ will contain a volume of threads that is proportional to $\Delta x \, \Delta\theta$.

What have we proved? We have proved that regardless of the initial probability distribution, we shall end up with a final distribution consisting of a chaotically tangled thread like structure that for practical purposes is equivalent to a uniform distribution. This has resulted from the instability of orbits, and this instability results from the extreme sensitivity of the angle of scattering to the initial position.

The advantages of this procedure over the usual one are:

(a) We make only a qualitative assumption that there is an arbitrary initial distribution, and obtain a quantitative result for the final distribution. This result is the starting point of the usual theory of probability.

(b) It is easy to verify experimentally that the initial conditions always fluctuate slightly, i.e., that they can never be reproduced accurately. This is the physical meaning of an arbitrary initial probability distribution. On the other hand, it is impossible in practice to verify the usual hypotheses about uniformity in phase space – it is too difficult to measure the positions of 10^{23} atoms, which is what is needed to do this.

(c) We have almost gotten rid of the concept of probability. Instead of assuming a certain ensemble, we assume an experimentally verified lack of reproducibility in initial conditions. Chaotic instability of trajectories then produces the usual ensemble. It is like the Cheshire Cat in Alice in Wonderland reduced to its grin.

Now can we get rid even of the grin? The answer is "probably". To do this, note first that the cause of fluctuation in initial conditions is chaotic lack of reproducibility in the surroundings, arising from the same instability of trajectories. A reasonable conjecture from a physical point of view is that the orbits of the universe not leading to chaotic fluctuations constitute a set of measure zero. (I can see how to prove it). Thus, we need only postulate that the universe is not in one of the states of measure zero in which there is no fluctuation in initial conditions for particular systems. This postulate certainly agrees with every experiment that has ever been done (i.e. there are always "errors")

I hope that you get the idea – I will send you more details later.

Love
Dave

Letter 79. Folder C119, dated: Rec Aug 14, 1952.

Number on photocopy: 26

Universidade de São Paulo
Faculdade de Filosofia, Ciências e Letras
Caixa Postal: 8105
São Paulo

Dearest Miriam

I got your letter from California, and I am just sending you a short letter. I have sent another one previously to Washington.

In your last letter, you state that "unlike George and me", you have a "phlegmatic" temperament and can work untroubled, without crises, etc. Everytime I think of you saying that, I cannot suppress a smile of amusement. You who worried yourself into an ulcer because of unhappiness, and who always write me of how unhappy you are, whenever you are alone, of how you don't like your work, etc. When it comes to withstanding loneliness, I have had more practice than you, and I think I can do it better. But by now, it's too much for me. I am going to try to find people here, but have no great confidence in being able to do so. Even the "left" here seems very superficial. No one here has ever, as far as I can see, done any thinking. But it is better to have the company of superficial people than none at all. Meanwhile, I must plan on how to get out of here. To return to the US is unthinkable. Not only the danger, but even worse, is the fact that I would have no job, except in some place like Louisville, Kentucky, which is 10 times as bad as São Paulo. But I think that things here are so hopelessly corrupt, inefficient, and demoralized that it would be ridiculous to plan to stay here and try to build a school of physics. The people here talk big, but very few of them want to do anything.

The first step is to see whether I get my passport back. I have asked my lawyer to make a request. If they won't do it, perhaps I can have him start a process in court. (I see that someone else in my position got a favorable decision in a passport case).

At the moment, I am in a bad mood. Now that vacation is over, I'll feel better, because at least I'll be occupied.

The vacation that you describe sounds interesting. I have crossed the continent so many times, but I shall probably never do it again. I don't feel any "roots" as you describe them in the U.S. I feel that it has turned into a madhouse. Once there were possibilities in the U.S. for good, but now it seems to me that everything rotten and false, which I have always feared, has grown enormously and has choked the good possibilities. Although I am sick of Brazil, I haven't the slightest desire to return to the U.S. I remember it as a sort of foreign country, because what it has become is foreign to its original ideals and possibilities. The fluorescent lamps and television screens have somehow come to symbolize for me the transformation of the U.S. into another country, because in their cold glaring light, the emptiness of life is somehow emphasized; i.e., one has superficial polish, cleanliness, and brilliance, but it is totally empty and meaningless. This is exactly what has happened to the middle class, with its modern kitchens, cars, etc.

Love
Dave

P.S. I have sent you a mimeographed manuscript on qu. theory, and I am curious to see whether you got it. (I have tried using the Brazilian mail for printed matter, but I am told that they often rob the stamps and throw the mail away.)

Letter 80. Folder C119, dated: Rec Aug 20, 1952.

Number on photocopy: 27

Universidade de São Paulo
Faculdade de Filosofia, Ciências e Letras
Caixa Postal: 8105
São Paulo

Dearest Miriam

I am sorry to delay answering for so long, but first I went for a brief vacation in the mountains and then we had our international Congress of Physics. The vacation (1 week) was very nice and refreshing, but the conference was on the whole, boring and discouraging. There were 8 American physicists, 8 Latin Americans, and a few Europeans. The Americans are clearly very competent in their own fields, but very naive and reactionary in other fields. I gave a talk on the causal interpretation, plus an additional informal seminar. I enjoyed giving the talks, because of the sense of struggle, and for a short time I felt alive again. But the reaction was on the whole very discouraging to me, because it shows that the majority of physicists cannot be reached by arguments. Rabi's opposition was typical. He said that it wasn't a new theory – it was based only on hopes. They don't seem to realize that the usual theory has produced no results for a long time and is also based on hopes, hopes which are rapidly failing. When someone does a very complicated meson calculation, he bases it on the hope that some day this will have some relation to experiment, even though the present mess in the theory makes such a possibility very implausible. There is a peculiar double standard here. For the usual theory, everyone says "We have hopes, even though not very bright ones, so let us work on it, discuss it, etc." For the causal theory they say "We aren't interested until you produce results".

It now appears to me that the problem is very difficult and may not produce "results" for a long time. But I am discouraged, and I don't even feel like writing anything because I have the feeling that at this great distance no one will understand, and even if they do, they won't be interested. Things may improve a bit when Schiller arrives, but I am certain that in the long run, I shall have to leave Brazil. Physics here is in a terrible mess and cannot really grow on a sound basis, so there is no point in putting a lot of effort into that. Also research won't produce any "results" for a very long time. Meanwhile, I still have no one that I can really talk to. I am sure that good people exist, but it is a complete mystery how to find them. I have tried in logical places, but it seems to be harder even than in the U.S. People are even more separated here than in the U.S. The language makes additional difficulties. As for Rio, I can't stand the city. Its hot climate combined with its really ***insane*** traffic exhaust me in about two days. Nobody can think in Rio. By evolution, everybody with any tendency to think has been run over by crazy bus drivers, truck drivers, or by just plain crazy motorists. This is really true. You can't cross the smallest street without looking several blocks to see if there is a truck or taxi coming which might decide abruptly to turn in without the slightest decrease in velocity, but with wheels skidding and barely holding the road. If someone is in the way, they just pay no attention, unless he is bigger than they are. If you go for a stroll in Rio, you come back exhausted, and with an illimitable hatred of all motorists, that fills your being and displaces all other thoughts. Also, the noise of traffic is so intense that you can't talk, and the smell of poorly burned oil of the buses and trucks so bad that you wish you could close your nostrils. Anyway, Rio is a city of really terrible poverty, which is adjacent to the most fantastic wealth (within a few yards of each other). The general impression that the city gives me is that of a natural setting of unmatched beauty, combined with a social order of unmatched insanity. I really dislike the place. São Paulo is similar in every way, except that the traffic and the poverty are a little less intense, and the climate a little cooler.

You are right – I really should find me a girl. This would help take my mind off the generally unpleasant situation here and in the world at large. My hopes for getting a passport are fading, but I cannot contemplate staying in this blind alley forever. But meanwhile, I should try to make my stay a bit more pleasant.

I hope that you enjoyed your vacation. Please write soon.

Love
Dave

Letter 81. Folder C119, dated: Mailed Aug 19, Rec Aug 26 [1952].

Number on photocopy: 28

Universidade de São Paulo
Faculdade de Filosofia, Ciências e Letras
Caixa Postal: 8105
São Paulo

Dearest Miriam

I have written you two other letters – one to Berkeley & one to Washington. Although you have written me five, two of them were just clippings, and don't count as letters. After all, I could send you clippings too, if you want me to. So the score is now 3 to 3.

I am not quite as depressed as I was when I wrote the other two letters, but the situation doesn't look so good. The lack of companionship is still a serious problem. Now that classes have started again, I am a bit more occupied, and therefore happier. The other problem is that I am very discouraged about being able to communicate my ideas to other people, especially over such a great distance. My experience with the Physics Conference in Rio brought home to me with a shock how little interested most physicists are in the questions that interest me. So a few articles or a book will hardly affect them at all. Even those who might be interested generally, do not understand what I am driving at except when I have a chance to talk with them for a long time. Thus, for example, Smith now understands this infinity of levels stuff pretty well. There was another strongly positivist young man, here, Anthony Wataghen, whom I converted completely away from positivism, by steady pressure. He even arrived at saying that it is important to publish these ideas on an infinity of levels, but I told him that nobody would believe it, or understand it, if it appeared as an article. The customary ideas are so thoroughly interwoven into most people's thinking that it takes days and days of discussion to lay the basis for a complete re-evaluation of these ideas. And each person has his own objections which must be answered differently. So I feel sort of thwarted. Similarly, on this probability problem, experience shows that it takes many discussions even to show people that the problem exists; i.e., that the present theory has something inadequate about it. Even with you, it has taken many letters to get to the point where we are perhaps ready to discuss the problem. Similarly, with the infinity of levels. Of course, the fact that it takes a month to write and get an answer is a big disadvantage too. In any case, I would appreciate having your response to what I have said thus far about the probability problem and the infinity of levels.

The difficulties of communication have an oppressive effect on my imagination, because I always feel that these ideas will necessitate a bitter struggle, even to get to the point where people will realize what the problem is that I am aiming at, or to admit that the problem even exists. And that is only the beginning, because after this, the problems themselves still remain to be solved. So that thinking about physics will bring mostly worries and tensions, and there is now very little else with which I can occupy myself.

I still hope to try to get better acquainted with people here. I know that there must be a large number of good people, [unreadable] who perhaps don't know very much, but whose hearts are, so to speak, in the right place. But how to get in touch with them? Most of the people I see here are as dull as the typical US bourgeois crowd. Smith, after $2\frac{1}{2}$ yrs, seems to have begun to get acquainted with a few interesting Brasilians, but thus far spends most of his spare time with secretaries from the US Consulate, of all things. He wants to fix me up with a secretary who he says is very liberal and all that, but I say that this would be pure insanity. He suggested in fun, that she might help me get my passport back, but I told him that she wouldn't be of much help unless she were Mrs Shipley, in which case the sacrifice would be too great to be worth it.

Seriously though, I have only very small hopes of recovering the passport. But the idea of remaining here year after year is frightening. By marrying a Brazilian, I could get citizenship here in 2 years, and then get a Brazilian passport, but this seems to be very roundabout. Also an appropriate Brazilian girl is not yet in sight, although there are several very wealthy young girls I know who would probably be interested in marrying me. Unfortunately, by a peculiar coincidence, they are not my type. In any case, the world situation is deteriorating so fast that in two years, it may no longer be interesting to travel to Europe.

I shall be very interested in hearing what kind of job you get. Also, I presume that you are refreshed by your vacation and ready for work. What are your ideas on Margenau's book?

I shall now answer one of your questions on the role of matrices in qu. theory. Personally, I dislike the matrix formulation of qu. theory, and regard it as excessively formal. For this reason, I gave only a brief treatment of the subject in my book, just enough to define the terminology, because the student will need to be able to recognize this terminology if he is to read much of the literature on qu. mechs. In the causal interpretation, the matrix formulation falls back to a minor role; for whenever the Schrodinger's eqn. is linear and homogeneous, one can, as for any other linear + homogeneous eqn., give it a matrix representation, if one chooses. But as soon as the equ. ceases to be linear, the matrix representation ceases to be useful. Moreover, in my point of view, the wave function does not describe all properties of the electron – there is also a particle with a definite position, and its properties cannot be represented by the matrix. Thus, my view is that the matrix formulation of qu. theory is a mathematical convenience which can be utilized in the restricted domain in which the present form of the quantum theory is valid, but which will prove inadequate in newer domains, where a less abstract formulation will be needed. [missing words] presents the superficial aspect of a generalization of the Schr. equations, this generalization can only be achieved at the expense of making the restriction (usually implicit) to a linear homogeneous equ. Like many other purely formal generalizations, it is more restrictive than broadening in its effect on our ideas; and is particularly dangerous because it restricts us under the guise of extending our point of view to cover a wider domain.

Please write me soon
Love
Dave

Letter 82. Folder C120, dated: Rec Sept 2, 52.

Number on photocopy: 29

Universidade de São Paulo
Faculdade de Filosofia, Ciências e Letras
Caixa Postal: 8105
São Paulo

Dearest Miriam

I got your last letter, and the delay in my writing was just due to a combination of factors. The main one was that I figured I would wait till you got back, since there was some doubt that the letters would arrive at the right place and the right time while you were on the move. I don't like the idea of letters getting lost on the way.

Your not knowing the meaning of the word "miserly" is particularly puzzling, because you used it correctly. You told me first that I hadn't paid Eugene the $100 that I owed him, and that I had a tendency to be miserly. However, now it develops that you think I am also a sort of old-maidish character.

Things look a bit brighter, now that I am busy. But the general state of the world, especially of everything in Brazil, is best described by the Portuguese word "porcaria", which has no exact English equivalent, but which is derived from "porco", meaning "pig". This word is a very useful word in Brasil, so it is an excellent description of all machinery, such as elevators, etc. and of anything done by the Brazilian gov't. I have discovered the general rule, which explains the functioning of machinery here, the sixth law of thermodynamics, valid only in Brasil. "Everything that is supposed to move is stationary, and everything that is supposed to be stationary moves." But seriously, the gov't is so hopelessly corrupt and incompetent that a serious crisis is developing here, which will be initiated by the "dollar shortage". There is a chronic food shortage, originating in the trek of workers to the cities, without a compensating rise in agricultural productivity, and riots are breaking out everywhere about the high cost of living. There is an electricity shortage, resulting from monkey wrenches thrown in by the Brazilian gov't, which for the last 5 years refused to approve the importation of generators, but insisted on Cadillacs instead. This power shortage is hitting industry now, creating unemployment. Meanwhile, because of the continuous inflation, capital is going into real estate speculation, and the construction of useless apartments, far beyond the reach of 95% of the people, in cost. This also wastes foreign exchange (Concrete - steel, copper, elevators, etc., must be bought abroad). The roads are in terrible condition, also the railroads, so that it would be impossible to bring enough food into the big cities (even if there were enough on the farms). There will be an oil shortage soon, resulting from the dollar shortage, which will further aggravate the already bad transportation problem. To top it all, there was a failure of the wheat crop in Argentina, which necessitates buying wheat from the U.S. with already scarce dollars. All they need now is a failure of the coffee crop – almost the only source of foreign exchange (or a drop in the price of coffee). Meanwhile investigations into the Bank of Brazil and other agencies show a complete rottenness and corruption., which suggests that the gov't neither knows what it is doing, nor cares. It is a case of each man filling his own pockets, even in the face of a national disaster, which is what Brazil faces in the next 10 years.

I'll write more later. I am particularly anxious to hear your ideas on the probability problem, and on Margenau's book.

Love
Dave

Chapter 24
Letters to Miriam Yevick, 1952, Part 4

Letter 83. Folder C120, dated: mailed Sept 13 [1952].

Number on photocopy: 30

Universidade de São Paulo
Faculdade de Filosofia, Ciências e Letras
Caixa Postal: 8105
São Paulo

Dearest Miriam

I was very glad to receive your last two letters, and am awaiting your next one on your ideas on Margenau et al. I am now a lot less discouraged, since Schiller has arrived. He is quite good, and his speciality, general relativity, is exactly what I need to extend the causal interpretation of qu. theory to gen. relativity. So I should be able to work here profitably with him for another year or so (if he isn't drafted).

By the way, did you get the mss. I sent you on the proof that probability approaches $|\psi|^2$ in causal interp. of qu. theory. I wish you would read it, especially Secs 1, 3, and 4 as it illustrates certain phases of the problem of probability in general.

I have made some progress on the probability problem. To do this, I abstract the model further. Consider a 2 dimensional gas of point particles, undergoing collisions with fixed obstacles distributed in a definite but disordered way.

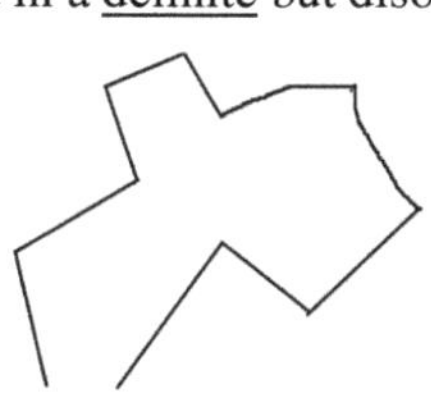

The particles will undergo a series of deflections, and you can see intuitively that a chaotic distribution will result. Our problem is to prove the development of chaos in individual cases without the use of hypotheses on probability.

C. Talbot (ed.), *David Bohm: Causality and Chance, Letters to Three Women*,
DOI 10.1007/978-3-319-55492-1_24

The essential aspect of a collision here is the extreme dependence of the deflection, $\Delta\theta$, in a collision, on the initial position of a particle.

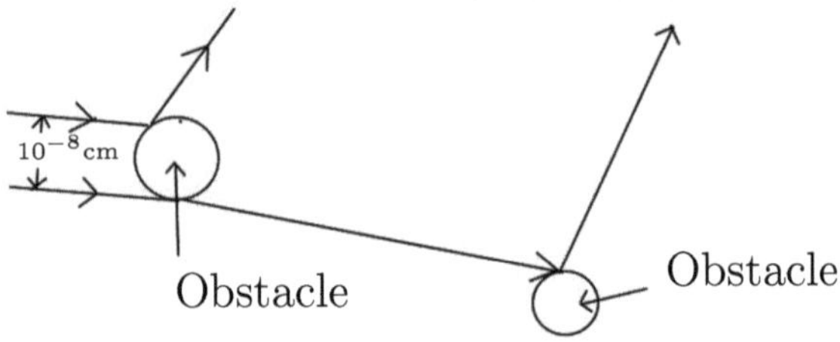

As you can see in the figure above, the change of angle may be anywhere between 0 and 2π, and a change of initial location of 10^{-8} cm. (the size of an atom) will change this angle from 0 to 2π. The future motion, even in one collision, is very sensitive to the initial location, and in a series of collisions, this sensitivity is accordingly multiplied. Thus, the motion of particles is very unstable.

The next step is to abstract the model still more. We assume that a particle always makes a new collision after a time, T, has elapsed relative to the previous collision, during which time it moves a distance $L = VT$, where V is the speed. In reality, there is a statistical distribution of times T, "free paths", L, and velocities, V, but we simplify the model by assuming that all particles have the same speed, V, and the same free path, L. (There is a distribution in directions of motion, θ, however).

The next step is to express analytically the extreme sensitivity of collision angle to initial position. To do this, we first adopt a center of polar coordinates at the point P, where the nth collision occurs.

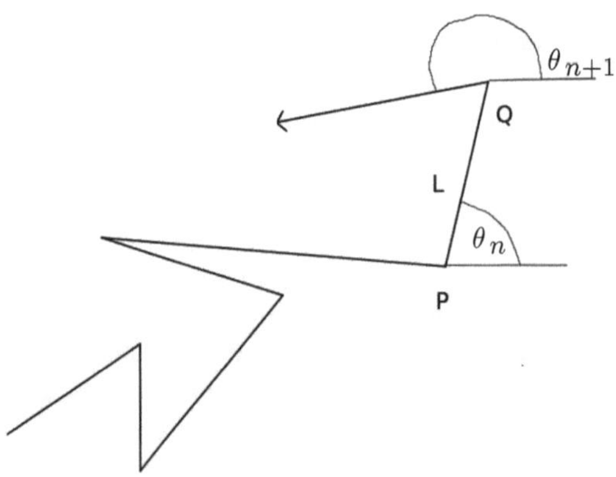

Let the angle of motion after the nth collision be θ_n and that after the $(n+1)^{th}$ collision be θ_{n+1}. We then assume that in a collision, we get

$$\theta_{n+1} = K\theta_n$$

where K is a <u>very big</u> constant (10^3 or more).

What does this mean? It means that when the position of the particle at the pt. Q changes by $L\Delta\theta_n$ the angle after collision changes by $KL\Delta\theta_n$. Thus when θ_n changes by

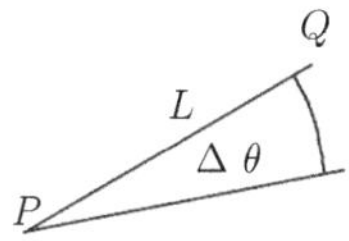

$\Delta\theta_n = \dfrac{2\pi}{KL}$, the angle θ_{n+1} changes by 2π. Changes of angle larger than 2π lead, however, to covering the same range of angle. In other words we must write for the physically significant angle,

$\phi^p = \left(\dfrac{\theta}{2\pi} - N\right) 2\pi$ – where N is the largest integer in $\dfrac{\theta}{2\pi}$

You can see that this hypothesis leads to a behavior that contains the essential element of the real physical problem; i.e. the extreme dependence of final angle on initial position.

We are now ready to draw some interesting conclusions.

After n collisions, we have

$$\theta_n = (K)^n\theta_0$$
$$\Delta\theta_n = (K)^n(\Delta\theta_0)$$

Thus 2 particles beginning with slightly different values of θ will end up in very different places, because K^n is so big. In fact, no matter how small $\Delta\theta_0$ is, we can always find an n so big that

$$\Delta\theta_n \gg 2\pi$$

This enables us to prove our first theorem:

Given an arbitrary initial distribution, $f_0(\theta)$, we shall always ultimately get a uniform probability distribution.

(I leave the proof to you – it is fairly obvious).

This theorem is already a big step forward, because instead of making the quantitative hypothesis of uniform a-priori probability, we make the qualitative hypothesis of arbitrary initial probability, and deduce the quantitative result of uniform probability after many collisions.

The distance moved by the particle between the nth and $(n+1)^{th}$ collisions is

$$(\Delta X)_n = L\cos\theta_n \qquad (\Delta Y)_n = L\sin\theta_n$$

Thus, we have

$$X_N = X_0 + L\sum_{n=1}^{N}\cos\theta_n \qquad Y_N = Y_0 + L\sum_{n=1}^{N}\sin\theta_n$$

I leave to you the proof of the statement that given arbitrary X_0 and Y_0, and an arbitrary initial distribution of angles, we shall ultimately approach a uniform distribution in X and Y.

We must now however go still further. For we have had to assume some kind of probability originally. We could have assumed, for example, that the initial form of the function $f(\theta_0)$ was a function that was zero everywhere except in a very small region of width $\Delta\theta_0$, near a given angle θ_0. This corresponds to experimental information usually available in a measurement of θ_0, since $\Delta\theta_0$ could represent the error in the measurement.

The usual hypothesis of "random errors" is the following:

Within the range, $\Delta\theta_0$, of error, θ will fluctuate chaotically from one measurement to the next (without any order).

This is what we mean by the statement that the probability is uniform within the region $\Delta\theta_0$. It could conceivably be non-uniform in such a way as to confine itself to a number of still smaller sub-regions. But our physical hypothesis is that such things do not occur. Such a hypothesis is very reasonable, and to base the theory of probability on it is much better than to, as is usually done, assume "uniform a-priori probability" in θ. But we now want to go further, and to get rid of all statistical hypotheses.

To do this, we consider a particular sequence, corresponding to a given value of θ_0. We assume that $\frac{K\theta_0}{2\pi}$ is not a rational number. Then it is clear on intuitive grounds, that the set of successive points

$$\left(\frac{\theta_n}{2\pi} - N\right) 2\pi \qquad \text{(where } N \text{ is the largest integer in } \frac{\theta}{2\pi}\text{)}$$

should form a dense set as $n \longrightarrow \infty$. The problem is to prove that they form a uniformly dense set, as $n \longrightarrow \infty$. If we could prove this, then we should have proved that the time average of θ_n is uniform for all θ_0, which is what we want to prove. Of course, there will be a set of points of measure zero in which this is not true, such that $\frac{K\theta_0}{2\pi}$ = a rational number.

Our only hypothesis is then that in the real world, no systems occupy these points of measure zero. Such a hypothesis is not only reasonable – it is almost inescapable. For every system is continually subject to slight disturbances from outside. Remember that because of instability of motion, the slightest disturbance is multiplied indefinitely with the passage of time. Thus even if there existed particles initially having θ_0 a member of the set of measure zero in which θ_n does not cover a dense set uniformly from 0 to 2π, there would surely be external disturbances which threw it out of this set of measure zero. Thus, we replace hypotheses on probability by a simple, qualitative, and practically inescapable hypothesis:

Certain values of θ_0 having a measure zero do not occur in reality.

I have asked mathematicians here about theorems in number theory which would prove that the set θ_n was uniformly dense. All agreed that the result was extremely plausible but knew of no way to prove it. In fact, they say it usually is easier to prove theorems in number theory starting with the theory of probability. But this problem is of crucial importance, for it furnishes the prototype of a whole new branch of mathematics, i.e., to deduce theorems on chaos (or randomness) starting only from causal laws. In the case of numbers, you have sequences generated by definite rules (i.e. causally), which nevertheless lead to randomness. This is the simplest example of the general problem that I wish to attack. As far as I can see, this is a sadly neglected branch of mathematics.

Well, I'll write more later

Love

Dave

Letter 84. Folder C120, dated: Rec Sept ? [1952].

[The date is obscured in the original photocopy].

Number on photocopy: 31

Universidade de São Paulo
Faculdade de Filosofia, Ciências e Letras
Caixa Postal: 8105
São Paulo

Dearest Miriam

I am sending you a brief supplement to the letter I sent a few days ago.

First I want to define the problem more carefully

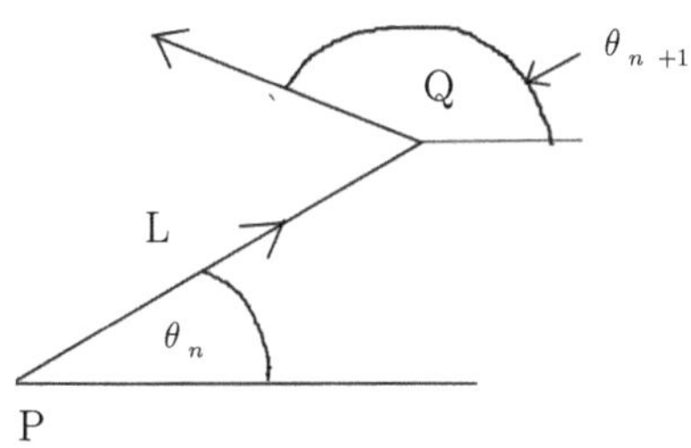

Let θ_n be the angle of the particle after the nth collision. We now assume
$\theta_{n+1} = K\theta_n$
But now I wish to define K more carefully than I did in the previous letter. I want K to be a large integer. The advantage of choosing K an integer is that it leads to a further simplification of the problem. For starting with θ_0 somewhere between 0 and 2π, we shall in general obtain $\theta_1 \gg 2\pi$.
But the physically significant part of θ_1 is
$\theta_1 = K\theta_0 - \left[\frac{K\theta_0}{2\pi}\right] 2\pi$ where $\left[\frac{K\theta_0}{2\pi}\right]$ is the largest integer in $\frac{K\theta_0}{2\pi}$. We now have for the physically significant part of θ_2
$\theta_2 = K^2\theta_0 - K\left[\frac{K\theta_0}{2\pi}\right] 2\pi - \left[K^2\theta_0 - K\left[\frac{K\theta_0}{2\pi}\right] 2\pi\right]$
But if K is an integer, the above reduces to

$\theta_2 = K^2\theta_0 - \left[K^2\theta_0\right] = K\theta_1 - [K\theta_1]$
Thus, the formulae are simplified.
Now our problem is to show that except for a set of values of θ_0 of measure zero, (in which $\frac{K\theta_0}{2\pi}$ is a rational number), the distribution of θ_n becomes uniformly dense as the number of collisions, N, increases without limit.

I think that I can indicate the direction in which one should seek a proof of this theorem. First, we note that in each collision, a region of width $\Delta\theta = \frac{2\pi}{K}$ expanded into a region of width 2π. In n collisions, a region of width $\frac{2\pi}{(K)^n}$ is expanded into a region of width 2π
Now, consider any value of θ_0' for which after N collisions, we do not have a uniform distribution of the θ_n, but instead a systematic tendency to cluster values of θ_n in definite regions. Now, we know that θ_N will change by 2π if we simply go to $\theta_0' \pm \frac{2\pi}{K^n}$. Thus, the measure of trajectories near the one in question which have the property of clustering in a given region must be of the order of $\frac{2\pi}{K^n}$. From this, we conclude that if there were only a finite number of such values of θ_n', their measure would have to approach zero as $N \to \infty$. But it would be conceivable that there were an infinite number of such points.

Let us now define a function $P_{N,\Delta\theta}(\theta, \theta_0)$ such that $P_{N,\Delta\theta}(\theta, \theta_0)\Delta\theta$ gives the number of values of θ_n lying in the range between θ and $\theta + \Delta\theta$ after N collisions for a particle which has started at $\theta = \theta_0$. Let us suppose that for a particular θ_0, $P_{N,\Delta\theta}(\theta, \theta_0)$ is not a constant, but is a function of θ. Then, as we have seen, there will be a region of width $\Delta\theta_0 \cong \frac{2\pi}{K^n}$, within which approximately the same functional form for $P_{N,\Delta\theta}(\theta, \theta_0)$ can be obtained, once the change of θ_n inside this region is negligible.

Let us now consider the $(N+1)^{th}$ collision. This will cause the region $\Delta\theta_0$ to decrease to $\frac{2\pi}{K^{N+1}}$ In other words, the measure of each region $\Delta\theta_0$ leading to a non-uniform distribution of θ_n decreases very rapidly with N. Our problem is to show that it decreases more rapidly than the new points, θ_0', can be generated, for which $P_{N,\Delta\theta}(\theta, \theta_0)$ is not a constant, (considered as a function of θ). It seems most plausible that this is so, but the problem is to prove it

Love
<u>Dave</u>

P.S. I have just thought of another method of proving the theorem. Suppose we consider the particles in a given region $\Delta\theta_n$, after the nth collision. There are K uniformly distributed regions of width $\Delta\theta_{n-1} = \frac{\Delta\theta_n}{K}$ from which the particles would have come near $\theta_{N-1} = \frac{\theta_N}{K} + \frac{2\pi m}{K}$,where m is an integer less

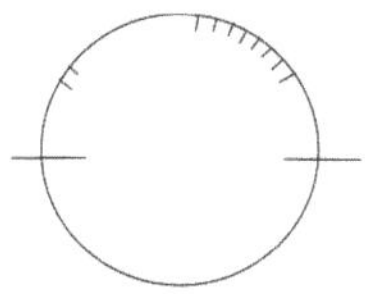

than K. Continuing the process back, each of the regions $\Delta\theta_{N-1}$ could have come from K regions of width $\Delta\theta_{N-2} = \frac{\Delta\theta_{N-1}}{K} = \frac{\Delta\theta_N}{K^2}$, located at $\theta_{N-2} = \frac{\theta_{N-1}}{K} + \frac{2\pi m'}{K}$

$$= \frac{\theta_N}{K^2} + \frac{(2\pi)^2 mm'}{K^2} + \frac{2\pi m'}{K}$$

By continuing this process backwards, we arrive at a practically uniformly distributed set of regions $\Delta\theta_0$, each of width $\frac{\Delta\theta_N}{K^N}$

We now consider a sequence of possible angles $(\theta_0, \theta_1, \theta_2, \ldots \theta_N)$. We ask. What is the measure in the space of θ_0 of the sequence? That is, what range of θ_0 could give rise to the sequence (admitting a uniform range) $\Delta\theta_0 = \Delta\theta_1 = \ldots = \Delta\theta_N$. It is easily shown that the measure of all trajectories passing thru a given $\Delta\theta_N$ is $\frac{\Delta\theta_N}{2\pi}$. If from these, we consider only those passing thru a given $\Delta\theta_{N-1}$, the measure is easily shown to be $\frac{\Delta\theta_N \Delta\theta_{N-1}}{(2\pi)^2}$. Thus the measure of a set of trajectories passing through a given set of θ_n's within a given range of $\Delta\theta$'s is $\frac{\Delta\theta_0 \ldots \Delta\theta_N}{(2\pi)^N}$.

We now define a new number $\rho(\theta)\Delta\theta$ which is the number of values of θ_n lying between θ and $\theta + \Delta\theta$. For any given function, $\rho(\theta)$, we can define the measure $W(\rho(\theta_1)\rho(\theta_2)\ldots\rho(\theta_s))$
where we have divided the range of angle θ into "s" equal regions of width $\Delta\theta$.

Now the distribution $\rho(\theta_1)\ldots\rho(\theta_s)$ can be achieved in many ways. Each way of achieving it has a measure of $\left(\frac{\Delta\theta}{2\pi}\right)^N$. But it is easily seen that the number of ways is, $\frac{N!}{N_1!N_2!\ldots N_s!}$ so that

$$W = \frac{N!}{N_1!N_2!\ldots N_s!}\left(\frac{\Delta\theta}{2\pi}\right)^N$$

From now on, it is a straight forward calculation to show that as $N \to \infty$, the measure of distributions $\rho(\theta_1)\ldots\rho(\theta_s)$ that are not uniform approaches zero. Thus, we show that:

The range of initial conditions (in θ_0) corresponding to a non uniform distribution $\rho(\theta)$ approaches a measure of zero as $N \to \infty$.

This is not a theorem in probability theory, for we do not assume a random connection between θ_n and θ_{n-1} but a definite connection.
PPS

Have you seen discussion of Causal interp. of qu. theory in one of later issues of French magazine, Pensée.

Letter 85. Folder C120, dated: Sept 23, 1952.

Number on photocopy: 29 (same as letter 82)

Universidade de São Paulo
Faculdade de Filosofia, Ciências e Letras
Caixa Postal: 8105
São Paulo

Dearest Miriam

I haven't heard from you for a long time – what's the trouble?

I am very busy now. Things are picking up, both in qu. theory + in theory of probability. I am sending you part 1 of a mss., of which part 2 is in the process of preparation. I have solved the problem of showing how chaos develops through the causal laws themselves, in simplified cases, and can see how the theory can be generalized. These will appear in part 2.

Love
Dave

Letter 86. Folder C120, dated: Rec Oct 13 [1952].

Number on photocopy: 32

Universidade de São Paulo
Faculdade de Filosofia, Ciências e Letras
Caixa Postal: 8105
São Paulo

Dearest Miriam

Thanks for your letter. I am curious to know of the reason for the long delay in your answering my letters.

As for me, I am busy with too ideas. Things are working out gradually on the causal interpretation of the quantum theory, and the possibility of a complete victory is at last in sight (But not the victory itself, as yet.)

First of all, let me say that the theorem on the equidistribution of the series, $K^n\theta_0 - [K^n\theta_0]$ is already proved by a theorem of Weyl, which is much more general than we need (Mathematische Annalen, 77, 1916, p 313). Also, I obtained a much simpler proof, based on a method whose meaning can easily be seen physically, which will be included in a brief paper that is being prepared, not for publication, but for preliminary distribution to those who are interested. I'll send you a copy as soon as it is ready.

I am afraid from the tone of your letter that you did not appreciate the idea that I am trying to get across. I do not wish to dispute the fact that set theory will probably be useful in calculations, from my point of view, as well as from the usually accepted point of view. However, I wish to make clear the difference between the two points of view.

Now, usually, one assumes an "a-priori" probability $f(\theta_0)$. This is equivalent physically to some sort of hypothesis about chaos or randomness. But in the case of the sequence

$$K^n\theta_0 - [K^n\theta_0]$$

we have proved that this distribution is chaotic. In other words, we have demonstrated that a causal sequence of numbers leads to chaos. This is as different from assuming chaos to begin [words missing] I am really anxious to be sure that you have grasped this point. All of the rest of this talk about whether set theory is or is not useful is just a side issue, from my point of view. In any case, there is no doubt that in the letter that you sent me several months ago, you assumed randomness to begin with, so that the results lead to nothing new.

As for the role of sets of measure zero, this is very different in my point of view from what it is in the usual point of view. Usually, one defines a probability function, $f(\theta_0)$, and asserts that sets of measure zero do not contribute to the total probability fn. any set of finite measure. But my point of view is something different. It is to follow an individual particle, starting at some initial angle, θ_0. We have shown that if we let N, the number of collisions become big enough, then chaos is sure to develop, except for a set of θ_0 of measure zero. But from experience in physics, we know that a set of measure zero cannot be realized in practice. For to realize it would require us to adjust exactly the value of some physical quantity, and we know that this is impossible.

Finally, let me say that not every causally defined sequence will lead to chaos, but in general, just those which involve instability of motion; i.e., a small change in initial conditions leads to a big change in final conditions. The discussion in the letter of several months ago was too general, in that it did not attempt to limit itself to the special types of sequences that lead to the generation of chaos.

To sum up, my point of view is that hypotheses on a-priori probability are unnecessary, at least in principle; and that if we know the laws of motion, we should be able to deduce the development of chaos, as well as the numerical value of the a-priori probability.

Love

Dave

Letter 87. Folder C120, dated October 16, '52.

Number on photocopy: 33

Universidade de São Paulo
Faculdade de Filosofia, Ciências e Letras

Caixa Postal: 8105
São Paulo

Dearest Miriam

I received your latest letter, in which you proved my conjecture. I was unjust to you in my previous letter; you do seem to understand my idea, but apparently you don't like it. Your first reaction was to try to prove my conjecture false, and then when this didn't succeed, your second reaction was to try to minimize its importance, by bringing up difficulties which can be resolved with a little thought. I don't know why you don't like the idea. My first reaction was to get angry at you, because I felt that you are acting like a typical little self-satisfied mathematician, who likes to imagine that his little world of fantasy already contains everything worth thinking about seriously. But then I felt that perhaps you are only angry at me for not paying enough attention to your offer to teach me about Kantor's theory of transfinite numbers. If this is what is bothering you, let me say that I am willing to learn about them. I must confess that I start out with a general dislike of the language he uses and of the ideas he expresses (insofar as I am able at this stage to appreciate his ideas). I am suspicious that it is really one big fraud. It seems to me that you, who always have to be shown that something is not "being put over on you" should tolerantly allow me to be a bit doubtful about a theory which treats infinity as if it were finite, and then gets into all sorts of paradoxical-sounding statements, by so doing. I like to regard the concept of infinity in the literal sense of the word as taken over from its Latin root; i.e., "without end". In other words, I like to think of a finite sequence without limits. But the theorems proved always refer to a finite but large sequence. Otherwise I don't know what you are talking about. The concept of "infinity summed" strikes me as being really metaphysical[1] (in the bad sense of the word). That is, it cannot be regarded as the approximate conceptual image of any property of a real object, or collection of objects. Perhaps I am mistaken in my notions as to what Kantor does. If so, please correct me.

As long as we are on the subject of metaphysics, I wish to say here that the concept of probability as it is currently used is very unsatisfactory. As I pointed out in the paper I sent you some time ago, there is no physical property of a real object or collection of objects of which the probability could be the approximate conceptual image. We can regard probability as representing information but then we get into the embarrassing question of why mere information should produce a chaotic distribution of real objects or events. On the other hand, I maintain that the concept of probability is applicable only when there is a real physical process that leads to chaos. When this happens, our information will be incomplete (as a rule). But loss of information is due to the same physical processes that are responsible for

[1]In fact, since you once spoke of the infinity of levels as a "religion", let me say here that Kantor's theorems have a mystical religious flavour, to my taste at least. I always regard the ∞ of levels as a series without limit, of which at each stage, you consider only a finite number, however large [probably 'large' but most of word is missing] The important idea, however, is that there are always more levels, beyond any given level.

chaos. Thus, the usual point of view is (like most modern bourgeois theories) upside down, since it seems to make matter depend on our information about it, instead of regarding our information as an approximate reflection of objective properties of matter.

In a few weeks, I shall send you part II of the paper mentioned above, in which these ideas are elaborated.

I shall now give a simple proof of my conjecture, which requires no "sophisticated" concepts of measure, but only the most primitive ideas, easily pictured physically.

Let us begin by dividing the unit interval into $\frac{K}{\Omega}$ intervals where K and Ω are integers, both large, but where $K \gg \Omega$. We denote the interval by X_Δ and its width by $\Delta X_\Delta = \frac{\Omega}{K}$

$$X_\Delta = \Delta \frac{\Omega}{K} \text{ where } \Delta \text{ is an integer}$$

We now consider our sequence of N terms, with

$X_n = K^n X_0 - [K^n X_0]$
where X_0 is the point at which our sequence starts
We now define the function
$Z_\Delta(N, X_0)$
which is the number of times that in a sequence having N terms, the point X_n falls within the interval between X_Δ and $X_\Delta + \frac{\Delta\Omega}{K}$. The function $Z_\Delta(N, X_0)$ gives the density of points in our sequence.

Our problem is to prove that except for a set of X_0 of measure zero, Z_Δ approaches a constant, independent of Δ and X_0 as $N \to \infty$.

We begin by considering any given sequence of points,
$X^0, X^1, X^2 \ldots X^n \ldots X^N$

We now ask. If a point, X^1 appears between X_0 and $X_0 + \frac{\Omega}{K}$, from what region of X_0 could it have come. You will easily convince yourself that it could have come from K regions each of width $\frac{\Omega}{K^2}$ that are uniformly distributed over the unit interval. Thus, the total measure of sequences for which X^1 lies in the region between X_0 and $X_0 + \frac{\Omega}{K}$ is $K\frac{\Omega}{K^2} = \frac{\Omega}{K}$.

We now consider those sequences which not only enter a given region between X_0 and $X_0 + \frac{\Omega}{K}$, after the first scattering process, but which enter any other given similar region after the second process. Any particle entering a given region must have come from a uniformly distributed set of regions in the space of X^1. But of those, only a fraction $\frac{\Omega}{K}$ will also have entered the given region of X^1 space. Thus, the measure of sequences passing through a given ΔX^1 and a given ΔX^2 is $\left(\frac{\Omega}{K}\right)^2$. Similarly, the measure of sequences of N terms passing successively through given $\Delta X_1, \Delta X_2, \ldots \Delta X_n$ is $\left(\frac{\Omega}{K}\right)^N$. In other words, the initial conditions in X_0 which can lead to any given type of sequence are uniformly distributed over X_0, and consist of $\frac{K}{\Omega}$ regions each of width $\left(\frac{\Omega}{K}\right)^N$.

Now to find the measure of a given density function $Z_\Delta(N, X_0)$, we must take the measure of each sequence that can lead to this function, and multiply by the number of such sequences. The number of such sequences is just the number of interchanges $\frac{N!}{(Z_1!)(Z_2!)\ldots(Z_\Delta!)}$. Since N and $Z_\Delta \to \infty$, we can use Stirling's approx. The logarithm of the measure of a given sequence is

$$\sum N \ln N - \sum_\Delta Z_\Delta \ln Z_\Delta + N \ln\left(\frac{\Omega}{K}\right)$$

You will easily verify that the sequence of greatest measure is the one with $\frac{Z_\Delta}{N} = \text{constant} = \frac{\Omega}{K}$.

Moreover, expansion with $\frac{Z_\Delta}{N} = \frac{\Omega}{K} + \delta f_\Delta$ yields for the measure a number proportional

$$e^{-\frac{\sum(\delta f_\Delta)^2}{2}N}\left(\frac{\Omega}{K}\right)^N$$

We easily see that as $N \to \infty$, the measure of trajectories for which δf_Δ differs appreciably from zero approaches zero. More accurately; given any set of $\delta f_\Delta^{(0)}$, however small, we can always find an N so large that the measure of those trajectories for which $\delta f_\Delta > \delta f_\Delta^{(0)}$ is as small as we please.

This concept of measure is very simple. The measure of any given sequence is $\left(\frac{\Omega}{K}\right)^N$, which approaches zero very rapidly as $N \to \infty$. This reflects the instability of trajectories, which requires more + more accurate "aim" to get a given sequence. Although the number of trajectories for which $f_\Delta > f_\Delta^{(0)}$ approaches infinity as $N \to \infty$, it approaches ∞ so slowly that the sum of the sizes of such regions approaches zero as $N \to \infty$.

Now, for the physical meaning of this result.

(1) As $N \to \infty$, almost all initial conditions lead to chaos, i.e. to an approx. uniform distribution of values of X_Δ. Thus, for almost all initial conditions, we predict chaos, instead of (as is usually done) assuming it.

(2) There are special conditions not leading to chaos. Let us consider one region

of width $\left(\frac{\Omega}{K}\right)^N$ of this type. As $N \to \infty$, this region gets narrower + narrower, so that to stay in this region indefinitely, the particle would have to have a certain exactly defined value of X_Δ.

If there were only a finite number of such regions, then our conclusion would be obvious; for we know that it is impossible at least by design, to have an exactly defined value of any physical quantity. It is possible by "accident". But we then simply make the postulate that such values do not occur in reality. This postulate is not only very plausible, but the fact that the laws of probability do work in systems having chaotically unstable motion is empirical proof of this postulate. For we know that if such special conditions existed, there would be no chaos, + the laws of probability would not work. The same reasoning holds when there are an infinite number of points not leading to chaotic motion, as long as the measure of these points is zero.

What are the advantages of this procedure over the usual one?

(1) It makes a qualitative postulate, and permits you to deduce the quantitative values of the "a-priori probability"

(2) It shows clearly what are the limitations on the applicability of the laws of probability – i.e. it can be applied only when the causal laws lead to chaotic motion, except for a set of initial conditions of measure zero.

(3) It enables you to give a clear physical meaning to a-priori probability. It is the mean calculated value of the ratio of favorable to total cases, for the cases where the motion is chaotic. Thus, it is an objective property of an individual system, and is not just a measure of lack of information. Yet, it does not have the disadvantages of the usual objective definition as the ratio of the actual number of favorable cases to total cases; for this definition is correct only for an infinite number of cases, and is therefore meaningless in any real problem.

From the math point of view, you may think that all this talk is unimportant – that only the development of formulas is really important. But I want to assure you that the relation of math concepts to reality is also an important point. You will have to come out of your ivory tower some day – why not now?

Love
Dave

Chapter 25
Letters to Miriam Yevick, 1952, Part 5

Letter 88. Folder C120, dated: October 16, '52.

Number on photocopy: 34

Universidade de São Paulo
Faculdade de Filosofia, Ciências e Letras
Caixa Postal: 8105
São Paulo

Dearest Miriam

I just thought I would write a few things other than this problem in the theory of probability.

First, it does not look as if I will get the passport back, my lawyer writes me that it is unlikely, although, of course, still possible. This means I will have to get used to living in Brazil for a while. Now that I have Ralph Schiller to work with, it doesn't look quite as bad as it used to, especially since the work has really promising possibilities. So I wouldn't mind spending another year or so here. It should be profitable, at least from a scientific point of view.

With regard to the book on dial. mat., I believe it is essential that I first do two things. (1) Carry the causal interpretation to a point where it definitely gives some new result. Otherwise, the arguments will lose much of their potential force (2) Carry this probability problem further. It is in just these two fields that current idealistic points of view are creating confusion that prevents further progress, as well as sowing seeds of discouragement, cynicism, etc., in the people who are forced to absorb the current point of view in their work. (I wish to emphasize here that the usual outlook of mathematicians, based, as it is, on the idea that they deal with a superior little world of their own, not connected with the nasty real world, is responsible for much of the disillusionment and cynicism that is characteristic even of the personal relations of most mathematicians. For they are led to look down on mere human beings, and

C. Talbot (ed.), *David Bohm: Causality and Chance, Letters to Three Women*,
DOI 10.1007/978-3-319-55492-1_25

eventually even on themselves.) (3) The concept of the infinity of levels must be an integral part of this book. It clears up endless sources of paradoxes, and makes dialectical mat. really understandable in common sense terms. But this ∞ of levels is intimately connected up, both with qu. theory, and with probability. For the causal laws at each level arise from the existence of chaos at lower levels. For example, the gas laws are valid at our scale, only because of chaotic molecular motions. Likewise, I believe that the causal laws governing the motion of electrons arise from the existence of chaotic motions of the particles of the substratum, out of which electrons are made, etc. ad infinitum.

Generally speaking, my purpose is not only to re-iterate existing ideas on dialectical mat, but to go further into new developments of the kind sketched above – also to show in detail how the orientation due to this pt. of view serves to develop new ideas, while the usual orientation suppresses such development, and leads to insoluble paradoxes, mysteries, and general confusion.

The ∞ of levels is important from another pt of view. Bourgeois scientists, even leftists (like you) have a tendency to feel that their general framework of ideas already includes all possibilities, and that only the "details" need to be filled in. This creates the illusion of "freedom" since they think they are "free" to do everything that can possibly be done. Actually they are confined by not being able to see the limitless possibilities that are visible as soon as one gets out of the customary conceptual framework. Mathematicians seem to be particularly bad, in this respect. For they seem to feel that because of the iron-clad logic of their proofs, nothing has been "put over on them". Thus, there is nothing (except for details) that they could possibly be overlooking. The task of mathematicians then reduces to an endless grinding out of detailed theorems, and the excellence of a mathematician is measured by his output of such theorems per year (Physicists are just as bad in their own way, however) The ∞ of levels, as applied to math concepts immediately shows the error of this point of view. For each set of concepts must be based on a still more fundamental set, etc. ad infinitum. Gödel's work demonstrates this, since he has shown that no complete self-consistent system of theorems is possible, which provides their own basis (i.e. their own proof of consistency).

To come to another subject, that of women. I am still in the same state as before – i.e. nothing new. The difficulties are appreciable, since I really cannot remain interested in a woman long enough to get to the point of going to bed, unless there is something more to her than there is to most of the girls I see here. As you know, some initiative + planning are needed to get anywhere, even in this field; and for this you need some enthusiasm. Perhaps if I got going, the enthusiasm would generate itself, but the vicious circle is that the field never seems promising enough to make efforts worth while. This is not entirely true. There was a rather pretty little girl in my 3rd year class who definitely attracted me a lot, but she went and got married almost immediately (Only about 20 years old) I can see that she is quite intelligent, is far to the left, etc. and moreover, there is this feeling of my being attracted strongly to her, which is usually absent with most women. It is difficult to define the cause of this attraction, but it is a sense that she still retains a child-like wonder and curiosity about the world, while at the same time being mature, intelligent, and cautious enough so that she doesn't let anything be "put over on her". To reach this degree of wisdom without cynicism and disillusionment is something that appeals to me in a most fundamental way in anyone, man or woman. Anyway, it is always a cheering sight when I am lecturing to look at her once in a while, but I am afraid that I embarrass her once in a while, because I sometimes feel as if I would like to grab her in my arms + kiss her; and this must reflect to some extent in my expression, even if I try to hide it.

Well, so much for the "woman question". Maybe we can get around to the book on dial.mat. in a year or two.

Love
Dave

P.S. Do you have any opinions about the mss. that I sent you on "proof that $P \rightarrow |\psi|^2$ in causal interp. of qu. theory". Also the mss. on probability.

Letter 89. Folder C121, dated: Rec Dec 1, Nov 22 1952.

[Nov 22 written below Dec 1, presumably date sent].

Number on photocopy: 35

Universidade de São Paulo
Faculdade de Filosofia, Ciências e Letras
Caixa Postal: 8105
São Paulo

Dearest Miriam

I received your last two letters. You are right. The time has come to stop irritating each other, as matters are becoming too serious for that. The Eisenhower victory was quite a blow to me. Even the more conservative of the Brazilians were disappointed, as they all preferred Stevenson. Eisenhower's election greatly increases the probability of war, and makes fascism in the US practically certain. Nevertheless, you should remember two things. Stevenson was defeated (1) because he supported Truman's domestic mistakes, and (2) because he gave no hope of ending the war. A lot of people voted for Eisenhower because he promised to go to Korea, and try to do something, while all that Stevenson said was "It's up to Moscow". It's true that the promise of Eisenhower was made tongue in cheek, but still, it was not entirely irrational for people who did not understand what is going on to prefer a man who promised some action to one who promised, at best, a continuation of the present hopeless stalemate in Korea. In any case, Stevenson's failure to alter Truman's home policies was a sign of lack of "greatness", and it was just this which probably added the finishing touches to his defeat. I think that Stevenson's greatness was more in words than in deeds, since he wasn't great enough to see that the central issue of the election was peace.

Please don't become depressed. The central duty of people who oppose the present state of affairs in the US is to stay alive. Remember that things will have to change, the worm turns, etc. The American people were twisted, turned, and corrupted by a definite process of high-pressure salesmanship, which was aided by the complete confusion and shallowness of the left. Now we shall all have to pay while one form or another of chaos develops. But our time will come. And when it comes, each person who has some understanding will act like a seed, a nucleus, transmitting his understanding to a large number of people. So, it is important for you to stay alive, keep a clear head, keep on developing, and above all, not to be overcome by panic. If you decide to leave, you must do it now. But it isn't clear to me where one should go. In Europe, only England is worth while, but England may be destroyed by war. Israel is a hopeless mess. Brazil is very doubtful. Its ties to the U.S. are getting closer every day. Brazil is now facing a crisis of unprecedented severity due to lack of dollars, and the gov't shows evidence of being hopelessly incompetent & corrupt in the face of this crisis. All that can come out of it is a complete dependence of Brazil on the U.S. The only good thing about Brazil is that the gov't is so hopelessly corrupt that it cannot even efficiently become a good colony of the U.S. My feeling is that wherever you are, you will be hit by what is coming. Brazil is safest from direct physical destruction, and if you are not too well known, or sought after, you can buy almost anything with a little money, as the gov't is completely completely corrupt. But if the U.S. asked for me, I'm sure Brasil would give me up. At least it's very probable.

I'm sorry I delayed answering for so long. I was on a vacation, and also was quite busy, just before going, as well as somewhat ill. The vacation was in the mountains (about a mile high). It was very nice, as I had a chance to rest. The mountains in Brazil are very empty of people, and give one an impression of timeless peacefulness, which is very good for the nerves these days. I suggest that you try some similar isolated place in the US without newspapers or radio. It will help you recuperate to get ready for what is coming. What one needs now is a "very long" view in which you can face the possibility of serious reverses and setbacks for the next 10 or 20 years. But the ultimate outcome will be worth any sacrifices that we may be forced to make, even if we personally don't live to see it. For unless one identifies oneself with this ultimate outcome, life will be worthless. Thus, the only hope of maintaining life is in a long run optimism, and since the alternative is a pessimism equivalent to death (and which would probably lead to death in one way or another) one has to root oneself more firmly than ever in the notion that humanity has limitless potentialities, which are being turned to bad ends these days. To sense these potentialities, it is necessary only to watch a young child, and to think of what this child will probably become, in the environment that exists now. You may recall a poem that Wordsworth once wrote to the effect that children seem to come from Heaven "trailing clouds of glory" and that they later lose this "Heavenly" character. He used this as an argument for immortality of the soul – a typical bourgeois confusion, which helps him avoid the responsibility of facing the fact that society turns these little "angels" into "devils". I merely cite this example to show how widespread is the consciousness of the great difference between what is promised in the young child, and what is realized in the young man or woman.

This brings me to the point that I have always emphasized – that it is ideological confusion that is responsible for the present mess, especially in the U.S. Obviously the people are acting against their own best interests. And why are they doing this? Because they are so mixed up that they don't know where their interests lie. But an important cause of this confusion is a certain view on life, which I shall call the "Bourgeois view", although a better name is perhaps the "general ruling class view". In this view the world is looked on as made up of fixed types of elements. "Freedom" consists only in being able to rearrange the pattern of the elements. Thus, there are rich, middle class & poor people. If a man can move from one class to another, he is said to be "free". Even the socialists tend to hold such views, since they essentially want to put everybody in the middle class. But nobody notices the fact that the entire system of elements and classification is itself not inevitable, and is indeed the product of the existing structure of society. Thus, people become cynical, because they see no way out of the same old merry-go-round. But in the dialectical method of thinking, we see that it is equally false to say that "a thing is what it is" as it is to say "it is not what it is". For everything is in a state of becoming, in which it comes into existence and passes away into something else. Thus, a thing cannot be exactly "what it is" since it is already becoming something else. In a sense, it already is "what it isn't". Thus, there is an opposition between that which exists and that which is coming into existence, and this is the normal state of all matter. When the process of change is slow, we may abstract the situation, and think of a thing as "existing" but remembering that this is only an approximation, because underneath, it is, even if only slowly, already becoming something else. The mistake of ideologies of all ruling classes is to imagine "that which is" is permanent. It may move around and change, but only in a superficial way, since in essence, the kinds of things that exist, and the laws governing their motion are conceived of as fixed. Dialectically, however, the very nature of things (including human beings) may change with conditions, and it is already a serious interference with human freedom not to have the opportunity to realize these changes, when they are necessary. Thus, in a capitalistic society, a person is not free to become what he is ready to become, and needs to become, in order to continue to exist as an integrated human being.

The illusion that "things are what they are" spreads to all phases of life, including science and mathematics. In math, it takes the form of making theories "static", in the sense that the mathematician axiomatically defines a "space" or a "set", and then imagines that all possibilities are covered by transformations within this set. In physics, a general framework of ideas is accepted, and all new ideas then are required to be bigger, better, or more ingenious combinations of things within this framework.

Of course, it is no mystery why any ruling class likes the idea that "things are what they are". This idea will disappear completely only after all traces of class distinction have been wiped out.

Incidentally, I did not wish to imply in my letters that all mathematicians are short-sighted. I had in mind only the majority, of the type of Chung and Doob and Co. Levy is doubtless much better. As you said once to me, he is an "intuitionist". I would like to say a few words about this brand of math. It is characteristic of bourgeois society that because it regards things as static, it is always getting stuck on one of the two opposing sides of a dialectical process. The "logical" mathematicians insist on cold deduction. Nothing is to be "put over on them", etc., etc. This is all very fine, but then where do new ideas come from? They just "come" from geniuses, and then a whole new generation of Chungs, Doobs, etc. works out their consequences. To avoid this one sided concept, the intuitionist admits ideas from general experience, from physical imagery, etc. This is an important correction to the opposite point of view, and I wish to admit its value here, once and for all. But it is also one-sided, because the ideas now come from a region beyond logical understanding, i.e., the "intuition". Now, in the dialectical point of view, one sees that both of these concepts are abstractions, limiting cases, of a broader concept, which synthesises the two opposing points of view. For logic is certainly an intrinsic part of the thinking process, and yet we admit that it is not all of the process, but only an abstraction. But it is not necessary to add another external process, a mysterious "intuition". One must instead remember that the thinking process is basically a process in which we deal with reality. There are two limiting ways of doing this. One is the narrowly practical way and the other is the extremely abstract way, that modern mathematicians pride themselves on. The bourgeois outlook tends to make a person think of these as two separate things. Actually they aren't, but are different aspects of the same thing. Now when one deals with reality, either through observation or through practice, one is continually forced to change one's ideas, in accordance with the things that happen that are not quite in accord with the theory. Thus, there is a continual impetus to new ideas. The "logical" mathematician insulates himself from this, and then the intuitionist to correct for this error, permits the effects of general experience to come in by the back door, through "intuition", which is after all only a set of vaguely sensed ideas which arise in response to our general experience with reality. But by bringing mathematics into closer touch with real problems, we wouldn't need so much of the vague stimulus due to intuition, since the imagination would be directly stimulated by having to work on a real problem.

As for the mss on probability, it will soon be ready, and I'll send it to you as soon as possible. But I have no extra copies, as yet, to send to other people.

Now that I have been back from the peace of the mountains for a day, I am getting scared a bit. Everything is so mixed up. What do you think of all these "revelations" against Slansky, et al, in Czechoslovakia? They sound a bit manufactured to me, because Gottwald comes out a saint, while most Jews appear to be American spies. It's a bit hard to believe, but what they say may of course, be true. I'm afraid that both sides are succumbing to hysteria. But these things cannot alter the fact that in the long run, things will come out O.K.

Love
Dave

Letter 90. Folder C121, dated: Dec 10, 1952.

Number on photocopy: 36

Universidade de São Paulo
Faculdade de Filosofia, Ciências e Letras
Caixa Postal: 8105
São Paulo

Dearest Miriam

Enclosed is a tentative copy of the paper I mentioned on probability. Note that I have revised Sec. 1, as well as added Sec. 2.

I would like to make a few comments on your previous letter. In this letter you mentioned a rotating die, and pointed out that a small range of initial conditions of this die would lead to a big range of final conditions, so that it would be impossible to predict the future behavior, and the probability would become uniform. This statement needs some more criticism and comment.

Let us first note that the final polar angles, θ, ϕ, of the die depend on the initial angles, θ_0 and ϕ_0, and on the initial angular velocities, $\dot{\phi}_0$ and $\dot{\theta}_0$. Now suppose a small initial uncertainty in the angular velocities $\Delta\dot{\theta}_0$ and $\Delta\dot{\phi}_0$. As a result of this, we get a big uncertainty in the final angles, if the time, t, is large

$$\Delta\theta = \Delta\dot{\theta}_0 t \quad , \quad \Delta\phi = \Delta\dot{\phi}_0 t$$

This is an example of how instability of motion causes information to be lost.

But let us look at this more carefully. In each individual case, the velocity $\dot{\theta}_0$ and $\dot{\phi}_0$ are fixed, so that a definite θ and ϕ will result after a time, t. If we define a function $P(\dot{\theta}_0, \dot{\phi}_0)\Delta\dot{\theta}_0\Delta\dot{\phi}_0 = dN$ = mean number of systems in region $\Delta\dot{\theta}_0\Delta\dot{\phi}_0$, then we see that after enough time, the probability density in θ and ϕ will become uniform. But notice that we haven't got rid of probabilistic assumptions yet . . .

[Page(s) missing]

be more likely to make him a better bourgeois. It ain't so easy to become a prosperous bourgeois! It needs long hard work, preparation, cultivation of the right people, writing the proper sort of papers, wearing the right clothes, saying the right things, etc. All of this takes energetic planning, and if George sets his mind to it, he can really make a go of it. Please warn him that I think he is in grave danger. Maybe becoming a proletarian is just what he needs, to liberate his potentialities for being a great agitator. Tell him to think of the future generations, who will read the polemics of George Yevick on Materialism and the Scientific Society, realizing that they played an important role in the transformation of 1973-75, which wrought such revolutionary changes in the structure of the world.

How about instinct v.s. rigor again. I said something about this problem in my previous letter, and I am going to more or less repeat it. I think that it is a characteristic bourgeois confusion to say that some people do things by "instinct" others by "rigor".

For just what does "instinct" mean? This is a word that does not explain what is happening, but just a means of hiding ignorance. Usually, when you say a man does something by "instinct" you don't mean that he was born with the ability (This would be the literal meaning, of course). Instead, you mean that a person is very familiar with the field, and can sense connections before they are evident from purely logical arguments. But all new ideas must be sensed vaguely, before they can be proved logically. For with pure logic, we can only start from definite hypotheses and end up with definite conclusions. The new idea consists in the new hypothesis. To make new hypotheses, you need in general to find a new system of logical classification. In other words, it is not enough to use the old classification, just replacing certain judgements of truth by falsity and vice versa. But how do you know what new classifications should be used? You say that this comes "instinctively" by "intuition". By this, you mean that you really don't know – i.e. it just happens somehow, in a miraculous but unintelligible way. But this is just the old bourgeois static mode of thinking, which is unable to get off the horns of the dilemma of logic v.s. so called "instinct". All real thinking processes involve a unity between logic and so-called "instinct". The "instinctive" aspect represents the ability to form new concepts on the basis of a general realization of how things are connected. It does not follow the rules of logic, but it is not an irrational process. It is based partly on connections suggested by general experience, sometimes in the same field, sometimes in other fields (in the latter case, we use so called analogical reasoning). But principally it is based on a continual imaginative search for a way of looking at things which unites them as an interconnected whole. It is not, however, separate from logic. (a) Because the ideas must be logically consistent to be acceptable (b) Because logic often plays a genuinely creative role in leading to unexpected conclusions, which force a re-examination of the old conceptual basis. "Carrying things to logical conclusions" is very important in this regard. For example, at first, people would not accept solutions of quadratic equations leading to numbers whose squares are negative. But the logically established contradiction between real numbers and the general solution of quadratic eqns. was resolved by a new concept, almost directly suggested by the logic – i.e., imaginary numbers. In my own experience, I showed that qu. mechs. could be expressed in a form very similar to classical mechanics, with the addition of a quantum potential. This suggested the idea of postulating a quantum potential. Additional postulates suggested themselves as necessary to lead to a consistent account of the theory from the new point of view. But "intuition" + logic were bound into an inseparable unit. "Intuition" alone leads mostly to empty speculations. As a matter of fact, I find it hard to see where "intuition" even entered, since I was guided largely by a materialist philosophy, which led me to look for just such things. In general, when you analyze such things, you discover that the mystery disappears. But when you don't know the factors in detail (i.e., when you are not guided by a consciously expressed philosophy) you don't see where the idea comes from, and therefore ascribe it to a mysterious intuition. This is all the more likely to happen when you try to understand how other people get ideas.

Now let me say that to think in this way is not automatic – it has to be learned. It happens that I like this way of thinking, and that is why I learned it. But it took a lot of work to do it, thinking day by day, and "living" with a long series of scientific problems. I do not believe that people are constitutionally determined as "intuitionists" or "logicists". But because of various factors they may be attracted in a certain direction which is very hard to change thereafter. I agree that some results can be obtained from the formal method. But I think of it this way. If a man looked at the world with a distorting mirror that exaggerated certain things, and suppressed other things, he could eventually learn in part to adjust himself, so that he could get along. He might even be able to deal with some of the exaggerated features very effectively. But it would be better to look at the world more directly. At times, it might even be useful to look through the distorting mirror, which would emphasize connections not easily visible by normal methods of seeing. But when he does this, he should be conscious that he is obtaining a one sided distorted image of the world, and not imagine (as many people do) that he is seeing everything as it really is.

With all of this stated, let me say that I would like to hear something about Kantor's theory of transfinite numbers. Also, can you send me a simple reference that gives some idea of what he has done, without a book full of math. symbols? You have ...

[Something missing, or pages out of order]

With regard to abstract paintings, they usually leave me just as cold as does formal elegance (by itself) in math. I can see their existence, but cannot understand why they are desirable in themselves. I like to see a formally elegant way of resolving a real problem, because the two go hand in hand. The greater the formal beauty, the more powerful, in general is the instrument for resolving the problem; and vice versa, a really effective solution of a problem will almost always turn out to have a great deal of formal beauty. Similarly, I think an artist should always try to communicate something in the most effective manner possible. If he succeeds, he will almost inevitably have created a thing of beauty. But just to create a thing of beauty which communicates nothing seems to be an empty activity. The mere fact that such empty activity is now socially acceptable is to me, [unreadable] additional sign of the decadence of bourgeois society (Both in math & painting). It is an unhealthy sign of a withdrawal from reality.

Let us come back to the problem of continuity. I can see that Cantor may have made an important contribution here, although he can hardly have solved the whole problem. The concept of the limit of a discrete countable set is clearly inadequate, as you say, to deal with continuity.

Let me say what I think about this problem next. I think it is inseparably connected with the infinity of levels. When we look at the world, we see both discrete and continuous aspects. Discrete objects seem to stand out in a continuous background (People, tables, houses, trees, and on a deeper level, atoms, electrons, etc.) But a more careful analysis shows that these objects cannot be entirely discrete, because (1) Their boundaries are continuous with the background. For example, a microscopic examination shows that an apparently sharp edge usually fades continuously but rapidly into the background. (2) At a more fundamental level, they are always made up of smaller elements which behave approximately continuously. (3) All objects come into existence continuously in time and go out of existence in the same way. Thus, the concept of discrete objects must be an approximation, an abstraction. But continuity is equally an abstraction. (1) Because even though objects do shade continuously into the background, etc., they have a unity and duration of existence which suggests that the concept of discrete object has some validity. (2) Because all continuous media have been discovered to be made up of discrete but small elements. Of course, these elements are probably made up of new continuous media which are made up of still smaller elements ad infinitum.

Thus, the concepts of continuity and discreteness are two opposing sides of a dilemma that cannot be resolved within the static conceptual framework that permeates a class society. As in the case of being and not being, which are resolved in the concept of becoming (see my last letter), we need a concept that resolves continuity and discreteness, and expresses them as limiting cases. Thus, we begin by saying that reality is neither continuous nor discrete, but instead that it is characterized by ever-changing aspects in space and time. Where the change is fairly abrupt in space, and where a certain unified behavior manifests itself over some time, we may discuss it as an abstraction, the appearance of discrete units. Those domains of space in which changes are not abrupt and in which there is no appreciable unity of behavior from one place to another we call continuous. But as is characteristic of dialectical opposites, each is continually reflecting the other and transforming into the other – This is just because the two do not really exist separately, but are "opposite sides of the same coin". Thus, continuous substances are analyzed in terms of discrete but small elements, while discrete elements are in turn analyzed as structures in a continuous background, structures which can come into being, transform, and go out of being.

In mathematics, we have found it convenient to analyze the continuous field of real numbers as a limit of a discrete countable set, such as the rationals. This does not mean that the reals are identical with the limits of the rationals. It merely means that the rationals can in this way be made to reflect and exhibit some of the relationships of the reals. Nevertheless, the reals are qualitatively new things, having new properties not contained in the rationals. This is characteristic however of dialectical processes in Nature and in thought, since a thing, as well as an idea, can lead to something new and qualitatively different from itself (e.g. a seed + soil + water + air leads to a tree. Similarly, rationals contain the seeds of the reals, etc.) It is important to remember that the logical application of an idea is not a tautology. But when by logical thinking we transform an idea into another form, the conclusion is in some ways qualitatively different from the starting point. The concept of logic as a tautology is another characteristically bourgeois static concept, in which all of the possible ideas are imagined and laid out in a "set" with static logical relationships between them. But if we consider the development of ideas in time, we see that one side of a logical statement may very naturally exhibit a connection with something else, which would be very obscure, in terms of our starting hypothesis. Thus, the two sides of a logical statement are qualitatively different, because each side has a different kind of connection with an infinity of other possible logical statements, as well as with the rest of the real world. Because these connections are infinite in number, it is not really correct to imagine them all laid out in a static set, but it is also necessary to consider their development in time, since it would take an infinite time to really explore the infinity of connections, each of which is qualitatively unique. It is this sort of thing which makes me uneasy about modern "set theory". Although such a theory has many uses, it creates the grave danger that people will imagine that the world is really built that way, so that it makes the world picture static, and therefore false. It must be recognized that set theory is an abstraction, and that mathematicians will eventually have to burst the bounds of this abstraction, so that the significance of set theory can be understood in its proper context. And to burst the bounds of such static abstractions, mathematicians will some day have to return to a consideration of new problems in reality. This has been the only source of new ideas in math thus far, and is likely to remain the only source.

Now about my "intolerance" for formalists. From my point of view, it is the formalists who are intolerant, as they (most of them) refuse to look at any idea outside the limited domain which they take for the whole world. For example, in physics, people refuse to consider the causal interpretation of the quantum theory, because in their point of view, "physics consists in taking numbers out of experiments and putting them into equations and vice-versa". If a study doesn't lead to an equation, it doesn't mean anything at all to these people. I can remember the people at the Institute for Adv. Studies, who are so sure that my work is of "no importance" despite the fact that the lengthy calculations of this group have hardly led in the past five years to a single result that can be applied to experiment. Yet, they are convinced that they already know everything that is "important".

Aside from my personal feeling against the formalists, there is the fact that they have been an important factor in turning theoretical physics into a grim dull business, in which only those with stomachs strong enough to go through 20 pages of ugly meaningless calculations are able to survive. Moreover, they have distorted the structure of physics and made it idealistic, thus helping spread the idea that the world is such an irrational place that not even the physicists are able to understand any of it rationally. Thus, science, which could have been a powerful instrument for giving people confidence in the possibility of a rational solution of social problems, has been turned into an instrument of obscurantism. Moreover, this point of view has led physicists to regard physics as that which is done by a small coterie of physicists, and even to lose the idea that there might be an important relation between their work and society, not only technically but also philosophically. Instead, people have been made cynical. I think formalism + cynicism go together (in art, science, music, or what have you). For the formalist sets up abstract standards of formal elegance, and judges people by how they meet these standards. Since few people meet them, he regards the rest of humanity (and ultimately even himself) as stupid sheep, not worthy of further concern. Instead of judging things objectively, he compares everything with arbitrary subjective standards set up by the small group, inside of which everybody is always busy determining the pecking order. I have seen this in math + physics already, and know enough artists to see how this type of insanity brings about the same reaction in art. So I think that in its long range effects, excessive formalism is a vicious thing, and I do not see why I should be tolerant of it, since it is the enemy of everything I have thought good.

Finally, what you said about throwing the die is just the sort of thing I have been trying to say for a long time. I shall send you an mss on the subject, as soon as it is typed.

Love
Dave

Letter 91. Folder C121, dated: Dec 23, 1952.

Number on photocopy: 37

Universidade de São Paulo
Faculdade de Filosofia, Ciências e Letras
Caixa Postal: 8105
São Paulo

Dearest Miriam

I was very glad to receive your letter of Dec 10. I notice that you are not answering my letters very often – an average of 1 out of 2.

I guess you are right that I should make some determined effort to get out of the state of bachelorhood. It is not very pleasant, especially at night, when there is no one to talk to, and one is too tired to work. But it is really hard to find a girl who is, so to speak, "better than nothing". In other words, with most of the women I can see here, I find that after a short time, I would feel an intolerable pressure to get out of their company because we shall by then have exhausted most of what we have in common. So there doesn't seem to be much of a solution, whichever way you look at it. I don't think my lack of initiative is due entirely, as you imply, to a fear of getting "hurt", but much more to a fear that I will be "trapped" with someone with whom I have no desire to be. I fear that by now I have no capacity for being happy with most women, because I cannot be satisfied with just "living" from day to day, as this is about all that one can do with them. I cannot remain close to another person for a long time, unless we are doing something together. Otherwise, the ties between us would tend to become tenuous and artificial, and pretty soon we would have nothing to say to each other, and I would begin to feel that I might as well be by myself. This has already happened to me several times in the past. Naturally people regard this as an excessive degree of self-centering, and it probably is. But I can't [missing word(s)] do anything about it. Also I don't think it is entirely self-centering, but is in part a genuine inadequacy of most women, as they are raised in this society. As a matter of fact, you are the only woman with whom I seem to have the possibility of inexhaustible conversation of mutual interest.

To change to the subject of politics, let me say that it is distinctly possible that they will kill a lot of the "seeds" who carry the beginnings of a new kind of society with them. But we must remember several things.

This is not certain, and until it is certain (i.e., until they are actually dead), these "seeds" must continue to struggle and act on the assumption that there is a possibility that they will have some effect. This is necessary
(a) Because, statistically speaking, a certain number are sure to escape, as long as all of them struggle. But if the struggle ceases, the number who will escape will be greatly reduced.
(b) Without this belief in the future, the "seeds" will cease to be seeds, i.e., they will decay and turn into cynics. Then, even if they do survive, it will mean nothing, either to themselves, or to society. Also, they will adversely affect other people, who are potential or actual "seeds".
(c) Objectively speaking, history never repeats itself precisely. The future always has unforeseen contingencies, and it is only those who prepare themselves by struggling continuously against great, and even apparently hopeless, odds who will be in a condition to take advantage of these unforeseen turns of events, when they come.

It is important to remember in analyzing processes of this kind, that the very opinions and concepts of the people who analyze may play a key role in what happens, so that these ideas are an inseparable part of objective reality, and must be taken into account in the analysis.

In your case, I am inclined to think that you and George should go to England, even if it should mean that George might become a "proletarian". You once told me that I am in greater danger of becoming a bourgeois than is George. But I think that this is not so, first because of course, I am in more difficulties than he is, but secondly, because the very energy that would make George a better radical also tends to make him a better bourgeois. Don't forget that it takes a lot of work to be a good bourgeois. You have to flatter the right people, wear the right clothes, write the proper sort of papers, say the right jokes, go to the right parties, have the right manners, etc. All of this takes a type of "drive" of which George has a lot more than I do. So tell him I think he is in grave peril unless he gets out of the trap right now. And if he becomes a "proletarian" this may stimulate him to do important work that he will never do as long as he stagnates in his present position. After all, when he looks back on his life at the age of 70 or 80, will he want to say "I lived a comfortable bourgeois life – kept my mouth shut, became a professor, a respected member of the community, etc" And maybe it is already too late to do it, so that he will commit himself to this life, and later be kicked out under conditions that are much less favorable for making the transition into an effective "proletarian".

As for Chung's second letter, I never got it. (I did get the first). Perhaps it came with your comments on my paper on "Proof that Probability approaches $|\psi|^2$ in Qu. Theory" which I also didn't get. Incidentally, I am still very anxious to hear your comments on this paper, also your comments on intuition vs. rigor.

I asked Gross to get me a sub to M.R. but he hasn't answered me. In case he hasn't done it, could you renew my sub. Also, I would like a sub to Stone's new weekly. Get the money from the "fund" I established with Gross.

It is very sad to hear what happened to Melba. She wrote that she had prospects of a few consulting jobs. I had already offered her money, but she said that she didn't need it yet. Maybe I'll just send her some anyway. She is a very courageous woman, and I feel sure that she will make out somehow.

As for your ideas on probability, they sound interesting, but these problems are very difficult analytically, unless you can find some neat approximation technique. It would be good for you to work on such problems for a while, whether you are successful or not.

As for me, I have made further progress and extended the proof of the quasi-ergodic hypothesis (which appears in the mss I sent you) to the case of real collision processes in 3 dimensions. Also, I have developed a new point of view on the role of stat. mech. in determining a distinction between past + future (i.e., a theory of irreversibility) of which I shall write you more in the next letter

Love
Dave

Chapter 26
Letters to Miriam Yevick, 1953, Part 1

Letter 92. Folder C121, dated: Jan 2, 1953.

Number on photocopy: 38

Universidade de São Paulo
Faculdade de Filosofia, Ciências e Letras
Caixa Postal: 8105
São Paulo

Dearest Miriam

I was glad to get your letter, but wish you could write more often.

As for me, I am sort of discouraged by having to stay in Brasil – the return of my passport was definitely refused. I really don't like the place much. It is a land of great misery for the many, and even the few who have money are not very happy. The government is hopelessly corrupt and incompetent, as is the direction of most enterprises. The university reflects the general corruption and incompetence. One has a feeling that to do something here (like helping build up a physics dep't) is like writing on sand, except that the difficulties in your way make it seem more like chiselling into granite. It is possible to live and work here as an individual, more or less disconnected from the society as a whole. But the degree of social cohesiveness is very small. Even the Brazilians do not seem to become really warm friends of each other. They associate at work, but seldom see each other. Each couple sort of lives in a shell of its own. At least this is in São Paulo. In Rio, there is more sociability and friendliness, but the general impression is that it is very superficial. One thing in the way of people getting together is the lack of good communications. Few people have telephones, and a long distance call is almost impossible, even if you do have one. The public transportation (buses, cars) is inadequate, unless breaking down (as is the water and electricity supply). For the poor people, things must really be intolerable, as the price of food has gone way up, and there is often no water for days in working

C. Talbot (ed.), *David Bohm: Causality and Chance, Letters to Three Women*,
DOI 10.1007/978-3-319-55492-1_26

class districts. The whole scene is depressing even to Brazilians, as one can see from the uniform pessimism of the articles that one sees in the press here. Also, the lack of people who really seem to be worth knowing is very depressing. Even when you meet such a person, the difficulty of establishing contact is great, not only because of the language, but also because of a generally different background. As for the girls in the university, they seem to be mostly "sweet young things". Incidentally, this girl I wrote you about is already pregnant. I wonder if it was an accident, or if they did it on purpose. The man is still getting his B.S.

I suppose I am in a bad mood now, because I haven't had enough sleep in weeks. They are building a new apartment across the street, and they start banging away at 7 every morning, while there's too much noise here to get to sleep before 12 at night. They are building at least 4 apartments per block in São Paulo. It is a building boom, fed by the super-profits that people are afraid to invest in industry. The building boom contributes, however, to the exchange shortage that is strangling the economy of the country.

My work is going at a slow pace. I have recently made some progress in understanding more clearly the irreversibility of development in statistical mechanics, without some of the mysticism that is usually associated with the question of increase of entropy. I shall try to explain it to you in a later letter, when I am less sleepy. However, my experience is that these things can be explained only in systematic articles (if at all). To get a new idea across is a discouraging job, as even people who would tend to favor it, usually oppose it, since they haven't understood it, and naturally keep on thinking in the old ways. To explain such things by mail is almost impossible, as it takes 6 months to clear up a single misunderstanding.

I am glad to hear that you are fighting against subservience to the "mature scholar". Keep it up!

As for the book on dialectics, etc., you have no idea of how hard such a thing would be. For myself, I do not feel ready for it, as I must clarify and develop my ideas some more. It won't do just to repeat the old stuff in new language – The problems must be carried to a new level. By working on qu. theory + theory of probability, I have become much clearer as to the new ideas that are needed. Even when a person is convinced of the general principles of dialectics, he still carries a great deal of positivistic and idealistic confusion in his detailed picture of the science in which he works, because when he learned the material, the two were presented as inseparable. It takes a lot of work to disentangle such things. My ideas are still developing along these lines, and it is too early to crystallize them in the form of a book. For example, on this question of intuition vs logic, I figured this out recently while on my vacation, as a result of thinking of that discussion we had several years ago, when we ate with George + Serge Kramansky [name not clear]. Whenever such questions are presented to me, I keep turning them over in my mind, often for several years. But naturally I had absorbed some of current misconceptions about "intuition" even though in my explicit philosophy, I already could not accept the "intuitionist" point of view 100%. But it takes long, detailed leisurely thinking to disentangle such a question, or even to start on the right track toward disentangling it.

Another thing to think of is that even the leftists will not readily accept the new ideas suggested in such a book, unless they are presented with extreme clarity and convincingness. I remember an argument with you and George on dialectics, which we had in my car as we were returning from Berwick, where the two of you violently attacked the notion that one should consider only two opposing forces in an analysis, instead of 3, 4, or an infinite number of "shades" of distinction. The idea of varying "shades", all stretched out statically besides each other, is a characteristic distortion and limitation of vision, which one picks up from bourgeois society. It is very similar to the idea of a "set" of classes of objects, all neatly classified separated from each other, determining a general scheme of things within which one moves by choosing one element or another (just as one moves between various pre-determined and pre-existing social classes); But the processes in which this very method of classification originate are not considered. Similarly, the various "shades" usually result dynamically from the specific balance between opposing forces or processes; and to change the shades (or more important, to change the very system of classification into "shades" that is appropriate), one must discover the opposing forces, and aid the force or process that produces the changes desired. In other words, the key to a real understanding of nature, which will make it possible for man to shape nature in new ways, is to find the dialectically opposite forces or processes that underly the bewildering variety of "shades" of differences, that present themselves to us, in a superficial first inspection of the problem. The bourgeois stress on various "shades" is then effectively a way of crystallizing the first superficial conclusions that follow from a preliminary view of the problem, so that one accepts the situation as it is now, and does not try to make a fundamental change, satisfying oneself with choosing one of the various "shades", that are available. But these different "shades" are basically just different expressions of the same forces, and therefore in the long run, it doesn't make much difference which "shade" you choose, if you don't try to find out the forces that underly the classification into shades, any more than it makes much difference which color clothes you happen to buy.

I read the "Brave New World" of Huxley many years ago. I agree with you that the ideas in it stink. Such ideas have been fashionable for a long time, and provide a convenient excuse for people to complain about the state of things, without them having to do anything about it. As you remarked, he shows the characteristic tendency, (implicit in bourgeois thinking) to crystallize all difficulties and contradictions found in existing societies, and to regard them as permanent and inherent in human nature. As I might remark, he did not try to discover the underlying but opposing forces as tendencies, which create this particular state of society, and which would provide the key to any changes that are to be made. There is certainly a real contradiction in existing society (especially in the West) between free development of the individual and the mechanical organization of production on a large scale. But this contradiction is not inherent and unavoidable – it is the result of processes which are conditional, and spring from present forms of society. To solve the problems, one must look deeper and find what are the forces that determine "human nature" itself, instead of metaphysically conceiving it as a fixed thing that eternally comes into conflict with "mechanism". As Engels said, we need a new situation, in which man + nature form a larger unity. Man must become "naturalized" and nature "humanized".

[... ends here]

Letter 93. Folder C121, dated: Jan 9, 1952.

[1952 written on original not correct – must be 1953].

Number on photocopy: 39

Universidade de São Paulo
Faculdade de Filosofia, Ciências e Letras
Caixa Postal: 8105
São Paulo

Dearest Miriam

Thanks a lot for your letter. It is difficult for me to tell how many of your letters arrived, as I do not keep a systematic file. Like all my papers, they are all mixed up. But I am certain that I never received any of your comments on the paper – Proof that $P \to |\psi|^2$, nor did I ever receive a second letter of Chung (to which you once referred).

I would like to comment on your idea that I am too "bitter". I do recognize that it is not good tactics to be this way in articles, and I modified the conclusion somewhat in the draft I sent to the Phys. Rev., in accordance with this idea. But you must remember that I feel these problems intensely. I have what you might call a passionate desire to fight this stupefying spirit of formalism, and pragmatism in physics (Only results count – thc ideas behind them are just "window-dressing"). And it hurts to see the lack of response to what seem to me important indications of what is wrong with our ideas, which explains why there have been such few results in the past 20 years. At times, such frustration cuts one's insides like a hot knife being twisted inside your heart, and it is hard then to avoid some "bitterness". However, I confess that I am a bit disappointed in the rather pompous attitude of Gene and George – banging on the table saying "Let him give us results". It's true that Schwinger + Feynmann have produced some "resultonhas" (Portuguese roughly translatable as "resultlets"), but these little mice are all that come out of 20 years of labor by the mountain of thousands of theoretical physicists.[1] Now, in one year, all by myself I am supposed to give them "results". Where, by the way are the "results" of George and Gene? I haven't seen any flood of papers from them recently. They apparently do not regard it as a "result" that I demonstrated that much of the conceptual structure of the usual interpretation is arbitrary, in comparison to that of the causal interpretation. They claim to have a materialist philosophy, but apparently that is meant only to be displayed, and not to be used. Thus, one need pay no attention to this philosophy but just go merrily working along with the usual interpretation, producing nothing, until someone drops

[1] Also, neither of them has done anything at all in the past 3 years worth mentioning.

some "results" in their laps. I can understand it when Rabi says such things, because it is his explicit philosophy. I know that it is George's real philosophy too; but I had thought that Eugene knew better. He should know that "results" do not come from a flash of genius, but require careful preparation, and a critical understanding of our conceptual background, in order to know what errors one has subtly absorbed that might be blocking the way. But all they can think of is that I may have offended a few nitwits, who need to be "attracted" into doing what they ought to have sense enough to do for their own good. As for Saul Epstein, I have not yet decided whether he would be more helpful on my side, as on the other side. He is so confused that I don't think he can accomplish much either way. But I can't help getting irritated when someone comes eager to be on my side, and then nicely uses all the arguments of the other side. At least I prefer that he openly admit he is on the other side, and be done with it. He can be compared to a man who shoots you in the back, and then begs pardon saying he was using a theorem from which he deduced that bullets come out from a gun at an angle of 90^0 relative to the barrel, and that he was therefore really shooting at someone else.

You may tell George + Gene that it's time to stop this pompous beating on tables and get down to the real issues. Also, as you say, Gene's little plan to sit around for 20 years and keep his heart pure while all his friends go into concentration camps is not very practicable. For one thing, I think he is lulling himself by imagining that the FBI pays no attention to him. The mere fact that he wrote papers with me + that I visited his house as a friend is doubtless on record somewhere; and this will cause them to look at him with suspicion.

All in all, I think Gene is much more in danger of "going to pot" than I am, unless something happens that will knock him out of the smug groove in which he moves. Also, if he thinks I will go to pot here in Brazil, let him get me a passport, and I'll gladly leave. Or perhaps he thinks I should go home and avoid "bitterness" by sitting in jail, or in a concentration camp, while he comfortably waits till the next 20 years of fascism are over.

Incidentally, you mentioned that Gene said some things that "sent shudders up your spine", in regard to my "shirking my duty" as a spiritual guide to Gene. Could you explain this in more detail? I have tried to maintain a correspondence with Gene, but he seldom answers, and when he does, he never comments on what I have said previously. So it is difficult to see what he expects. Anyway, by now he is old enough to try to guide himself once in a while. I will be glad to discuss anything he wants to, provided that he responds to what I say once in a while; but some of the initiative is surely up to him. In view of the fact that he + Sonya will not do political work, it is especially important that they get out of the groove for a while, and go, for example, to England, where they will be exposed to new ideas, perhaps returning after a year or two, if they decide not to stay longer.

As for me, I would return to the States only if they dragged me back. I have always been completely out of sympathy with the American spirit, although I think I understand it pretty well. This extreme pragmatism – immediate results are all that count – is what led the people to their present sad state. I have hated this spirit, even as a child, especially since it was so strongly embodied in my father. I once used to think of America as a land of hope and freedom, but since then I have seen that this America was a dream country, that never existed and never was even a potentiality. The correct emblem for the US should be, not the eagle, but a pig with his nose in the pork-barrel representing each individual taking care of his own career, and careful not to let his eyes see more than 6 inches in front of him (also symbolic of his ultimate fate after he is "fattened" up).

I hope I have not offended you by this letter, but some of the things you said in your previous letter made me angry. I recognize the tactical need of avoiding bitterness, but I felt that you were concentrating on superficial points, by not commenting on what you thought, objectively, of what was said in the paper. And I would still like to hear those comments.

There is no question that life will be frustrating no matter what you do. The situation is not comparable to the fight of Israel against the Arabs. There the fight was tough, but the social spirit was well integrated. Here, we are faced with a real disintegration of the culture, comparable to that which occurred (but much more slowly) in Roman times. And the only way of keeping going now is to keep your eyes firmly on the society that is sure to come. In Portuguese, they have a word for it, " porvir" meaning "that which has to come to be"; I have been reading an abolitionist poet Castro Alves, who played an important role in freeing the slaves here. There is a phrase which sound magnificent in Portuguese, but not so good in English. It was a poem that ends up roughly by saying that "The future (porvir) confronts the past; freedom against slavery". But I always get a lift out of the phrase "O porvir enfronte o passado", and think of the new society that has to come to be, which in America exists only in the hearts and minds of a small minority, confronting all the organized stupidity and viciousness that is already decaying and preparing to go out of existence.

Love
Dave

Letter 94. Folder C121, dated: Jan 13, '53.

Number on photocopy: 40

Universidade de São Paulo
Faculdade de Filosofia, Ciências e Letras
Caixa Postal: 8105
São Paulo

Dearest Miriam

I hope you weren't too badly offended by my previous letter. I was really shocked and hurt by your letter, because it showed the development of an astonishing degree of misunderstanding among people on whom I had been counting, as well as certain tendencies that are even more disquieting.

Let me begin by admitting that I see the futility of bitterness. It eats into your soul, and gives nothing in return. But even when one sees this, it isn't always so easy to stop being bitter. In the past few months, I had been getting over this bitterness, but the shock of your letter brought it back, and now it may take me several weeks to get back where I had been.

I want to explain why your letter was so shocking. In general, I had begun to accept the idea that little was to be expected of Western Physics these days (especially in America). It is too corrupt and militarized to be very much interested in fundamental questions, in the best of circumstances. In addition, the general cynicism and cowardice prevalent in the culture leads young physicists to accept without protest whatever the "big shots" tell them to, even though many find unpalatable some of the consequences, like working on 18th order calculations, because there is nothing else in sight. I had nevertheless hoped that there were a few people who understood the importance of analyzing the ideas underlying the "results", instead of acting like the primitive medicine man, who accepts the "result" that seeds planted with the appropriate prayers will sprout – so that the prayers were obviously needed to produce "results". Similarly, modern physicists, especially in the West, have accepted the idea that quantum theory with its a-causal philosophy produced "results". Therefore, we need the a-causal philosophy, until some new witch-doctor (perhaps me) produces the new magic rites that will replace those of Schrödinger, Heisenberg and Dirac. The analysis of concepts to see what is important and what is not is of no importance to them, because it does not produce immediate "results". But the problems before us are complicated, and at best could only be resolved in a favorable environment of discussion, circulation of new ideas, etc. (which we do not have in the West) in perhaps 5 or 10 years. In our environment, they might never be solved, until there is a fundamental change of outlook. I notice that a vigorous criticism of the foundations of quantum mechanics is going on in the "East", and in 5 to 10 years this may bear fruit, as a large number of people discuss the ideas, counter-discuss them, propose solutions, criticize them, etc. But here, according to the pragmatic American view, I alone am supposed in a year or two to produce a scientific revolution comparable to that of Newton, Einstein, Schrödinger + Dirac, all rolled into one. If I do this, then our broad-minded American theoretical physicists will condescend to listen to me. Otherwise, it is not necessary for them to think about these problems. (Naturally, not since there are so many more lucrative problems to think about, such as H-bombs).

Now, I had thought that George + Gene (especially Gene) had developed far enough to appreciate these points. But I suddenly found out they hadn't, and I was left in serious doubt as to whether you had appreciated them either. This made me wonder. For how many people am I writing? Perhaps even for an infinitesimal fraction of the small number that I had in mind already. You mention various discussions of my papers. Those few that I have heard of have been carried out at such an inadequate level that they don't amount to much. Also, as for the literature, we have a letter of Halpern (hardly worth mentioning), a few childish remarks by Epstein, and a paper in the Japanese Progress of Theoretical Physics, which raises all of the standard positivistic objections to the theory. However, the author of the latter article appears intelligent, and for this reason, I have taken the trouble to prepare a careful answer to him, which will be published in the same journal. (In general, Japanese theoretical physics is carried out at a higher level than American, though it still suffers from excessive formalism).

In any case, life is already unpleasant enough in Brazil, and to feel that one's efforts are being wasted just makes things that much tougher to bear. The fact that points as simple as those in my paper can be misunderstood by people of the type to whom the projected book is to be addressed also makes me wonder. For if the book contains any new ideas, the numbers of people who will be able to resist environmental pressure sufficiently to grasp them will be minute. This experience only re-enforces my feelings about your first reaction of ridiculing the infinity of levels, as well as what I feel is a desire on your part not to comment on the paper that I sent you (if I am wrong on this latter point, please correct me). Also, your suggestion that I try to make my ideas attractive is tough, because the situation is so unpleasant that I am almost afraid of a new idea. It just means

[. . . missing page(s). This last section probably comes from an earlier letter as the paper referred to at the end was submitted to Physical Review June 25, 1952.]

one I suggested to you originally. The problem would then be to show directly that the average number of particles in a given element of volume is a definite number, except for a set of trajectories of measure zero.

I haven't yet had time to study your mss carefully, but it is now being studied by Prof Farah, of the math dep't, who will report on it later in a seminar. I am interested mainly in the general problem of sequences, and of measures of trajectories in that sequence, of course, and not so much in the specific problem under discussion.

I am sorry that I mistook your meaning about my "being so far away". I was very busy + distracted by many things at the time. But the way you wrote it, it was somewhat ambiguous, because you implied that I was trying to "explain the universe" and then said that "I was so far away". Incidentally, I am a bit angry at you for calling the limitless number of levels a "religion". I regard it as typical of the supercilious way in which modern mathematicians sneer at all new ideas not in the comfortable sphere which they have become accustomed to accept as "deep". I really get angry when mathematicians say that they want to be really sure that their work is solidly founded, no nonsense about it, etc, and then proceed to base themselves on the most abstract postulates, having nothing to do with anything real. Then when by their own methods, they show that their mathematics has no foundation (logically speaking), they stew around a little worrying about these "deep" problems and then do not hesitate to sneer at new ideas which might remedy this situation as a "religion", forgetting that their own postulates are hardly more than a religion either.

Thanks very much for sending the books. Stone's book is particularly enlightening.

I have been busy writing up a paper showing that as a result of collisions, an arbitrary probability distribution approaches $|\psi(\vec{x}, \vec{t})|^2$ as $t \rightarrow \infty$, in the causal interpretation of quantum theory. This is an important point, as it really is necessary for the logical completion of this interpretation. You should get a copy in a month or two.

Will write more about probability soon

Love
Dave

Letter 95. Folder C121, dated: Feb 24, '53.

["Must be 1952 because of refs to Gross" written on original – not correct must be 1953].

Number on photocopy: 41

Universidade de São Paulo
Faculdade de Filosofia, Ciências e Letras
Caixa Postal: 8105
São Paulo

Dearest Miriam

I just arrived from São Jose dos Campas, where I was visiting for a week or two, and found your letter. Now that I have had a chance to get out of São Paulo, all this fuss seems to be much less important. I am very easily irritated in São Paulo, partly because it is too noisy to get much sleep, partly because the food makes me sick, and partly because the traffic is such a a mad confusion that you risk your life every time you cross the street, which is full of recklessly speeding cars, and buses which give off a dense cloud of black smoke, because their engines are not adjusted properly. Anyway, with a nice rest in the country, and with good food, I became much less irritable. But already, after being back only a day or two, I am becoming tired + sick again. São Paulo is worse than N.Y. Only one city I know is still worse – and that is Rio.

So I hope we can forget this little incident, which is partly due to misunderstanding. You did not write clearly what Gross and George said. In fact, you gave the wrong impression altogether. It is true that I am somewhat touchy about my "baby", but perhaps you might remember that there is no lack of people to criticize it. So when I hear what seem[s] to be completely unjustified criticism from people who I expected to be friendly, it tends to disturb me. So please try to be more accurate in the future, and I'll try to be less touchy.

As to you and George's coming here, I might be able to locate enough money to cover your expenses in Brazil. It would be practically impossible to get the trip paid, however. It is already late, and we must do something rapidly, if this is to be done. For the Brazilian way of doing things is so full of delays that there will be a lot of. . .

[. . . missing page(s)]

. . . another struggle to get people to listen to it, and to understand it. And if I don't try to do this, then the whole business seems so futile that it isn't worth bothering with. It isn't so pleasant to have to "sell" your ideas, by putting them in attractive packages, as if they were drugs or cigarettes.

You might tell me that I am exaggerating these things, and doubtless I am. But it is hard not to do so, in this isolation. And one of the few consolations I had – that at least some people would understand these things and back me up – is now open to serious doubt. What you tell me about Eugene's plan to go into a shell for 20 years is particularly disquieting. Also, I am bothered by the slowness of your reaction to it. You should have jumped on him, and torn him to pieces. Instead, even in your letter to me, you merely said that you "doubted" it, and later admitted that you were sure it wouldn't work. It was your duty to tell Eugene + Sonya on the spur of the moment that it wouldn't work, because I am sure that you knew it even then, since we had been discussing just this point for the last 6 months. In any case, Eugene's going into a shell apparently also involves accepting the prevailing attitude in science, which is not surprising. And this also disappointed me, and presented me with a cruel problem. I could, of course, resolve it by just saying "O well, – just another American physicist who is following the well-worn track to its inevitable conclusion." But I find it tough to do so, because it makes me very sore inside, and even bitterer. But I don't know what to do about it. Experience suggests that writing to him would be a dubious thing, since the written word is so easily misunderstood.

Finally, your mentioning that Eugene said things about me that "chilled your blood" is disquieting. To my way of thinking, something that "chills the blood" is pretty bad – it calls up pictures of massacres, etc. You probably should never have mentioned this business to me in the first place, but now that you've done it, you had better let me have the facts, as the damage is already done. Let me say a few things frankly about Eugene, since he seems to have talked without reserve to you. There are several serious faults that he has not yet corrected. One is a tendency (along with Sonya) to "take their friends to pieces" whenever they get off by themselves. When I visited them, they did this so consistently that I became bored and depressed by it. I am sure that they do the same to you and George.

I interpret this as a form of compensation, with which they try to make up for their own dissatisfaction with their futile life in a middle class environment closed to new ideas, by convincing themselves that everyone else is even worse off still. So I wouldn't take Gene's complaints too seriously. Secondly, Gene himself lacks initiative in exploring new ideas, and in applying his philosophical principles. He continually asks for "guidance" and needs to be infused with enthusiasm. Also, his papers, while good from a mathematical point of view, impress me as being too mathematical, since they show little tendency to enter into the physical significance of things, despite the fact that he accepts a philosophy which implies the need to do this. He refrains from making a really sharp analysis of a problem, even though I am convinced that his abilities would permit it. I often have a feeling when I talk to him that there is a "fog" over his mind, that makes concepts hazy, like a drowsy summer's day. Part of this comes from a formal absorption of these concepts, without an effort to digest them through sharp analysis. Finally, I think he tends to "browbeat" people who are less familiar with a subject than he is, giving them lectures with which he hopes to "devastate" them, instead of trying to understand the problem.

Now that you have started this business, you had better carry it to a finish. I don't think it wise you that you talk too much about it to Gene – perhaps mention a few general points. But please let me have more details as to what is up, and on the basis of these, I shall try to write a letter to him, re-opening the discussion. He seems to be pretty angry at present, as he sent me an order for reprints for a paper of which he is reading the proofs, without even a note inside. You should, however, tell him frankly what you think of his ideas of drawing into a shell for 20 years.

Love
Dave

P.S. As far as bitterness is concerned, I often detected such a note in you references to Chung, etc. I feel that the struggle in which I am engaged is even bitterer, from a scientific point of view, as well as because I have to carry it out in isolation in Brazil, and cannot even travel to discuss my ideas with sympathetic people, who might be able to help.

P.P.S. In order to illustrate the philosophical points mentioned in this letter, I would like to tell you what an experimental physicist friend of mine wrote. He is no leftist nor has he studied philosophy, but on the basis of simple common sense, he notes that the "big shots" made a big fuss about the fact that qu. mech. seemed to prove that causality needs to be given up. Now that I have proved that that is not necessary, they suddenly find the entire question "unimportant", and say that we must consider only "results". In my opinion, it should be clear that the whole reaction is nothing but a set of tactics to throw dust into people's eyes. Their basic purpose is to get across an idealist philosophy. When it suits their purpose, they frankly stress the importance of philosophy. When it doesn't, they forget philosophy and stress "results".

I find that generally speaking, experimental physicists tend to favor the causal interpretation, whereas theoretical physicists (even leftists who have studied philosophy like Phil Morrison) are opposed to it, because it produces no "new" results. I think this is an indication that theoretical physicists, who are one stage further removed from reality than experimentalists, are more easily swayed by the confusing and corrupting environmental pressure in their ideology. Thus, it develops that I do have some support, but it comes from somewhat unexpected quarters; whereas the kind of support I had expected from some theoretical physicists has failed to materialize. Well, if this is the case, then to hell with the leftist theoretical physicists! In the long run, it will be so much the worse for them.

Another point I can't help mentioning. You have doubtless read Charles Lamb's essay on "Roast Pork". In this essay, he gives a fanciful history of the origin of Roast Pork, saying that it was first made by accident when a Chinese boy lit a fire that burned down the house, inside which had been a pig. After returning to the ashes, they discovered the succulent roast pork. For several years thereafter, everyone made roast pork by burning down his house, until some bright fellow figured out that one could do it with a fire, without burning down a house. Similarly, qu. mech, was introduced by people who threw out causality, and for 25 years thereafter, people never did qu. mech. without throwing out causality. Now I show that this is not necessary. But modern American theoretical physicists are not as clear-headed as were the Chinese, who immediately saw the advantage of not burning down their houses, when it wasn't necessary. Instead, the physicists want to throw out still more of causality in connection with a "fundamental length". It is fortunate that the ancient Chinese were not governed by theoretical physicists, or else they would have suggested that since burning down a house made good roast pork, still better roast pork could be made by burning down something bigger, like a forest and to anyone who asked that we analyze just what it is that is responsible for roasting the pork, they would answer "That sir, is pure metaphysics. Roast pork is roast pork in any case. We refuse to listen until you produce a new kind of roast pork." And meanwhile, since we are used to making our roast pork by burning down houses, we will continue to do so, until you show us that your suggestion of roasting it over a fire makes better roast pork. Then banging on the table, they would all say "Produce results, that's our motto. Bring your results to the free market-place of ideas; we will inspect them, and if we like them, we'll buy them, maybe giving you an associate professorship if they are really good".

Letter 96. Folder C122, dated: Feb 28, 1952.

[1952 written on original not correct – must be 1953].

Number on photocopy: 42

Dearest Miriam

I just received your 2nd letter after the "storm". I am really sorry that I caused you so much unhappiness, and wish I had not written those letters. Next time, let us be more careful. When it takes 15 days or more to get an answer, a hard word can cause a lot of trouble.

As for the question of "feelings" I have to guard myself about getting too excited about what I am doing, or else, these excessively intense feelings will destroy me. I suppose that I learned the same reaction with people a long time ago, so that by now it comes without thinking. Anyway, I have been steadily revising downward all my estimates of what I might hope to accomplish, and of the importance of the work. You are quite right in saying that I am in a demoralizing position by sitting here and working, while you people at home face real trouble. I suppose also that one becomes demoralized sitting in the U.S. surrounded by people who have blinded themselves to what is really happening. And I suppose also that it would be demoralizing to sit in a concentration camp. The chief characteristic of the modern world seems to be that no matter you do, it is demoralizing. This is a sign that the entire society is deeply infected with rottenness from top to bottom, from right to left.

I am reaching the point where I don't really care too much what happens to me, now that it has become clear that I can't really be happy in this society no matter what I do. Your suggestion that I marry someone just to "share life" probably would not work. After 2 or 3 days, we would exhaust all topics of conversation, and then contact between us would gradually disappear. I am afraid I am so constituted that the one thing that is more killing than anything else is to "just live". The futility and meaninglessness of the thing always has made this intolerable to me; and if that is all that a person is doing, I cannot become very interested in her (or him), unless I feel that there is a potentiality in that person for something more. That is why I don't think I have ever really enjoyed myself after the age of 12 or 13. For in this society, this is the age when people become "practical", and accept things as they are. When I was a child, I used to feel a dividing line near this age. With people under this age, one could establish contact, but with people over, I felt that there was no possibility of real understanding. So I have a constitutional necessity to be oriented to something more than "what is". This is why dialectics has such a strong appeal to me. Now, in Brazil, the hopeless acceptance of things as they are is even deeper than in the U.S. (at least in the educated classes). Thus, it seems hard for me to establish much contact with the people here, even the leftists. They seem to have no hopes, no dreams, but only the "sensible" desire to make the best of things and live on top of a volcano without any real effort to do much about it, or even to think much about it.

I suppose that you are right in that I should try to find some girl who would have some interests that would give us at least a little in common. I don't feel at all optimistic about the possibilities, however.

As far as my hate for America is concerned, you must remember that this is because I once loved it very much, as the land where there was freedom from the ties of the past that bound Europe, and where there was hope for the future. People like Melba are the [missing words] such people are the sparks in a soggy mass of wood. The time [missing] "sparks" will catch fire, but this, I fear, is a long way off.

[missing] as the "book" is concerned, I feel that I am now experimenting with the [missing] of some of the ideas that will appear in it. For example, I am giving 4 [missing] lectures on causality, discussing the history of the concept, its role in classical physics, its role in qu. mech., usually interpreted, then causally interpreted, and then a philosophical conclusion on the limitless number of levels. Also, some of the articles I am writing serve to clarify my concepts. An important problem will be to get clear on a large number of details, which are brought up when you really start to write about the subject. I feel that a number of articles (perhaps to be published in a British philosophical journal) will help lay the basis for starting the book.

I have begun to feel that perhaps we could have made a go of it in Brazil. There would have been some difficulties, on both our parts. You would probably have felt the isolation more than I.

Now to discuss some of the math points that you raised

(1) My opinion is that <u>all</u> probability distributions should be derivable from causal laws, as a result of chaotic motion or development. The example I sent you was only illustrative. If there were forces, the final chaotic distribution would be more complex, but it is in principle determined by the forces, which determine the law of motion.

In other words, I think that all chaos comes from causality. In the case of statistical mechanics, the particular aspect of causality that leads to chaos is instability of motion. The usual trick of defining a "random variable" is in my opinion limited to being a useful artifice. It is useful because it helps us dodge the difficult question of "<u>Why</u> is the variable random?" But it permits us to dodge this question only in those cases where we know by experiment or experience that it is random. Thus,

it is a way of expressing mathematically certain empirically observed properties of the real distribution. But in my opinion, these empirically observed properties of "randomness" should also be deducible as necessary consequences of the causal laws applying in each particular case.
(2) The real mathematical problem is precisely to generalize the type of reasoning I went through to a wider class of problems. There is no question from the physical point of view that chaos is really produced by the actions of the causal laws. The fact that we do not at present know how to solve the corresponding mathematical problem should be regarded by mathematicians as a challenge. This does not mean that the theory of random variables is useless. Far from it the theory [words missing] by experiment, or by theory, we show that in the problem under consideration, the events in question really do fluctuate "at random" (i.e. chaotically). But to limit yourself to accepting the random fluctuations as a hypothesis is to confine unnecessarily the development of math. For the next step is to prove that the causal laws themselves are what cause the events to fluctuate chaotically. This will give you a theoretical criterion for when the theory of probability can correctly be applied.

Naturally, you can't come out in 1 day with a complete generalization of the theory of the development of chaos through the causal laws. My suggested program is, however, to start in stat. mech., where the causal laws are all known, and where the problem at least seems feasible. A first step might be to try to take into account in detail how collisions among hard spheres would lead to chaos.

Another aspect of the problem is how chaos in turn leads to new causal laws at a higher level. Thus, the molecules bouncing chaotically off a wall produce an almost constant pressure. The pressure is constant just because the distribution of molecules is chaotic, so that the fluctuations in the rate at which the molecules strike the wall almost cancel out.

I have a feeling that people often say "The problem is very difficult, and it is much easier just to postulate a set of random variables. Why all this bother, just to get the same answer?" But then this is just an example of the pragmatist philosophy that stultifies science + math. Our objective is not only to get "results", but to understand reality. And if in reality, chaos comes from causality, we are blinding ourselves to a real problem, if we avoid the problem just because of mathematical difficulty. Instead, the mathematical difficulty should be a sign that we need to develop new forms of math. to handle the problem more easily. Of course, if your objective is to maximize the rate of production of publishable "results", such considerations will carry no weight. But if your objective is to try to understand a little more about how the world really works, then such problems should be of interest.

Please don't be angry or sad about all this fuss that was made over a matter of little importance. I think that everything is all cleared up now.

Love
Dave

Letter 97. Folder C122, dated: March 9, 1952.

[1952 written on original not correct – must be 1953].

Dearest Miriam

I have just received your third letter after all this trouble. By now I hope that you have received my previous two letters, and are less angry. Why do you always put the worst possible construction on what I do? When I didn't answer, didn't it occur to you that I might be on a vacation (or that your letter might have been lost – always a possibility).

Your third letter was not very pleasant, but it doesn't hurt me, since I know that it is just your reaction to the delay in my answering your other two letters. I would like to explain a few points, however. I was not, as you say, ready to "abandon my friends" in favor of science. The problem presented itself differently to me. It was as follows:

(a) I knew of Eugene's class philosophical position in the past.

(b) I heard from you that he pounded on the table and asked for "results" – apparently a reversal of his old position.

(c) If this had been really true, it would have been very serious. It would have meant that Eugene was absorbing the current rotten point of view in physics – where it is easiest to remain honest. What then would have been the probability of his remaining honest in other fields, where it is more dangerous?

(d) If Eugene had really reversed his philosophical position, this would have meant that I could no longer have trusted him as a friend. For a reversal of scientific philosophy merely to be comfortable would have meant the possibility of a similar reversal in any other connection. With regard to George, the situation did not seem to be so alarming, since he has never held a clear materialist philosophy anyway. Such an attitude would therefore not have been very inconsistent, in his pragmatic point of view.

(e) All of these things seemed to be confirmed by the other things you said.

(1) That Eugene & Sonya were going to withdraw and remain safe.

(2) That he had told you things about me that "made your blood run cold". What these could have been, I still can't imagine.

Thus, you can see that you presented a pretty consistent picture, in which even one of my best friends seemed to be turning away from me, and running with the tide. This is one reason why I was so hurt. Of course, it turns out that the real picture is quite different, since Eugene was merely saying that I needed results to make an impression on American physicists in general. I should have realized that this was the correct interpretation. In fact, I did think of this interpretation, but I was already very over-wrought by other things, and half sick; and your final statement about Eugene telling you things that "made your blood run cold" led me to adopt the first interpretation. . . . [words missing] Thus this misunderstanding: You must remember also that I have seen example after example of complete misunderstanding of what I have written, and this adds to the irritating character of the whole situation.

Now to comment on a few other points. With regard to the "discipline" in the USSR, I don't see how they can survive without it. Thus industrial production is not yet over $\frac{1}{4}$ of that which can be arrayed against them. Hence, without maximum utilization of resources, they would perish. But this is possible only with a rigid self-discipline on the part of each person, coupled with an understanding of the need for such discipline. I read in the papers that by 1960, their industrial production will equal that of the rest of Europe (and that the population of the two areas will be about the same also). This will be a sort of turning point. But the only sensible attitude in the face of collective peril is rigid discipline. What would you suggest? All of these bourgeois virtues of relaxation, etc. are possible only when there is a large surplus. Up till now, only a few people have been able to enjoy such relaxation, and these on the backs of the majority. When the surplus gets big enough, and when there is real security, however, then everybody can afford to relax somewhat and live a more pleasant life. You are asking for a miracle – a relaxed life under material conditions which are not adequate to provide it. There are plenty of things wrong with the USSR, but as I see it, the excessive discipline is not one of them. The main things that are wrong are: (1) Excessive authority in the leaders – lack of facilities for critizing leadership (2) Lack of real thought on the part of the majority of people in connection with fundamental issues – i.e., a passive acceptance of doctrine coming from the top (3) Identification of criticism of policy with treason (4) A tendency to believe that mere decrees and doctrine are able to overcome natural difficulties – not paying enough attention to a careful study of the objective situation, which alone can provide the key to solving the problem in general (Example – Lysenko's excessively dogmatic presentation of his basically correct point of view). All of these difficulties arise in part out of a lack of widespread diffusion of knowledge and understanding concerning those problems treated by Engels, Marx, Lenin, etc. In other words, there is a rather mechanical understanding of the real meaning of Marxism. Hence, what is crucial at this stage is a critical re-examination of such fundamental questions in the light of new knowledge and experience.

With regard to life in Brazil, I am beginning to consider the idea (not very seriously yet) of returning to the US when my contract is finished (provided that it is not obvious that this would mean jail or concentration camp). However, one problem is to locate a possible job, preferably in some metropolitan area. I think Brazil is in a pretty hopeless mess, and as long as I can use English a lot better than I'll ever be able to use Portuguese, I may as well go back. England is still a possibility, of course. I got an offer of a professorship in Australia, but that is too isolated for me.

With regard to the paper of Yang + Lee, it looks like good work, but I haven't yet read it carefully. It is really quite far from my main interests at the moment.

Love

Dave

Letter 98. Folder C126, dated: rec'd Mar 17 [1953].

Dearest Miriam

I received your letter of March 2. Many of your letters are being lost. I do not believe that this is due to malice, but rather to unreliability of mail deliveries in Brazil. In general, I have lost no letters addressed to the University address (Caixa Postal 8105) probably because the mail is delivered directly from the Post Office Box by a man from the University. Thus, it might be better to send the mail to the University, even though it may be a day or so slower at times (especially over weekends and holidays).

I am glad that the storm between us is over. Your last letters made me feel much better. Please don't imagine, however, that I am a feelingless monster. The feelings are usually deeply buried, and the reason that I bury them is probably that they would run away with me otherwise (witness the recent explosion).

With regard to George, it is likely that I could get the National Research Council to cover the cost of his stay here. It is no use buying tickets in cruzeiros, however, as the cruzeiro has effectively been devalued already. I shall look into the possibility as soon as possible. However, I am afraid that I shall have little time to discuss [unreadable] with George this July-August (winter here) not only because of duties, but because 2 other people are already coming under similar arrangements. One is a fellow from Argentina, and the other is Vigier from Paris. (Vigier thinks that with de Broglie's influence, the French gov't will pay his trip). From Vigier, we hope to find out what they are doing there in France. Nevertheless, George is always welcome here, even if we won't get much science done. He may find it interesting, but perhaps not worth the expensive cost of the fare.

Now about the state of politics in the US. I really don't want to sit here and urge people to take risks or not to take risks. About the middle class, I am afraid that I am right. It is not that they are bad people; but with their conditioning, 99% of them cannot do otherwise than they are doing. Really, I am sorry for them, rather than angry at them; because most of them live such simply meaningless lives, and don't know why their lives lack fullness – all they feel is vaguely that something is missing. But I am afraid that nothing much can be done for them, except for the small percentage who will, for one reason or another, begin to see through the situation that is trapping them even more surely than it is trapping the working class. This small percentage is important, however, because from it must come a large fraction of the technical skills, etc., which are needed in any real movement to get at the root of our problems.

Now about your work, the method of stable distributions seems to be a very promising way of looking at the problem of condensation. At least, it seems to be the only possible way now known for framing rigorously the problem of "clusters" and condensation. The work of Yang + Lee is as you say, very schematic. I would compare it to my simplified example of the development of chaos. It is not the real problem, but there [words missing] less than 3 months.

I think that things may improve a little. I have hopes of George meeting people from various places – Argentina, England, Switzerland. We may eventually be able to build up a good group here.

I am sorry about Epstein. He is doubtless a fine fellow with good intentions. His ideas are so confused, however, that I just get a feeling of hopelessness any time I get a letter from him. If this is what the "friends" of the causal interpretation are thinking, then I feel "Heaven help me!"

As to what's going on in the U.S., it is quite terrible, although of course only what was to be expected. (I got a copy of the "Columbia Spectator" that you sent me). I wouldn't put much hope in the struggle of middle-class liberals for "civil liberties" however. They have already admitted that Communism is the real danger, so that they don't have a leg to stand on, when they protest against what McCarthy is doing. Their position is that while they agree with McCarthy on fundamentals, they don't like the way he goes about doing things. But McCarthy can always answer "We've got an emergency here, and stern measures are needed". the only way to resist McCarthy is to point out that the entire Cold War is a fake, and that the real danger comes from within our society, and not from without. But can you imagine our liberals saying this?

I would say the same about your "leftists" friends who visited Yugoslavia. In general, little is to be expected from most middle class leftists. For one thing, their leftism often stems from a protest against the way in which the present social system interferes with their individual lives. They dream of Socialism as a fine way to do as they please in return for 4 hours a day of work. But this just cannot be. Especially now, in the early stages, socialism cannot succeed without a determined "all-out" effort. The world is full of nice people who only want to be left alone to their little plots of land, careers, or what have you, to raise their little children, etc. Because of their conditioning this dream seems reasonable to them. But the sum of hundreds of millions of such people is to create an environment in which socialism is impossible. Such a point of view can only be combated by the most energetic of measures. Some of these measures may seem to be more energetic than is absolutely necessary, but the penalty for failure is the collapse of the entire society. So you can understand if things are done with excessive crudeness. If there were plenty of time, the transition could take place more gradually. But there really isn't so very much time, you know.

It's no use getting discouraged about the attitudes of the "leftists". Middle class leftists have always sold out when the going got tough, and 99% of them always will. It is a necessary result of their ambiguous attitude toward the whole social struggle.

A more serious problem, I think, is that of being able to criticise the leadership. From my observations about what is happening in European Communist parties, and from what I could see of what happened when they got rid of Browder here, the rule is so autocratic that they have no way to criticise a leader before it is too late. As a result, the leader may make serious mistakes, and then they have to have an internal crisis to get rid of him. When they finally succeed, everybody suddenly reverses his point of view, and calls the former leaders "traitors", spies, renegades, etc.
They are thrown out of the party, and hanged if the gov't is Communistic. This is a serious weakness, which could be prevented only if the leaders themselves were

always subject to free criticism, instead of having to wait until the situation gets so bad that the situation is that of a crisis.

As to your idea of a transformation to a set of "normal random variables", my feeling is that such a possibility will hold only when the deviation from "randomness" is small, so that a linear approximation is possible. This is what happens in the plasma theory.

As for my own personal life, I am now quite busy writing all sorts of articles, etc., giving talks on causality and preparing courses. At the moment, however, I am not feeling well, because of lack of sleep and the effects of something I ate, so I'll end the letter now. Just one more thing, however, I was waiting for an answer to 3 letters, so it was your business to "breakdown" and reply.

Love
Dave

Chapter 27
Letters to Miriam Yevick, 1953, Part 2

Letter 99. Folder C122, dated: April 2, 1953.

Universidade de São Paulo
Faculdade de Filosofia, Ciências e Letras
Caixa Postal: 8105
São Paulo

Dearest Miriam

I received two long and two short letters from you (The latest dated Mar 23). I find it hard to follow what you say about the distribution of particles, so I shall take your advice and wait until you have a fairly definite treatment before commenting further.

As far as George is concerned, I never got his letter about his coming to Brazil. I shall look into the question of getting him money. Only I am afraid that my ability to get money here is approaching exhaustion, partly because I have already asked for a lot, and partly because they are running out of money because of the approaching economic chaos in Brazil. My plan is to ask for 25,000 cruzeiros to cover the expenses of both of you for a two months stay in Brazil. If only George should come, then only half of the money would be needed. But I wouldn't count on this [unreadable]. I have my doubts that they will have the money to pay Vigier. In any case, I haven't heard from Vigier, so he is probably having trouble getting money on the French side. I hope that both of you can come, and will write George as soon as more definite arrangements can be made.

Right now, I am taking a few days rest at São José dos Campos. My apartment has become unliveable because they are constructing another ap't across the street, starting 6AM every morning including Sundays. São Paulo is crazy with ap't building. There are 3 going up on every block. I have been trying without success to find a quiet place at a reasonable distance from the university. Meanwhile, I am slowly being killed by lack of sleep. It really takes everything away from you.

C. Talbot (ed.), *David Bohm: Causality and Chance, Letters to Three Women*,
DOI 10.1007/978-3-319-55492-1_27

Brazil is going to pieces. Prices have doubled [unreadable] $\frac{1}{2}$ the workers of São Paulo are on strike with brutal repression by the police, and everyone is talking openly of [unreadable] dictatorship. (With the devaluation of the Cruzeiro. I am no longer very rich, as my salary is $400 a month at the free market rate, perhaps $600 a month in purchasing power). And prices are still going up every day. I am awaiting a collapse here. Meanwhile the situation in the physics dep't. is becoming impossible. One of the professors is slowly going crazy and spends all the time poisoning the atmosphere with intrigue, another is a dirty rat and a thief of the lowest type. [unreadable] who has just returned from Europe is a Communist but the strangest type of Communist I ever imagined. He really would pass better as a sort of Jewish businessman. He engages in all the dirty intrigues here in a very short-sighted way, and in physics, he is a pure formalist and idealist, admiring Pauli as the ideal in [a] theoretical physicist. All this goes under the name of the purest dialectical materialism. (His name is Mario Schonberg, in case you want to know). There had been some hope of straightening out things here in the physics dep't, but with his return, the situation looks dark. The only hope is that he may soon leave, as he doesn't like Brazil, and would prefer to live in Europe.

I received an offer from Australia about 6 months ago, which I refused. But I am beginning to seriously reconsider, especially if they can give me some kind of passport. Because I am afraid that things will soon become impossible here. There are innumerable other things which look bad – for example that the physics group in Rio seem already to be under the control of the U.S. Embassy and the Atomic Energy Commission. (Please don't spread this as no good would be achieved by so doing). The gov't is so completely corrupt and short-sighted that one can only look on with amazement. The middle class is really totally demoralized, and lives from one day to the next. In the Northeast of Brazil, there has been drought for many years, and people are starving, trying to leave, and living in tremendous misery. Even in S.Paulo, the average workingman can no longer earn enough to avoid hunger. Profiteers are making 200% profit on foods. Rice, for example, is three times as expensive in the cities as in the farming regions a few hundred miles away. Yet no one knows who is responsible or what can be done. It is just utter chaos. Near the [unreadable] there is as yet no [unreadable] movement among the working people, although the beginnings of such a movement are barely visible in São Paulo. In the North East it seems to have spread quite far, however. In São Paulo and Rio, the Communist Party, for example seems to be made mostly of middle class Jews like Schonberg, who have just as much a cynical disregard for the people as do the rest of the middle class. They say, however, that it is different in the Northeast, which is, in a sense, the real Brazil, São Paulo & Rio being mostly populated by foreigners and immigrants.

The Jews of Brazil are a much more unpleasant lot than those in the US. Here, they openly assert their superiority to the so called "monkeys" who inhabit this unhappy land. Even the leftists often refer to the Brazilians as "monkeys". Meanwhile, they make enormous profits and live at a level that they could not achieve anywhere else. They do not assimilate, but remain a closed community, with the narrow short-sighted point of view, reminiscent of the Middle Ages. All they want is to "live well" and to see that their children marry and "live well".

To come back to more pleasant subjects, I agree with your self-analysis that in some of your traits you take after you mother. She somehow typifies the "Jewish Woman" to me, kind hearted and considerate in her own narrow spheres, but completely short-sighted and insensitive to anything outside of it, judging only by surface appearances, and tending to act as if what she did not already know could not possibly be of importance. Please don't get offended – I may be completely misjudging your mother. In any case, I don't want to imply that you yourself have all of these faults. But, for example, I had the impression that your first reaction to my suggestion about an infinity of levels was the sort of thing that might have been expected from your mother if she could understand the problem. Of course, in you, the traits of your mother are an inconsistency, rather than the fundamental characteristics.

Incidentally, I was very thrilled to see that you have begun to really "catch on" to the need for the concept of the limitless number of levels. I was very frustrated by your first reaction to this idea. Could you please explain to me in detail why you found it implausible at first, and what made you change your ideas. This is important to me, as I need to learn how to present the idea with the maximum [missing word(s)]

I detect a note of surrender in your attitude toward math. It is all very well to objectively recognise your own limitations. But you seem to be going too far. Are you sure that in your new humility, there is not an element of "delicious passivity"; i.e. a desire to let things just happen (in math, that is) without the need for struggle on your part? Actually, you do have a lot of ability in math, and with an orientation toward real problems you are in a position to supply something that most modern mathematicians lack, even though by tremendous concentration they may get ahead of you in the purely formal aspects of the subject. However, it is also true as you say, that we have to get closer to the people, and to try to give them hope. As yet, we do not have any definite plans, but the mere idea that something better is possible is important; for without it, people become cynical and hopeless. It is necessary somehow to integrate our scientific work with this other problem. For the hope of humanity lies basically in science, which can liberate people from slavery to bare necessity. But without the right kind of society, this will not happen. Thus, people must be made to appreciate the new possibilities opened up by science, and with this, the necessity of social changes to bring these new possibilities into being. Without science, social change would be useless and meaningless. With science, it opens up limitless possibilities for fruitful purposeful, and even "adventurous living", which goes beyond the hum-drum level to which most people of all classes are now condemned. We should stress not only that people will be more comfortable, but that this new level of material comfort and security will be the foundation for a new struggle of humanity with nature, in which the life of the everyone will take on purpose and meaning. We should not simply imagine socialism as a way by which everyone can obtain the meaningless "good living" of the bourgeoisie. For such a society would be doomed to decay before it started. In addition, a person who aims only at "good living" for all will easily sell out to anyone who promises the same thing in an easy way. It is important to stress that "good living" is only a means and not an end in itself. In this way, one may hope to capture the kind of enthusiasm that is now misdirected into religious ends, a kind of enthusiasm that can carry people through Hell if necessary to attain the ultimate goal. In this connection, the concept of nature as limitless plays an important role; so that you may not have been entirely wrong in comparing it to a "religion". What offended me was the supercilious way in which you made the comparison. As you know, I have always felt the feelings connected with religion are valid and necessary. It will be a sad day indeed when people cease to look at the world with a certain child-like wonder, and cease to see that there are still unresolved mysteries. The trouble is that under religion, this sense is twisted because the wonder and mystery are projected onto an unreal supernatural world beyond human control. The exciting prospect opened up by science is just that these things are in this world, and subject to our understanding and control, provided that we work together in a rational way. Thus, the sense of the infinite which people projected onto God is in reality a faint foreboding of the powers that we feel can be developed in ourselves. We can feel these new things trying to come into being in us. But religion takes away from us the courage to face our own potentialities. Science, properly understood, helps give us that courage.[1] In this sense, it is necessary in a properly organized society that science become an integral part of most people, at least in a general way. For the society will depend on science not only for its immediate sustenance, but also for its long range goals, which help nourish what we usually call the "spirit". A society in which everyone depends on science, and in which most people pay no attention to it, but regard it as a form of magic carried out by a small number of "wizards" or "geniuses" will also be doomed to fail. In this sense, I would agree with George that the next society must be scientific. But I do not believe that the scientists of today are capable of bringing this society into being. The most that they can do is to transmit to the most advanced sections of the people in general the true character of science, and thus not only counteract the idealist distortions of science which make it seem to be the enemy of the people, but even more, to show how science can belong to everyone (both directly and ideologically) instead of just to a small esoteric group of "geniuses" who are looked on with suspicion by everyone else. In this connection, it is necessary to develop science (and with it, math) in a rational way that can be transmitted to the people. For in its present form, science seems more like a lot of hocus-pocus, than like something that can be understood by everyone, and that even more can become a part of the being of everyone.

In this regard, to deny the infinite would be to leave this sphere for religion. For everybody senses that something corresponding to the infinite exists, something which can never quite be captured. To restrict science to a finite domain invites a sort of division of labor, in which science takes care of the hum-drum aspects of the workaday world, while religion takes care of the things of the "spirit", which escape the power of science, which deals only with the finite and the limited. We must recognize that the infinite exists in Nature, and that it can be dealt with by science. In doing this, we bring science to a higher level, for the theories begin to resemble "real life" a bit more. In this regard, the infinity of levels is only a beginning. For

[1]Science positivistically understood, denies these possibilities and thus plays the same role as religion.

I think that the limitless character of Nature will manifest itself to us in time in an unlimited number of ways.

To change the subject a bit, thanks a lot for the books that you sent me. "Man and his Gods" was very nice and "Woman the Second Sex" is interesting + instructive, even if perhaps distorted in some ways. The distortion is a matter of emphasis. First there is not enough stress on the fact that men and women are part of a society which forms both. This is stated, to be sure, but then most of the time, the world of women is treated in great detail, with man a being from outside, who interferes with her world, controls her etc., but who himself is not controlled by factors originating in this world of women. In other words, I think that such a book is basically wrong in concept. The correct book to write would be "Man and Woman, The First and Second Sex in a bourgeois Society".

Finally, I would like to mention an idea about defining chaos and randomness more precisely. You may recall that I regard randomness as "potential chaos"; for randomness refers only to an abstract infinite sequence, while chaos refers to a finite sequence. Now one way to get a "random" sequence is to take an irrational number x ($x < 1$) and multiply it by a series of integers, n. Then the sequence $s_n = nx - \lfloor nx \rfloor$ (where $\lfloor nx \rfloor$ is the largest integer $< nx$) is random; for if x is irrational, it can be shown that in the long run, all values of s_n between 0 and 1 occur with equal relative frequency.

Now, the question of the irrationality of x is connected with the question of whether there is a regular pattern or order in the expression of the number x, in a series of decimals (for example to the base 2). If we take an irrational number, then the various terms in the decimal have no order, so that as the number of terms n in the decimal approaches infinity, we should get a "random" distribution.

Now consider a decimal carried to a finite number, N, of terms. We would like a measure of chaos in the terms in the <u>finite</u> decimal. To the base 2, a typical number would be, for example

$$.1011010100$$

We would like to get a measure of the degree of chaos in the distribution of zeros and ones. I propose the following idea: Suppose that the function after N terms are repeated indefinitely as a periodic series, e.g.

$$.1011010\ 1011010\ 1011010\ 1011010\ \ldots\ldots$$

This series will give some <u>rational</u> number, y. Let us express y in terms of the lowest common denominator, n

$$y = m/n$$

where m and n are integers.

Then the size of the integer n would be a measure of the degree of chaos. For it is clear on intuitive grounds that a very regular distribution

.1111111

would lead to a rational function having a small n, while a complicated + irregular decimal such as .101101 would lead to a function having a larger value of n.

Could you comment on this idea and perhaps help me carry it further?

Love
Dave

P.S. Could you please explain to me Kac's ideas on Feynman's theory?

Letter 100. Folder C122, not dated.

[refers to March 30, but allow for delay in mail. The long letter referred to could be 99.]

Number on photocopy: 46

Dearest Miriam

I just received your letter of Mar. 30. I must say that I was very happy to be so thoroughly vindicated. I wrote you a long letter yesterday, and now I just want to add a few points.

(1) The confusion about probability is in my opinion a deeper and more pernicious thing than the confusion about the quantum theory because it spreads itself through so many different fields, including social science & statistics, as well as the natural sciences. It will be much harder to combat, because so much complicated math. has been written on the subject. Every time you say something, you will be referred for example to an article by Dr W.or Prof Z., who showed that by means of a super Hilbert space having special non associative properties, whatever you may be talking about has been successfully treated in the usual point of view. For this reason, a few mathematicians who have gone thru this stuff in detail and can say with authority "No, it just complicates the problem but doesn't solve it" will be able to play a very important role in this struggle. For example, in discussing stuff with Schönberg, I always have this problem. He has read a fantastic quantity of stuff of this kind without digesting it; but whatever you do, he throws it at you and leaves you to disentangle yourself, as if from a sticky mass of tar so that no time is left to really get down to the problem.

(2) I am not too surprised at the opposition to an ∞ of levels. Remember that you yourself were opposed to it at first! But young people find it a very attractive point of view, and it therefore has a good chance of spreading. If we can get some articles or books out on the subject, a big step forward can be made. For the usual point of view is pretty repulsive after you analyze it, from the standpoint of a better pt. of view, like the ∞ of levels. The idea that "they can <u>think</u> of only a finite number of levels" is pure idealism. For behind it is the notion that their present conceptual scheme of

things determines all that can possibly exist. The essential property of matter though is that it is always full of new things, many of which are surprises, thus always demonstrating anew that we had not really fully captured it by our ideas. Thus, we must always assume that the world is full of things that we don't yet know how to think about objectively and independently of our being aware of them. (Incidentally, you might ask these people how they manage to think of an infinite series. The ∞ of levels is basically no harder – it is defined logically even though the definition is never complete and explicit, just like a series that has no definite limit). But people have always found new ideas tough going. [Crossed out paragraph on an example of a chaotic process]

I changed my mind about discussing chaos in more detail right now. Will write more later.

Love
Dave

Letter 101. Folder C122, dated: April 21, 1953.

Number on photocopy: 47

Universidade de São Paulo
Faculdade de Filosofia, Ciências e Letras
Caixa Postal: 8105
São Paulo

Dearest Miriam

I just received your letter of April 13 yesterday. I am anxiously awaiting your "book" of which you sent me the table of contents.

I have sent a letter to George stating that we can offer to cover the cost of the stay of either him or both of you for 2 months. If he comes by himself, I can pay 15,000 cruzeiros; if both of you come 26,000 cruzeiros (for 2 months). This is on the theory that two can live more cheaply than one. We can get you a tolerable double room for not over 1600 cruzeiros a month. Food will cost you about 30 cruzeiros a meal (for each), 150 cruzeiros a day will cover food including breakfast, for the both of you. Thus, you need to spend 9000 cruzeiros on food, 3200 on a room, leaving you a surplus of 14,000 cruzeiros for travelling in Brazil, etc. You can buy dollars at about 45 cruzeiros to the dollar. Thus, you may find that you have a few hundred dollars left over, to cover part of the cost of your travel. Please tell George to write me soon, so that I can make final arrangements.

What you say about your mother is true. I probably did not judge her fairly. She is much better than the average bourgeois.

About the ∞ of levels and "religion", I want to amplify what I said in my previous letter. At present, there is a tendency to "share the work" between science & religion, science treating of the everyday affairs of the "material" world while religion treats of the intimate and more "interesting" affairs of the "spiritual" world. But there is a third class of people, which includes most scientists and many intellectuals, which prides itself on being so "hard-headed" that it does not need religion, and feels that it is basing its ideas only on solidly grounded facts and reasoning in science. This class also feels all of these vague things commonly referred to as "spiritual", but tends to relegate them to a private and individual world, a world of poetry perhaps, but a world having nothing to do with objective reality and with the prosaic material world. Such a trend is especially noticeable in the literature and poetry and art of the time, which is full of a very personal sort of symbolism, which represents the inner feelings of the individual or at most of a small esoteric group of such individuals. Thus, the world is split into the prosaic material world of everyday life and a magic spiritual world of purely personal fantasy. The latter is supposed to have no significance apart from the secret life of the individual himself. In my opinion, this is a distortion, having dual roots; first, in a distorted view of nature and objective reality, as being identical with the cold, dull, prosaic, and limited view of it which has been developed, and second, in a distorted view of our own fantasies and "spiritual" aspirations as belonging only to us as private individuals, neglecting their origin in our material structure, in our connections with society, and in our vague presentiments of what we might come to be under better conditions. This dualism is characteristic of the modern intellectual, who has developed a positivist view on life. For on the one hand, he likes to pride himself on how hard-headed he is, how he bases everything on observed facts and logic, and on the other hand, all of these other vague dissatisfactions, loneliness, sense of futility, longings and presentiments continue to surge up in his private thinking. To the latter, he ascribes the role of irrelevant personal fantasies, which he may indulge in according to "taste". In my opinion, this dualism must be eliminated. The concept of the ∞ of levels plays a role in this, for it shows that nature is not so cold dull & limited as we think of it. A similar concept applies to human nature. Moreover, we are led to think of our apparently private fantasies, longings, "spiritual" aspirations, etc. as a reflection of our material structure, of our true needs. At present, this reflection is distorted, but it nevertheless demonstrates indirectly a deeper level of reality that is trying to manifest at the level of conscious action and social organization. The current dualism leads people to ascribe their dissatisfaction with life to a mysterious struggle between the "id and the super-ego", between the "individual & society", etc., which is simply postulated to exist without further explanation. But the general concept of levels suggests that we should analyze how this struggle comes into being. Once we understand this, then we understand how to put an end to it. But to the average bourgeois, this struggle seems inexorable and unavoidable, so that he must confront life in a basically hopeless way. The general idea that everything in existence comes into being as a result of forces operating at other levels (in this case social) shows that we do not have to accept any properties of "human nature" as eternal, but instead, such properties are only conditional. When we change the conditions sufficiently, we can bring new properties into being. But in this regard, human nature is no different from Nature in general; for according to the ∞ of levels, <u>all</u> properties can be altered with sufficient changes in conditions. Thus, the ∞ of levels is an integral part of a better view of Nature in general, and of human nature in particular.

In this regard, religious feelings play an important role. These feelings are now directed at the supernatural. But such feelings themselves are natural and as experience shows they are an integral part of what it means to be a human being. For everyone wants to feel connected with something that goes beyond his immediate personal life. Such a desire is healthy, and reflects the fact that we can already feel within us the vague indications as to what is possible both in humanity and in nature. But to connect these things with the supernatural is to disconnect them from reality and to make them ineffective. It is necessary that the enormous forces which religious types of feelings can call forth be harnessed to real social problems.

In this way, both society and the individual will benefit. But to do this, a new philosophical view is needed. The view must be based firmly and rigorously in logic and in facts. But like a tree that is nourished from the soil, but becomes something different from the soil, this philosophy should not restrict itself to the cold, hard, prosaic facts which make it resistant and strong, but it should go beyond and connect up with the realm of the emotions, with the "spiritual" aspirations of humanity, with those vague moods and presentiments which cannot yet be defined, but which reflect the real humanity that is just beginning to come into being in each of us.

The object of a pure "religious" feeling should not be "self-denial" but rather self-fulfilment through work on common social problems that go beyond the individual. This is the trouble with most of the Christian religion, which creates a great chasm between the "self" and "God". Thus, the individual can "save" himself only by negating himself and devoting himself to "God". But this is a distorted view. In reality, the contradiction is between the "self" that has been developed in our insane society, and the "self" that can come into being, and that has come into being, if the individual is not to deny his most fundamental nature. But the self that has to come into being is one which identifies its interests with those of society. In our present distorted view, however, we think of the "self" as an eternal, given, thing which therefore cannot be transformed in a rational way, but must be destroyed, and replaced by a mystical new personality that is suffused with the "Grace" of God. Thus, the religious person feels that he is sacrificing his will to the "will of God". In reality, he is setting aside the most human part of himself, and viewing this part fantastically as another being – i.e. – God. Thus, when he does what is good for him, and for society together, he looks at it in a distorted way as a "sacrifice" of himself to an external being. In reality, he is transforming his "self" and a new "self" is coming into being, while simultaneously, the old "self" is going out of being. But this takes time and is never perfect. Thus, the new "self" is continually struggling with the old "self". Actually there is only one "self" of course, but we may analyze this self abstractly in terms of two struggling trends, one representing the past, the other representing the new personality that is coming into being. In reality, the true self is a result of the union of these two opposing trends, and it is this opposition which generates the change. But the change has roots not only inside the individual, but also in society. For the needs that bring about these changes, as well as the concepts which help direct them, generally originate in a large measure in our social relations.

All of this is by way of raising questions that may be important in the "book". I hope that both of you can come, so that we can have many interesting discussions. It appears that Vigier will be here also, early in June.

Love
Dave

Letter 102. Folder C122, not dated.

[refers to April 15 letter so placed after 101 which refers to April 13. Also must follow 99 on Schonberg.]

Universidade de São Paulo
Faculdade de Filosofia, Ciências e Letras
Caixa Postal: 8105
São Paulo

Dearest Miriam

Thanks for your letter of April 15. Your statements about how to "convert" people to ∞ of levels will be very helpful. What we really need then is a critical analysis of the weak points of existing ideas, which people tend to cover up, because they are afraid that they will discover that what they are doing has no solid foundation. This attitude is clearly taken over from the way one is taught to look at social problems. For here too one is supposed just to work in one's own field and pay no attention to the general foundations – e.g., we must regard capitalism as "given" once and for all, similarly with regard to "human nature" "masculinity and femininity", etc. Of course, the entire attitude is irrational, for wherever there are weaknesses (in science, society or anywhere else) it is to our profit to discover them and correct them. For the weaknesses will be there whether we are aware of them or not, and will some day destroy us unless we correct them first. The current positivist attitude is based on the idea that "what we don't know can't hurt us." But clearly it can!

I have been thinking about your friend, Mr [unrecognisable name], of the subjective probabilities. He said that statistical mechanical probabilities are the result of various subjective estimates of physicists, based on the Liouville theorem – quasi-ergodic conjecture, etc. I have just thought of an extension of this idea. As we all know, according to qu. mech., all laws of physics are basically probabilistic, and stem from Schrödinger's equ. + its probability interpretation. Even Newton's laws of motion at the classical level are merely statements of very high probability. Now, clearly the estimate of probability is subjective, being based on the fact that physicists are thinking of Schrödinger's equ. and Born's probability interpretation. In fact, the present laws of probability are more or less the result of an informal agreement among physicists. Thus, the prediction that it is very probable that an apple released from a tree will fall downward is only a subjective statement. If only all of the physicists would think differently, it might become very probable that it would fall upward, or sideward. Since there may be disagreement among physicists on this question, I suggest that the question be decided by a democratic vote. This is really the only appropriate method in the Free World. Of course, behind the Iron Curtain, it would have to be decided by orders from the Central Committee. In this case, it is quite possible that in the Free World, apples would fall to the right, while among the enslaved people of the East, they might fall to the left. Really, I don't think that physicists realise how much power they could have if only they liberated themselves from outworn metaphysical notions such as objective reality. The possibilities of modern physics go far beyond those conceived of by the ancient magicians and soothsayers.

Now about Schonberg. It may be true, as you suggest that I am a bit overhostile to him. But there are objective factors which make it difficult for me to feel free with him. First of all, there is his very formalistic point of view in science, which is in direct contradiction to the implications of his political position. He practically worships Pauli, for example, and thinks that Dyson is the greatest of the younger theoretical physicists, being far ahead of Feynman. He knows that Pauli is completely idealistic in his point of view in physics, and yet he not only does not resist Pauli's influence, but instead tries to spread it. His inconsistency between his scientific + his political philosophy is matched only by a similar inconsistency between his tactics in the physics Dep't and his political philosophy. From the first, he has supported the two stinkers who have been destroying the Dep't, both of whom are scientifically incompetent, and rats otherwise. I know this personally as I have been the subject of some red-baiting from them. Schonberg knows it too, but does not seem to think it important. Moreover, I know that Schonberg has cooperated with these rats in the past, in some very dirty and unprincipled deals. He seems to think that because he holds a Marxist philosophy, he has no moral obligation to anyone. Moreover, he has a tendency to surround himself with very shady characters, such as this man Ochialini, whom you mentioned in your letter. Ochialini was in São Paulo for several years, and from what I can hear, he really wreaked havoc. He is a completely Machiavellian character, and in my opinion, only partially sane. For example, when I first met him, we began to discuss the troubles in the Dep't. His suggestion was "Why don't you go to Adharas de Barras (a notorious demagogue and thief) and denounce them all as Communists". This friend of Mario Schonberg illustrated his Machiavellian twist of mind, when he warned me against rumors. He said that the trouble with Brazil is that there are too many wild rumors. You just can't trust rumours, he said. For example, there is a rumor that a certain assistant of mine is a homosexual. Of course, this is false, he said. It is true that he is a bit "delicate" but certainly not a homosexual. This just shows you how you must be careful with rumors, . . . etc., etc. Another thing that our friend Ochialini has done for us in his visit to Brazil is to go to the Italian Colony of São Paulo and to try to dissuade them from building a physics building that they had been planning to donate to us. This is only what I know of Ochialini from direct contact. The stories I hear from other people re-inforce the same impressions, but contain some really dirty things.

So in sum, I find Schonberg a difficult person to like. I shall certainly find a "modus vivendi" but personally he is such a colloidal character that I cannot really feel at ease with him.

Now about what you say about [unclear name] Kaufmann. This only confirms my own vague impressions. I have never felt that she and I had anything really to talk about, and for this reason I never said much to her. Usually, I get a sense with certain people that they are emotionally "dead", i.e., they have accepted things as they are, and are satisfied to "just live". With such people, I feel an inability to exchange ideas – or more accurately a lack of desire to do so. Generally, my reaction to them is extreme boredom. I feel that Hal Lewis is entering the same category, but that when I left, this was still largely superficial, since it was still possible to discuss many things with him.

Of course, [unclear]'s ideas are ridiculous, even childish. She seems to be more interested in avoiding responsibility for suffering than in helping people to eliminate suffering. As far as she is concerned, 2 billion people can live for centuries in misery and semi-slavery, but she will do nothing until these people are ready in a neat orderly and peaceful way to take over their factories. Of course, you are right in pointing out the impossibility and utopian character of such a dream. Even if it were possible, it would leave society basically unchanged, so that at best, workers would be turned into little businessmen. But in reality, suppose that every factory was owned by its workers. We all know, however, that all factories are interdependent. Thus the car factories need steel, the steel factories need coal + machinery. A well-balanced mechanism of exchange + supply needs to be worked out. Who will do it? Whoever does it will inevitably gain tremendous power. How is his power to be controlled, in this disjointed world of little businessmen, whose only pre-occupations will be presumably to return to their modern little houses and to cultivate their little gardens, the more advanced among them perhaps to do a bit of mathematics? Society needs to be organized in such a way that these problems are faced explicitly from the start. And the only way by which such tremendous power can be set up without disaster is to change the nature of human beings in such a way that each person identifies his own interests with those of humanity as a whole. In this way, we could conquer the tendency of each person to aggrandise himself for a person would see no advantage in doing so, since he would be much happier by forgetting himself and working for the common good.

This brings us to the subject of human nature, both masculine and feminine. I am <u>not</u> in complete agreement with some of the things that you say and imply about femininity. You state that a woman's strongest emotional need is to "give herself to a man". For example, Sonia Gross wants to do Eugene's integrals for him, etc. The error in what you say is in attributing to women <u>only</u> a characteristic that belongs to all of humanity (including men). I doubt that there is an man alive who would not like to encounter a superior being to whom he could devote himself. The fanatical devotion of an army to a good leader is a case in point. The devotion of the masses to leaders like Hitler + Stalin is another (I don't want to imply that the two are equal, but only that there was a similar mass devotion in both cases). Now many writers, noting this devotion, have said that the masses are "feminine". But this is very wrong. One could equally erroneously say that women are "collective". The fact is that all humanity has a desire to find something above itself to which it can devote itself. This desire is in itself good. But the unintelligent choice of object of devotion can lead to disaster. Now if we imagine that we have a suitable object of devotion, a tremendous weight of responsibility falls off your shoulders. We feel that our life is so to speak "validated", our future path is certain, and all that remains is for us faithfully to serve the worshipped being. Now in the case of man, the opportunity for this sort of thing is rather limited under the present social system, for men are not encouraged to worship each other. At best, they may find a leader to whom they can attribute these magical powers, but this is rare, and involves at most a limited phase of life. But women are given every encouragement to worship and serve a man (or men) in every phase of life. Considering the great barriers in the way of their acting independently, the temptation is almost irresistible to give in, and to try to obtain a vicarious contact with life by subordinating themselves to the man, serving him worshipfully. There are however several reasons why this doesn't work very well. First of all, no man (or woman either) is worthy of this kind of worship and devotion. The woman will have to try to fool herself into thinking that he is, but after a little married life, she will quickly see that he isn't. This leads to serious troubles. Secondly no woman (or man) really wants to give up independent action. The implicit meaning underlying devotion is that if (A) devotes herself (or himself) to the more powerful being (B), she (or he) establishes a claim on this being which will result in the effective control of B by A (at least in part). Thus B will end up carrying out the will of A. This is roughly what is in most people's minds when they devote themselves to a God, and it is in most women's minds when they devote themselves to a man. But in reality, B usually pays no attention to A's devotion, and carries out his own plans, thus making A very unhappy. In addition, this kind of devotion is bad for B, since it makes him too conceited, and he forgets his human limitations, only to be called to account later by the resulting mistakes, which suddenly bring him back to the human level in a humiliating way.

In sum then, the kind of devotion that women think they want to give to men does not really satisfy anybody for very long, either men or women. It is really a form of play-acting, which comes out most strongly in sexual intercourse. The woman wants the man to dominate her symbolically. That is, for a few minutes, the woman wants to represent, not herself, but Woman in general, while her partner is to represent Man in general, the virile aggressive masculine element to which she must submit herself as the weak submissive feminine element. But this type of play acting doesn't satisfy anyone for very long. The woman doesn't really want to be dominated, but rather to dominate her lover by her feminine charm and submissiveness. The man senses this vaguely, and also realises within himself that he is no knight in armor, but just another human being. If he is not very perceptive, he may succeed, like a child, in imagining for 10 minutes that he is a great conqueror. But after it is over he will certainly lose interest in the whole artificial drama, which is a sort of ceremony dictated by society to represent the relative roles which it has assigned to man and woman.

In reality, every person has a healthy desire to subordinate himself to something bigger than he is. As far as I can see, the only thing worthy of such submission is the future of humanity as a whole, in its struggle with nature, and with disruptive trends within humanity itself. But such a goal does not permit one to shed responsibility. For this future is not yet in existence. It must be brought into being by our own efforts. Thus, instead of permitting us to shed responsibility, it requires us to enhance our responsibility, and to take the initiative. For this reason, it permits a healthy synthesis of our two most basic needs. (1) To take independent aggressive action and to struggle. (2) To devote ourselves to something beyond ourselves. For in this way, we are by our own actions, always in the process of creating that to which we are devoting ourselves, and vice versa, what we are creating gives us a goal which gives meaning to our lives.

The various forms of symbolic submission and domination implied in the current sexual practices in my opinion serve an important social purpose. They help deviate humanity from its natural goal, and channel certain needs to subordinate oneself to something bigger into an illusory sexual outlet. It is for this reason that our civilisation has been so preoccupied with sex and in such a confused way. For sex has gotten so mixed up with the process of oppression of man by man (generically speaking) that its real meaning has by now been lost.

Now there are clear differences between men + women. But I do not think that they have anything to do with "masculinity" and "femininity" as these are now understood. The visible differences are these: women are generally a bit smaller than men, not quite as strong physically, but more subject to emotional disturbances because of glandular fluctuations associated with menstruation, etc., and most important of all, they bear children. There is no evidence that any of these characteristics lead necessarily to what we usually call "femininity". All evidence is that "femininity" is a response conditioned by society. Even the pains in childbirth and menstruation are 90% the result of social conditioning which upsets the delicate functioning of the reproductive mechanism by all kinds of strains + worries, as well as by a lack of physical exercise which would build up its strength. (Proof – in primitive peoples, these things are much less severe). De Beauvoir shows in detail also how social attitudes towards menstruation as something dirty, dangerous, mysterious, help make it much more painful than it has to be. In a better society, and with some intelligent study of this problem, 90% of these troubles would disappear. The remaining 10% should be the subject of intensive research, aimed at stabilizing the chemical balance not only in women's blood, but also in men's, since men too suffer from similar problems of nervousness in lesser degree.

In most modern life, the difference in physical strength is not important. It could also be corrected in large measure if women exercised adequately. As for childbirth, this is different, since it implies that the woman must devote some time to her children, at least during the first year or two. As I said before, I think that in a proper society, and with some research, most of the pain + danger of childbirth could be eliminated. If we didn't have this insane competition, the loss of a few years from a career would not be a serious thing. In fact, even now, many women succeed in combining a career with raising children. Also, after the 1st year or so, the father can and should play an equal role in the raising of children with the mother. Finally, with properly organised nursery schools, collective restaurants, etc., much of the mechanics of running a home would be simplified to the point where it would be almost as much a relaxation from work as a job; especially if it were shared by father + mother. So as I see it, we must accept the consequences of certain objective physical differences between men and women; as for the rest, the evidence is that they are mostly of social origin. Small physical differences have been magnified by society into a tremendous chasm. Both men and women will benefit when they can "be themselves" and not the symbolic and mythical beings which society demands that they be. If after a thorough change of society and adequate research, some differences between men + women are seen to persist on the mental level, these should be recognised, and each sex given the chance to develop its own peculiar possibilities to the fullest. But I extremely doubt that either sex has a greater natural tendency to dominate or to be dominated.

So I think that you should resist this insidious tendency to submission, this traitorous desire which lies hidden underneath your feeling that women "naturally" want to "give herself" to a man. A woman wants to "give herself" but so does a man, and both must "give themselves" to a common end, which is worthy of such an action.

Love
Dave

P.S. On re-reading this letter, it occurred to me that I had not said what I thought of the role of sex in this problem. Now it is clear that sexual intercourse has as its origin the process of reproduction. But nothing is ever identical with that from which it has originated. It always has something new, characteristic of its own. A new feature of sexual intercourse is its tendency to unify and mobilise the entire personality, especially on an emotional level, thus enhancing and intensifying certain characteristics which may ordinarily be hidden, hardly visible, or even suppressed. But what is intensified or enhanced depends largely on social conditioning and past experience of the individuals in question. Thus, many people enhance, intensify, and symbolise in the sex act their desire for masculine domination (or the parallel desire for feminine submission). That this has nothing to do with the intrinsic structure of the people concerned is shown by the fact that perfectly normal males (physically) often take at one time the submissive role in homosexual relations, and the same person may also at another time take the aggressive role. (This also frequently happens in heterosexual relations). With such conditioning, sex is often the means by which inhibited tendencies toward cruelty and masochism are exhibited. The entire process of sexual excitation of the personality becomes connected with manifestations of domination, cruelty, and submissiveness. But another possible effect of the enhancement and intensification of the emotional processes occurring during sexual intercourse is to enhance feelings of generosity, love, happiness, well being, and good feelings, directed not only at the other person, but also toward humanity in general. This reaction is quite natural, but to develop to its full degree, it too needs some conditioning. We may as well recognise that conditioning of this kind is inevitable, so that to avoid bad conditioning, both society + the individual must consciously aim at good conditioning. With good conditioning, sexual relationships not only bring two people closer, but also help each to disclose to himself and the other certain facets of character, potentialities, etc. which are usually lost in the confusion of every day life. Once these have been exposed to view, this is an important step by which the individual transforms his character in accordance with the necessities of what has to come into being within him. The same role as that of sexual intercourse can be played by any intense experience which unites and mobilises the whole person, in such a way as to show him to himself and to others in a new light. In a future society, and effort should be made to multiply both the number, and the kind of such experiences, so that sexual intercourse would be only one among many things directed at the same goal.[2] But in the meantime, it will have to be normally one of the most important of such things, since life as it is now lived will very rarely afford us other experiences that can move us deeply.

Letter 103. Folder C126, dated: 1953.

[There is only the year on this letter and its date is not clear. I have placed it after 102 because of the "woman question" reference.]

Universidade de São Paulo
Faculdade de Filosofia, Ciências e Letras
Caixa Postal: 8105
São Paulo

Dearest Miriam

I just got your letter in which you reply to my "abstract" answer on the "woman question". I am afraid you misunderstand me completely if you think I don't enjoy "relaxing". I can even beat Lloyd Motz at it. Thus I sleep for a half hour every day after lunch, and I read for a half hour every day after breakfast. There is nothing I like to do better than to lie on a grassy hilltop and dream idly for hours while I listened to the wind blow and watch the clouds go by. The idea of planned amusement revolts me. The only reason that I didn't mention this need for relaxation in my previous letters is that I thought it was so obvious that it went without saying just like food, air and water. But to make relaxation a central goal in life would be just as mistaken as to make eating or breathing into such a goal. All of these things are not only essential, but quite enjoyable in themselves. But they acquire meaning only insofar as they fit into something larger. Nothing could be more futile than a life devoted solely to enjoyment and relaxation. In fact, these things become pleasant largely in the measure that there is something from which you are relaxing, and for which you are re-building your strength by relaxing. And this is what I dislike about the American (and bourgeois in general) stress on "good living". To me, there is nothing more boring and suffocating than pure unmitigated "good living" without a goal. The other aspect of the problem is that such things are costly in effort. To obtain them, most people have to struggle so hard that they have no time for anything else. Thus, the bourgeois divides his time into the work needed so that he can "live well" and the "good living" which he tries to enjoy, with incomplete success because he knows that it can continue to be possible only if he struggles in the never ending rat-race that produces no real results, beyond the day-to-day struggle and "good living".

---

[2] A few possibilities would be plays, dramas, collective singing + dancing and ceremonies, aimed at exalting the state of mind of each individual, and at making him feel his unity with the whole of humanity.

Aside from this, there is the long-run insecurity that follows because the economic and social system are so unstable that he does not know how long he can count on his "good living", especially for his children, if not for himself. Finally, the few people who do not have to devote their lives to this struggle to "live well" usually do so at the cost of the labor of others, purchased by inheritance of money, or by an easy job, which is a source of corruption, since if the person loses his easy job, his "good living" will be lost.

So I am not telling you not to "enjoy life". What I object to is to have to stress such a thing, instead of making it a by-product of a fuller life, as well as one of the conditions (like food and water) out of which a fuller life can develop. I think that this talk about enjoying life has some truth in it, but that in times like these especially, it can be very treacherous, just because it is so unrealistic. 99% of the people of the world are in no position to enjoy life at present, and the social upheavals which threaten because of this fact is making the enjoyment of life by the remaining 1% so insecure that for most of them, such enjoyment has already lost much of its real spice, and will lose still more in the future as the situation grows more chaotic. If you don't watch out very carefully, you may find that this desire to be a "little animal" for a while (which in better times might be quite harmless) could entrap you in such a way that there wouldn't be much enjoyment of life left, as it has done to the countless other people who are just a little less fortunate than you are. For example, if you got a few children, and if George lost his job, there would be little opportunity left to "feel like a little animal". I am not saying that you should not take these risks, but I am just trying to remind you that such risks exist.

As for enjoying life in Brazil, this is much harder than in the States. Even the newspapers are always saying how unpleasant life has become here. Everything is breaking down (transportation, electricity, water, etc.) the future is completely uncertain. In addition São Paulo & Rio are just plain very unpleasant cities to live at. There are no parks to speak of, and to get to a pleasant place requires taking overcrowded buses for hours, standing in the hot sun, etc., so that it is hardly worth it. When you get there, there is no safe place to eat and nor can safe drinking water be obtained. Then you must think of getting back. The city streets are nightmares of chaotic traffic moving at insane speeds amidst clouds of oily smoke from badly adjusted bus and truck engines. There is practically nothing to do here. On Sundays, most Brazilians stand in line for 2 hours to go to a mediocre movie. To buy a car would not help. There are few passable roads, and few places to buy food, etc., or to sleep over, where you can hope to go without coming down with diarrhoea that will make life a misery for a week or two. In the cities and on the few good roads that there are, traffic moves in such confusion and at such break-neck speed that it exhausts your nerves to cross the street on foot. To enter it in a car would be no relaxation at all, but just a nightmare. Also, you would have a good chance of having your car smashed, and being injured or killed.

As for me, I think this grim Brazilian life is killing me. Even in the States it was hard enough to find people with whom I could have any real contact. And in Brazil it is much harder, because Brazil is a primitive country with very few educated people, and most of these people, even those on the left, are pretty shallow. I don't think

Australia would be any better. It is even smaller than Brazil and even more primitive (culturally speaking). It is hard enough to find a woman here with whom I have any real common ground. It would probably be even harder in Australia, because from what I hear, women are kept down there even more than here. I don't really think it would be worth my while to go to Australia. I don't have the faintest desire to do so. If I can't get to Europe, I think it would be better for me eventually to return to the States, even taking into account the brutal repression that may happen there.

I am really beginning to be worn down by life in Brazil, as well as by the general situation. If I could get some satisfaction out of social relationships with the average sort of person, life would be better for me; but no matter how I try, I can't do it. I usually come out of a party or a discussion with nothing but a sense of extreme frustration. And the same goes for most of the girls I meet. My trouble is this. If I once start to take part in social life I get some mild enjoyment, or more accurately, a cessation of the annoyance which comes from being lonely. But it is such a mild pleasure (rather like that which comes when a dull tooth-ache stops) that it doesn't really give me a motive to make the effort needed to get it. I am capable of quite strong efforts when I feel that there is a worth-while goal. But it is very hard for me to do anything to obtain a "luke-warm" goal. Unfortunately, modern life affords to the individual only lukewarm goals. I would probably be better off if, like everybody else, I made a life and death struggle to obtain such goals, but it goes against my constitution so strongly that I cannot maintain such a struggle for a very long. And I wish to repeat that to obtain these typical luke-warm satisfactions that come from the available type of social life, quite a strong effort is needed. In my case, the effort is even stronger, because I must overcome a persistent feeling, which is by now very deeply rooted, that there is no real hope that I, as an individual, can get any real enjoyment in the present situation. Thus, the only goal left to me is the long range one, but this is of such a long-range that it leaves a big feeling of dissatisfaction in day to day life.

This brings me to the "woman question". I realised when I wrote my answer to you that it was abstract and general. But I had a particular purpose in mind. Your previous letters had indicated that you were at least toying with the idea that it is woman's "nature" to submit to the kind of life that the existing society requires of them. You didn't say this explicitly, but there was an undercurrent in which this idea was sort of half-implied. I merely wanted to remind you that the existing evidence is all against this idea. It is not "utopian" to analyse the role that women "ought to have" or will come to have in a better society. For a proper understanding of this role is needed to avoid, as far as possible, the traps which the present society prepares for everybody, but especially for women. So if you are to be trapped, at least you don't have to cooperate by imagining that you are merely fulfilling the role that Nature has decreed for you. It is true that such an analysis does not solve the immediate problem of an individual woman, but it helps provide a framework within which such a solution can more intelligently be sought.

As for the problem of individual happiness, my own idea (perhaps somewhat distorted) is that in the present disintegrating type of society, this is not a real possibility. It is a "will o' the wisp" which can only lead those who chase it to disaster.

This is, I feel, the trouble with all of us bourgeois. Instead of centering our lives on the needs of society, we keep on chasing personal happiness. Even in more stable times, this would be bad in the long run; but now, with everything coming down around our ears, such a search is just plain unrealistic. Of course, each individual should try, within the possibilities at his disposal, to be as happy as possible. But this should not be his central concern. It has always been a well-known truth that those who seek happiness with excessive intensity do not find it. To some extent, it must come from "losing yourself" in something bigger. In these days, this is especially cogent, because all of the usual props of personal happiness (home, family, children, careers, job, etc.) are threatened with destruction. One can only have secure happiness if one has the feeling of being connected with something that will survive the possible destruction of all of the conventional foundations of life. In the past, some people found this in religion. Today, however, religion can no longer serve, and we must connect ourselves up with something real, instead of with fantasies supposed to exist in another world.

Finally with regard to the "book". I think it would be interesting if you tried to write something. You may find it much harder than you ever imagined, however.

One more point. It would make your letters easier to read if you would put page 2 on the back of page 1, etc.

Love
Dave

P.S. My letter may be unusually gloomy, because I have a sort of cold, and feel weak and run-down.

Letter 104. Folder C122, dated: May 12, 1953.

Number on photocopy: 49

Universidade de São Paulo
Faculdade de Filosofia, Ciências e Letras
Caixa Postal: 8105
São Paulo

Dearest Miriam

I got your letter asking whether I want you to come to Brazil. The answer is, yes! I am very anxious to see you again. I was worried about our complicated 3-way relationship and for this reason didn't answer you for a while. But regardless of what happens, it would be good to see you again. I think we can raise the money a bit to 30,000 cruzeiros. It will cost the two of you about 10,000 cruzeiros a month for expenses. In 2 months, this will leave 10,000 cruzeiros left over, about $250, which is about $\frac{1}{3}$ the plane fare to Brazil and back.

What you say about people giving in to committees is disturbing. I read in the paper that Phil Morrison testified recently. What did he say and what is happening to him? Also what about Furry? What is happening to him?

I am making progress on the causal interpretation of the quantum theory and on the problem of the development of chaos. I am very interested in hearing about the progress that you have made. It would be a lot easier to talk about it if you came.

The new philosophical point of view is slowly falling into shape, everything fitting together very beautifully. We may even have a little scientific revolution on the side, with a bit of luck. But I am afraid that my first love is the philosophical problem and not the detailed dry scientific problems. Yet it is the latter which supply the means by which the philosophical point of view comes close to reality and demonstrates its fruitfulness.

As for correspondence with Eugene, I plan to start up any day now. It's hard to start up once it has died out.

Perhaps I am getting old, but I have noticed a gradual loss of interest in things sexual, not only a drop in what is usually called "desire", but also a drop in interest in general: Part of it is a sort of disillusionment, a feeling that it seems to be a rather childish thing to make such a big fuss about, but part maybe due to the operation of the curve which Kinsey reports for men. Fortunately, I still retain my physical strength and keenness of mind, so that I am not yet approaching senility. In fact, I have the feeling that one can do other things better when one is not disturbed by this desire that no one ever can really satisfy (However it may well be that such desires are only sleeping and not dead).

I am going to read Hopf's article soon. It does look as if he has ideas that are related to mine.

I have some interesting new ideas on the irreversibility of physical processes in time. It is hard to discuss them by mail at this early stage, but they throw light on the problem of how new things can come into being. I'll write them to you later (or better yet tell you about them when you arrive).

Love
Dave

P.S. I think you have gained an exaggerated idea about my intelligence, apparently partly in connection with my having solved the problem of the uniform distribution of angles in the case of a particle whose angle is always multiplied by K. Let me say that I had two unfair advantages over you:
(1) I was more familiar with the physical background of the problem, and thus was guided more efficiently into what we may call the essential character of the problem.
(2) I was not familiar with a whole host of theorems with which you were familiar. Thus, you tended to try to solve the problem by applying ready made theorems, instead of trying to understand its essential character. I could see you becoming bogged down in the categories created by the theorems that you tried to apply. It is an advantage not to be so familiar with so many theorems, which have arisen in the effort to solve other problems, usually much more abstract. It would be better still to absorb the essential method or idea behind the theorem and to keep it in reserve, to be used when an appropriate problem arises. But I am convinced that the theorems themselves are seldom likely to be of use in a new problem. Thus, I suggest that in your future work, you try to work out your own method of solving new problems,

even if you think or know that theorems exist that will do the job. This will not only give you practice in analyzing a real problem, but it may also lead to superior methods, which have grown out of the problem at hand, rather than having been grafted on to the problem from outside.

As for people like Doob + Co., it is obvious that they are very intelligent and capable. Only I think that they would get a lot further if they were a little less conceited, so that they could, without condescension, work on problems a bit closer to reality.

Chapter 28
Letters to Miriam Yevick, 1953, Part 3

Letter 105. Folder C123, dated: Received June 15, '53.

Universidade de São Paulo
Faculdade de Filosofia, Ciências e Letras
Caixa Postal: 8105
São Paulo

Dearest Miriam

Thanks a lot for your two letters. I have been feeling kind of tired lately and didn't answer you until now. I seem to have some kind of persistent infection that resembles a cold, but just won't quite leave me, so I feel low most of the time. Combined with the grayness of life here, this gives me a sort of feeling of approaching old age, which is I hope premature. Part of it is surely psychological, as I no longer feel that it is likely that I personally will be able to arrange a happy life, and the concept of a happy life for the next generation is perhaps a little too vicarious a form of satisfaction to get me out of the depressed state of tiredness in which I find myself. Oh well – I guess I will recover in time.

I have the feeling that both of us are being a bit silly in our discussion of the "woman question". As for me, I am the last person to want women to be identical with men. I believe that there is, and should be, a considerable difference between the two sexes. Only in the present society, the real differences are so hidden by irrelevant stereotypes that it is hard to know exactly what the real difference is. In any case, the current stereotype of "femininity" leaves me cold. For I am not interested in a creature whose main virtue is supposed to be passivity, dependency, daintiness, in other words, a sort of counter-foil to man's strength, independence, aggressiveness and roughness. I don't believe that any woman really finds satisfaction in this role. At best, she may enjoy using this role to gain control of men. But I don't want to be controlled in this way, although I sometimes feel that a man, in the present society, seems to have a certain moral obligation to pretend that he is being dominated by the

C. Talbot (ed.), *David Bohm: Causality and Chance, Letters to Three Women*,
DOI 10.1007/978-3-319-55492-1_28

"wiles" of a woman, in order to prevent the woman's dissatisfaction with life from becoming too great. Thus, very often, perfectly capable men develop a complete incapacity to deal with simple every-day problems, and leave such details to their wives. This is not only convenient for the men, but serves to give the women a purpose in life. For the woman likes to feel that basically, this man, he is a child [words missing] helpless in life as a [words missing] real contribution, by doing things that her man could [words missing] sometimes wonder if the man doesn't sense that family life will be a lot smoother if he develops this helplessness which serves, in part, to validate the woman's life.

The fact that we have nothing to argue about is shown when you say that you cannot really accept the current role of women in society. Of course, you want to enjoy what you call your "animal" tendency but under present conditions, you realise that you have to be a bit careful with such things, as the present society uses them as traps. I don't say that you can't enjoy such things, but I do say that many people who have tried it have paid much more than the enjoyment was really worth.

My own idea of the problem is that a woman is, in some respects, sure to have a personality different from a man. But this difference has no particular tendency to be more passive than a man. Indeed, after sexual intercourse a man is generally more exhausted than a woman. Thus, if it is true, as you say, that frequent satisfactory intercourse leaves a great many women in a passive satisfied animal-like state, it should do so even more to the men. The fact that this doesn't happen is merely a sign that socially conditioned attitudes in the man are continually driving [word missing] back toward aggressiveness. On top of this, we must add the form of conquest which many men have in intercourse.[1] This feeling is socially conditioned. Such a feeling is exhilarating and tends to release energies of all kinds. But the opposite side of the coin [missing] the man can feel that he has conquered only to the extent that the woman shows herself as having been conquered. But [missing word] will react on her psychology, and make her less aggressive, less [missing] etc., more satisfied to let the man take the initiative.

Now what I feel is that a woman should energetically [missing] aggressively be herself. For example, there is, for various reasons, some [missing word] in the personalities of men & women, some difference in orientation, in the [missing] which each is sensitive. This difference may originate in part in the physical structure and in part in the different relation to society (which will always have to be somewhat different). Thus, the two sexes can, to some extent, complement each other. I feel that a part of the attractiveness of a woman for a man is that she is somewhat different, and there is a feeling that in this difference, one may find something that has thus far been lacking. Perhaps women feel the same way about men. But I feel that the difference must be in qualities which are expressed energetically, and with a full force of the person's character; and should not be expressed as a passivity. For insofar as a woman is passive, she becomes a mere reflection of the man's desire, and is therefore, not a person who has fascinating similarities and differences; but instead just a woman who

[1] Which explains why conquering armies tend to express their sense [missing words] by raping the women of defeated populations.

reflects back his own will and desires. (This point is treated nicely in S. de Beauvoir's book). Thus, men generally tire of a passive wife, and seek the other aspect of women in adventures outside the family (without, however, much real success).

As for me, I must confess that the only women who have any real sexual attraction for me are those who I feel, have some independent personality of their own. What I most enjoyed is seeing a woman who is quite "womanly" and who still has a life of her own, a force of her own, opinions of her own. On the other hand, a woman who yields in her opinions, desires, etc. gives me a feeling of a warm, limp rag. Of course, I may be annoyed by some of the things that an independent-minded woman may insist on doing, but in the long run, it is better to suffer this annoyance than the monotony of a person who will always yield. On the other side, however, I could not stand a person who tried to order me around.

I feel, as you say, that there is a certain natural understanding between us, that does not exist between you and George, who always likes to put things directly in black and white, without considering the intermediate process by which black and white are related. Thus, his way of doing things strikes me as somewhat "jagged", except perhaps in his art work, which is, to me, a surprising phase of his character, indicating a greater real sensitivity to things than he cares to admit in his normal life. Thus, perhaps his "jaggedness" is like the stereotyped "masculinity" of most men, a response that he learned in the crude and rough society in which he was raised.

However, to return to our relations, there is a certain barrier between us, raised perhaps in part by my difficulty in maintaining you in a proper perspective. At times, this barrier goes down. For example, I remember once in Trenton when we were waiting for your train to Washington, and I was soon to leave for Brazil. When I looked into your eyes, I saw a totally different person, and I felt as if I were diving into an infinite sea, with infinite depths, in which were things that were perhaps infinitely promising, or perhaps infinitely dangerous.[2] But I became timid, and did not pursue this dive, perhaps because of the difficulty in reconciling this new view of your personality with my customary view. Perhaps if I had been less timid, things would have happened differently. But as you now say so frequently, things will develop to some extent by their own logic. It is still possible that we shall come to see each other during the next few years, as I think it is not outside the range of things to be hoped for that I may eventually succeed in getting to England or France (perhaps, even in 6 months to a year, but don't count on this too much). But you were probably right in deciding not to come now, as conditions would have been horribly mixed up for all of us. Meanwhile, both of us must arrange our own lives as best we can, reconciled to the fact that the possibilities for happiness are limited, especially under present conditions. It is difficult for me to advise you at this distance. You have to somehow realise your possibilities, both as a "woman" and as a person. This cannot be done very satisfactorily in the present society, so you will have to be satisfied with a partial realisation of both. This will require an intense struggle on your part, and also some careful cool thinking about your plans. You will have to avoid what I have called "insidious" tendencies toward surrender, but this doesn't

[2]This is the first time I have had such a feeling since I was very very young.

mean that you should avoid the risks of life. If you really want a baby, then you should have it in the same way that if you really want to do math, you must struggle to try to do it. But in both cases, the struggle will not be a magic resolution of the difficulties inherent in living in our insane society. The struggle to raise the baby will cost you some of your achievements in math, but doesn't have to mean the loss of all possibility of work, provided that you fight hard enough. But if you give in to the feelings by which the current society traps people, then it will be hopeless.

Well, so much for my sage advice.

Love

Dave

Letter 106. Folder C123, dated: Sept 8, 1953.

Universidade de São Paulo
Faculdade de Filosofia, Ciências e Letras
Caixa Postal: 8105
São Paulo

Dearest Miriam

Well, I guess it's about time that we renewed our correspondence after this long delay.

It seems to me the time has come for some plain and honest statements. My interpretation is that you are clearly in love with George, and have decided to stay with him. So let's leave it that way, and get back to at least an occasional exchange of ideas. As you said, events have a way of deciding such things, with the passage of time. So please let me know how you are, what you are doing, and what is happening; for it would be very bad if we were to permanently lose touch with each other.

As for me, I am feeling discouraged. First, it seems that progress in the causal interpretation is very difficult. Secondly, no one will pay much attention to it without further progress. Thirdly, I am discouraged by the fact that even those who are favourable have shown evidence of misunderstanding my articles. I have an oppressive feeling that it's no use working on new ideas, as nobody will even understand them, much less adopt them.

The situation in Brazil adds to my depression. There is almost no one to talk to. In this respect, the visit of George and of Vigier was like a life-giving breath of fresh air, but of course, much too short in duration. The situation in Brazil is bad in every way; culturally, and socially. On top of this, we have a very ugly situation in the dep't of physics here. I have been told by people who have previously been reliable that the two rats in the dep't (Stammreich + Souza Santos) have combined with Schonberg to try to start a long-range plan to get rid of me. Of course I have no proof of this, but what makes it plausible is:

(1) These three are known to have done precisely this to Wataghin, who occupied my chair before me.

(2) Schonberg + Stammreich are going around saying that my work is too "abstract" (the work on probability) and that they need somebody who will do calculations

on nuclear physics. This is somewhat of a joke, as Schonberg is working on a very abstract theory that has no conceivable use at all (as you can find out by asking George). Schonberg is also against me because he thinks that my stress on the causal interpretation has created an excessively "polemical" attitude among the students. Schonberg is 100% against the causal interpretation, especially against the idea of trying to form a conceptual image of what is happening. He believes that the true dialectical method is to seek a new form of mathematics, the more "subtle" the better, and try to solve the crisis in physics in this way. As for explaining chance in terms of causality, he believes this to be "reactionary" and "undialectical". He believes instead that the true dialectical approach is to assume "pure chance" which may propagate from level to level, but which is never explained in any way, except in terms of itself.

In general, Schonberg's actions disturb me, as he and the dep't rats did exactly the same thing in Wataghin's case, going around saying that his work was unsuitable, etc.

You may ask how he does such things in spite of his political position. He is however, a very complex, devious, and inconsistent person. I never was able to trust him from the beginning, after becoming aware of the extreme inconsistency between his political & philosophical point of view, and his actual behaviour in scientific research and in his relations with other people in the dep't.

All in all, everything points to the advisability of my getting out of here. But the problem is, how. (Please remind George to inquire about the passport business for me). Also I am very lonely here, with little prospect of meeting people with whom I can feel really close. If I stay here much longer, I am afraid that I shall lose most of my interest in life.

The international situation looks dangerous, with the overwhelming support Adenauer obtained in the German elections. The crucial issue is whether France will resist the European Alliance or not. We are far from being safe from another war, as yet. But the outlook is not entirely without hope.

Best regards to all

From your old friend
Dave

Letter 107. Folder C123, dated: Fall 53.

[There are two letters headed "Fall 53". This is the first because of reference to Miriam renewing correspondence].

Universidade de São Paulo
Faculdade de Filosofia, Ciências e Letras
Caixa Postal: 8105
São Paulo

Dearest Miriam

I was very glad that you have decided to renew our correspondence [unreadable words] of our relations unexplained in full detail for the time being. We still have a lot of things to discuss.

First of all, as to your suggestions about the passport. I do not see how to carry them out, nor do I think there is much chance of success in that direction. Do you have an appropriate lawyer in mind? Don't forget that they might ask me to sign some sort of anti-Communist affidavit. How would it look if I refused to do it, after making such a big fuss? As for returning to the U.S. at present [unreadable] doubtful procedure. I very much doubt that I would [unreadable] once I returned.

As far as which [unreadable] is not clear. It seems difficult to obtain a job in England without actually being there. Also, England looks like a doubtful place to me at present. Reaction seems to be pretty firmly in control there. Vigier says he can find a job in France. So this seems more likely. Of course, there is always the possibility of Israel, but from what I have heard of Israel, I can't see that it is so much better than Brazil. I had an offer from Australia, but Australia is so isolated that this would be the end of all chance to influence people on the causal interpretation. Also, from what I have heard, Australia is, socially speaking, a very backward place.

With regard to the paper on causality, I am afraid that I don't agree completely with your opinion. I think that it does need a lot of revision, with some pruning, and some additional parts. But my experience here suggests that there are a lot of people of university level in the sciences and having philosophical interests who would be able to read it. It seems difficult to see how one could make the subject a great deal less technical without multiplying the length by 3 or 4.
You must remember that Q.M. is a difficult subject to explain, not only because it is so complicated, but also because it is so unfamiliar. I might

[the next pages were attached to Letter 97, Oct 24, 1953 but seem better here, and the Weissberg comment predates the reference to Weissberg in Oct 24]
however, [unreadable] write it, stressing the Q.M. and the infinity of levels, developing the latter concept some more.

With regards to your offer of free advice on love-making, thanks. However I have the feeling that a woman who insists on certain technical details may already have a general orientation that would make life with her rather uninteresting to me. One of my main difficulties is the feeling that I cannot get interested in most women, because I feel that they are so limited, superficial, and shallow. Of course, this is due to the way they are educated in our society. But once they have been thus conditioned, there isn't much to be done about it. Of course, one can have sexual intercourse with them, but this by itself seems hardly worth all the trouble involved in reaching that point. I think that if I really had some confidence that it is worth doing, I could probably make stronger efforts in this direction.

I hope that you will excuse the somewhat discouraged tone of this letter. However, I am sick today with diarrhoea, as usual, and the weather is very hot and insipid. I certainly hope that it will rain soon and cool off. At the moment, I have neither ideas nor desires.

What you say about Soviet society is interesting. I haven't yet received any books by Alex Weissberg. Am I supposed to? Certainly there are many grave defects in Soviet society. Among these are the development, as you say, of a servile careerist point of view in many people. The mechanisms for enforcing a uniform point of view are very strong. A proof that something is wrong is furnished by the fact that a skunk like Beria wielded absolute power for so many years, without anything ever appearing in the press except praise for him. However, we must consider the alternatives. In America there is absolutely no hope for any kind of solution. There is no possibility but the development of a servile shallow population, amused and drugged by television, movies, advertisements, etc., all hating each other in reality, and living in an economy based on oppressing a large fraction of the people in the US as well as billions of people in colonial nations.

The ideology includes such irrational things as racial superiority, religion, belief in a non-existent "free enterprise" combined with an [missing words] pressure for everyone to conform. Until the economy ceases to be based on oppression of man by man, there is no possibility of a real improvement. And 99% of the American people have been conditioned to regard socialism, which is the only solution of the problem, as the most evil concept imaginable. All I can say is that if America comes to dominate the world, as it is now trying to do (and is in fact driven to try) then there will be centuries of misery deeper than anything that we have ever imagined. As for the Soviet Union, it is true that there are many unhealthy currents in it. But if they can once achieve a high standard of living + security from the constant threat of military invasion, they will be able to create the conditions in which more freedom is possible. Such freedom will not come without a struggle on the part of the people. But I believe that after the dangers which justify the present dictatorship have passed, it will be difficult to continue the present methods. It is true that many people have been turned into servile careerists. But as long as the economy is not based intrinsically on the oppression of one class by another, there is no reason why contradictions resulting from the functioning of a bureaucratic oligarchy will not eventually create the possibility for those who are dissatisfied to bring about a fundamental change. The fact that such contradictions do occur is proved by the fact that a man like Beria was finally kicked out. Perhaps those who replace him have many of his defects. But such fights inside the bureaucracy are always bound to occur in time. At present they cannot lead to fundamental changes, because the objective conditions don't permit it. It would take a really unbalanced person to suggest that police restrictions be lifted completely, as this would create a field day for an army of American agents. The fact is that it takes a long time for the ideal of socialism to be implanted so deeply in the people that all of them can be trusted to resist the temptation of a promise of an immediate improvement in conditions at the expense of a long run betrayal of the possibility for human happiness. When conditions have risen so high that American money is no temptation, then there will be a qualitative change of the forces involved. Meanwhile, however, I fear that the present generation, and probably the next, are going to have a very hard time of it.

Well, I guess that's all I have the energy to say today. Next time, I'll say more about my scientific ideas, such as they are. I am becoming somewhat discouraged about the possibility of ever explaining any such ideas to people who are far away, as my conversations with Vigier & with George, as well as with correspondence with many people, suggests that something new is almost impossible to explain by articles or letters.

Now that we have defined our relations, I can sign our letters with the greeting

Your old friend
Dave

Letter 108. Folder C123, dated: Oct 24, 1953.

Universidade de São Paulo
Faculdade de Filosofia, Ciências e Letras
Caixa Postal: 8105
São Paulo

Dearest Miriam

Thanks a lot for your encouraging letter. You are right, I must start to move, and in fact I had already independently started to do so. Let us hope that something can be worked out. The time has come for a new adventure and to get out of this dull place. The time wasn't entirely lost here, and I learned many things that I wouldn't have learned elsewhere. But things are now overdue for a change.

You are right about my attitude toward women. I shall have to try to care more for the next one. I have had enough of this loneliness. Lately I have tended toward being depressed, partly because I was sick but now that the decision is taken, I feel better. I am only a bit scared that things won't work out.

This immunity business looks bad. I am scared also for Melba and other people.

Firstly, I have read all the 3 books, including Kinsey. Kinsey's book I liked, as I think it helped me understand women a bit better. Knowing some of the facts in there, I can see how I can be more understanding in the future than I have been.

The other two books impressed me a lot, and also depressed me a lot. As far as I can see, a great deal of what Weissberg said must be true. Tremendous mistakes were made in the USSR and will doubtless be made in the future. The general calibre of the present-day leaders does not seem to compare even with Stalin, much less with Lenin. A great deal of what Deutscher writes seems also to be true. However, the net effect depresses me very much. My own feeling is that independent thinking and courage are much more lacking among the American people (right, left and centre) than they were in the Russian revolutionaries. I am sure that an American revolution (such as might happen in 50 years) would be much bloodier and more doctrinaire than the Russian was. On the whole, I have little confidence in the American people. Their education is so shallow that it is worse than useless, their high standard of living mainly serves to corrupt them. Of course, it must be remembered that the existence of a highly industrialised society would, in one respect, simplify the problem of constructing socialism. But the existence of such a large [missing words] businessmen and clerical and [missing]

[The next 3 pages attached to the above seem to be better attached to Letter 96, Fall 53 (1st letter)]

Letter 109. Folder C123, dated: Nov 10, 1953.

Universidade de São Paulo
Faculdade de Filosofia, Ciências e Letras
Caixa Postal: 8105
São Paulo

Dearest Miriam

I received your letter in which you say that you are going to have a baby. Although I was disappointed by this in one sense (as a sign of the final closing of a kind of door) I think that it will be a very good thing for you. After all, we can't sit around for ever in the same state, and the passage of time carries us forward irresistibly, each to his own peculiar course. You will need all your courage and energy to bring a child into the world in these troubled times, but it is definitely worth doing. It will be a difficult problem educating the child properly, when he will be exposed to all the influences of a rotten society, before he is old enough to understand what is happening to him. The least that you can do is to try to arouse in him a spirit of courage and love for the possibilities in human beings as individuals, and in humanity as a whole. But to raise such children is a very important task, which helps lay the foundation for a better future world. It is especially important that you should not give way to natural but selfish desires that your child should avoid the risks which follow inevitably if he develops in an honest way. He ought to be ready to face these risks, whatever they may be, rather than become rotten and corrupt inside, as so many of our modern liberals are (and as all of us are to some extent, because we have been nurtured in a rotten and corrupt society). It is better to live a full honest life for 10, 20, or 30 years than to survive to 70 or 80 in the way that most people do nowadays, hopelessly compromising all that might be good in them in order to remain comfortable, or to have the satisfaction of a fruitful career.

Well, so much for advice. You cannot expect to do as much scientific work in the next few years as you have done, but try to do as much as you can. But above all, don't give up hope of returning to your work. Your child is an important contribution, but by itself, it cannot be enough to justify you. You must try to do what you are able to do, in every direction that is open to you. It is necessary to avoid the panicky feeling that you are falling behind hopelessly in your work, because this is an exaggeration. If you work with enough desire later on, when there will be time, you can recuperate what is lost. After all, other people have done it, and you too can do it, if you don't lose hope. It may sound trite, but one should remember that the spirit in which one works is just as important as are the objective results, especially in times like these, when every person who maintains his spirit and avoids cynicism constitutes a living example of revolt against the decadent spirit characteristic of a decaying form of society.

As for politics, I am afraid that we are too far apart to carry on a discussion by mail. I have the impression that in proposing Elliott Montroll for president, you are pulling my leg (or someone else's perhaps). In any case, you asked how you can cheer me up. I had a very good laugh at this proposal, and if you can think of anything equally funny in the future, please let me hear of it. Seriously though, I am afraid that you are going off the deep end in politics, and are losing touch with reality. To dream that you could fool the bourgeoisie into giving up their power to a man like Elliott Montroll is pure fantasy. And to dream that Elliott Montroll would have the faintest idea of what must be done is even farther in the realm of fantasy. The critical questions are being decided outside the U.S. these days. The American people have given up the possibility of controlling their own destiny, and they will not get it back before events in other parts of the world have decided the issue pretty well. At the moment, France is in a key position. If they reject the European treaty, there is a good chance that they will start a movement leading to the isolation of Germany and to social change over the whole of Europe. For the necessary consequences of such a rejection would be to break the alliance with the U.S. Since they cannot stand alone, they would have to orient themselves towards the other side. With France outside the U.S.– Germany orbit, other countries like Italy, and eventually England would follow in time. Eventually, the U.S.would be isolated, and would have no choice but to face the internal problems that will in time develop. But if France goes along, Germany will take over Europe, and in 5 years, the only way out will be war if the Alliance is not destroyed in some other way in the meantime. At this moment, the French Socialists play a crucial role. Many of their leaders (such as Guy Mollet) are anxious to sell out, but there is some doubt that the rank and file would follow. I think that the behaviour of Socialists in supporting (or not strongly opposing) German armament is treasonable. It could easily lead to a disastrous war, in which 300,000,000 people are murdered. No punishment could be too severe for such behaviour.

I have been thinking of some of your criticisms of the behaviour of the left. In this regard, a man like Schonberg offers much food for thought. Obviously, he is a man of high intelligence, comprehensive political understanding, and has some ability to act. Yet he is hopelessly confused about a large number of questions. In this confusion, he reflects the twisted ideas of our bourgeois society. For example, in science, he advocates extreme idealism and formalism, such as that of Pauli, Heisenberg, etc. In art he is in favour of modern abstractionism. He is against the industrialization of society, and has certain desires to return to a handicraft and artisan society. As for women, he believes that a great many of them (if not most) would be happiest raising children, and making life pleasant for some man. In the physics dep't, he cooperates with the lowest most degenerate group of thieves that you could even imagine, justifying this by various [words missing] (you have to try to be kind to them, because they are very unhappy and insecure). As a result, the dep't is ready to fall apart, and a very dangerous situation is developing here for me. I hope that I can manage to leave before it develops.

In any case it is clear that just because a man is a Marxist, this is no proof against his absorbing a great deal of the rottenness and confusion of our society. The same must be true in every country in the world. Such people can have a very destructive effect, after the development of socialism, and they probably have had such effects in many countries.

Does this mean, however, that Communists are especially subject to such disadvantages? No! For if you look at people in other groups, you will see that such rottenness and confusion is the normal pattern. Hence no matter which group takes power, there will be a great deal of confusion, rottenness and dirty work.

But then can one adopt your advice and wait for "2 depressions and 3 new deals". Again, this won't work, because the world is too united. Those who don't seize the power when the opportunity presents itself will be smashed later. Hence, there is no choice but to act, with the unsatisfactory material at hand, when the opportunity arises. The process will be very messy, but not as messy as what will happen if you do nothing, or if you act inadequately. It is the only way in which a good society can eventually emerge. And if you put off action when the opportunity arises, you will only later be confronted by an even messier and more dangerous situation, requiring even sterner and more ruthless measures for its resolution. You may take the US as an example. American liberals caved in in 1946 when action would have been easy and safe. Now they are threatened with persecution, under conditions of such demoralisation that they can do nothing.

Sometimes I wonder if you would look with such equanimity on "3 new deals + 2 depressions" if you and George were making $2000 a year instead of $10,000 a year. Or if you were a miserable Brazilian worker earning $50 a month, living in a shack without water and in a slum without sewage facilities. Do you realise that a depression destroys lives just as effectively as a purge but on a much greater scale? People not only die of disease and hunger, but their spirit is broken, even if they do manage to live. And as for the average person living even in the US in "prosperous" homes, what kind of life is it, with no hope, nothing to live for except the next day, nothing better to do except watch the inane television programs at night. I think that you are underestimating the destructive effects of the present society on the people who live in it. These must be understood in striking a balance. And in the case of Russia and China, the destruction was much higher as the lives of 99% of the people were utterly miserable, and without hope (except in the "life to come" promised by religion). Would you advise the Chinese, the Moroccans, the Persians to go through perhaps 7 depressions, 5 reactionary gov'ts and 9 new deals to finally arrive at Socialism?

I think that if you strike a balance, the contribution of the Russians + Chinese Socialists is distinctly positive, particularly if you take into account the way in which the people of the world have to live in the colonial countries. [unreadable] to live for a year or more in a colonial country to see what "Modern Civilization" is doing to 90% of the people now alive on the Earth. And as you well know, Colonialism is an integral part of this "Civilization". When I see a socialist gov't give up its colonies then I will [unreadable] believe that they mean business. Otherwise it's clear that it's mostly pious talk.

[missing page(s)]

did, one can always find a way out of one's difficulties. For example if I had testified, I might still be working in Princeton, (whatever that would be worth I don't know). But I am surprised how easily a man like Levinson justifies everything in his conscience. Also I'm surprised at your tolerance of such a man. Only a few months ago, you wrote me that he took the "easy" way out in order to continue his present life, cottage at the beach, etc. Really, I am afraid that your whole point of view is being twisted. I don't know by what, but perhaps by what I shall call, for want of a better word, the "Weisskopf crowd". I pick [unreadable, presumably Weisskopf] because I happen to know him a bit. But the [unreadable] weaknesses in his character, and in general I do not trust even though he is perhaps well intentioned. My first impression of Weisskopf was very bad. We were discussing his work, and after I told him what interested me, he answered in a very virtuous way "I am working on much more modest things". He exhibits what may be called an "immodest modesty". That is, under the guise of modesty, he manages to center the problem of how virtuous he is in not doing things that are beyond him (instead of centering his attention on the problem itself rather than [unreadable]). He is the same in politics. For example, once in the old days, when Oppenheimer was still [unreadable], he said "These things are beyond my ability. I let myself be guided in political matters by Oppenheimer as he really understands these things". My impression of Weisskopf is that he is at best a sort of namby-pamby, not to be taken seriously. He has no courage whatever, and he adopts opinions which he thinks are relatively safe while still permitting him to pose as a liberal. What the rest of this crowd is like I don't know but apparently Levinson is not so very different.

[missing page(s)]
his problem must be long run only by temporarily sacrificing a part of his standard of living based on the slave labour of other peoples.

My own feeling now is that human beings growing up in a capitalist country or even in the early stages of a socialist society will for the most part not be able to solve complex problems of social adjustment very well, partly because they look at things "instinctively" in the wrong way. Until the time comes that (a) there is no objective need for one man to fear another and (b) the motivation of man is basically to satisfy his desire for individual fulfilment by working for the society, most people will, except in times of great emergency tend to place careerist interests first. As a result, bureaucracy will tend to develop, and there are bound to be many mistakes, and much injustice. But if the basic goals of the society are kept in sight, in time the change of conditions will create [unreadable] who will react differently. Then a much simpler and more effective organisation of society will become possible. But in grim times such as these, the role of few leaders who keep this [unreadable] is crucial if, as the masses tend to become tired and want to receive the fruit of their labours now. This is natural, but if they are allowed to give in too soon, the result may be disastrous.

This brings me back to my own personal problems. The point you mentioned about "caring" for another person has become very difficult for me. It is hard to know how this difficulty started but it probably originated in systematic disappointment in people throughout my childhood and youth, perhaps the discovery that nobody was really trustworthy and reliable, in spite of appearances. So I have tended to develop a protective habit of not taking people very seriously as individuals. Unfortunately the habit grows as I discovered that in all fields, one finds disenchantment and disillusionment. Thus, in physics, I see a majority of the physicists supporting the reactionary atomic policy of the gov't and most of those who don't support it still remain reactionary in their attitude toward physical theory itself. The two books that you sent me have tended to give me a very wary attitude toward the Communists, as I can see how through various forms of weakness, they may be led to betray what they are fighting for. The weaknesses arise from a bureaucratic desire for advancement, as well as from lack of courage to question the decisions of those higher up. As for the Socialists and liberals, I have long since ceased to believe that the convictions of most of them are strong enough to lead them to take actions that would inconvenience their personal lives. In any case, most of them have a pretty shallow view of the whole problem, which precludes their ever doing anything that would carry us very far in the direction that we have to go. As for the bourgeois, the less said about them the better. It is difficult to imagine myself caring what happens to such stupid, treacherous and distorted people. I suppose the working class and poorer people are still only partially spoiled and there is so much difference in way of life for me to come close to them. So I am afraid sometimes that present-day humanity (including myself) is too spoiled for me to be able to take them seriously as individuals. Their main present value in my point of view is that they constitute the soil so to speak, out of which the future humanity can and must grow. In small children, one can see the possibilities of what people can and will be in time. But as they grow up, they are inevitably twisted into the caricatures of human beings who are the adults today. Eventually, I am beginning to lose the ability to be very concerned about people, about physics, or other such things. The way I feel now, I could "care" only for the people of the future (or perhaps for the children of the present, if I had some of them to care for).

I think you are right in saying that I must become integrated more with people, with their struggles, etc. This would be easier if I could believe that they were struggling whole-heartedly for something good. But even among the best, you will find that some very limited interest predominates, and that outside the protection of the group, most people cave in. Thus one must become party to tremendous dishonesty, and eventually this dishonesty becomes an [unreadable] part of one's life. If one is a Communist, one must overlook all the weaknesses of the party, etc., as we have so often discussed. But if one is a Socialist or a left-wing liberal, one must overlook the ever more serious weaknesses in these groups. And the same for any other group, Zionist, Catholic, or what have you. If one does not turn a blind eye to these weaknesses it is impossible to continue to believe in the goal, and to work efficiently. And if you don't ally yourself with a group, it is still no use. Because then you must try to imagine that you yourself are good enough to be able to struggle entirely by yourself – also an impossibility, and therefore to be believed only by being dishonest with yourself.

[Ends here]

Letter 110. Folder C123, dated: Fall 53.

[The second of the "Fall 53" letters. Seems to follow 109 because of the criticism of "superficiality" arising from the American environment, rather than the "Weisskopf crowd"].

Universidade de São Paulo
Faculdade de Filosofia, Ciências e Letras
Caixa Postal: 8105
São Paulo

Dearest Miriam

Your latest letter was a sort of a shock to me. I did not realise that I was writing to you in an unfriendly tone. I merely felt that you had developed some ideas which seemed pretty silly to me, and I told you what I thought of them. After all, I do still have the right to have ideas different from yours, don't I? But don't think that this means I am no longer your friend.

Our ideas are by now so different that I don't think it is really possible for us to discuss politics by mail without hurting each other too much. Perhaps it would be better not to try it.

As for your criticisms about my relations with people, I accept them. I have to try to love people as they are, to work with them, and to have some relationship to them, as difficult as this task will [missing words]
during the [missing] politics much better than they do now. [missing] of the decadence that is overtaking American life. Don't be so superficial, Miriam. In previous letters, you have admitted this decadence and corruption. Have you ever asked yourself where it comes from? You seem to imply that it is due to the industrial organisation of our society.[3] This is a characteristically superficial evasion, worthy of a man like Orwell, who understood no politics, and not worthy of you. It comes because the "West" is following a political and economic policy of anti Communism, which everybody senses is hopeless, even those people who don't realise that the next step must be Socialism. People are bound to become decadent and corrupt when they see no way out of a hopeless situation. Decadence + corruption are not new things which arise with an industrial society.

They were well known in ancient Greece + Rome, where similarly [missing words.]

[The following pages are assumed to be correctly attached, but it is not certain].

had been [unreadable]

Finally, in answer to your question about [unreadable – possibly Vigier] and the Bunges, they are very nice people, who help make Brazil a more tolerable place (for me at least). Unfortunately the Bunges have returned to Argentina.

George did not behave too badly in Brazil. His main bad point was that he was too frank in criticising Brazil, before the Brazilians. "Let him who is without blemish cast the first stone". [unreadable words] than you do.
(3) To come back to your superficiality, I have been surprised by the low character of most of the books that you have sent me, since about the time George visited Brazil.

[3]Note, presumably made by Yevick: When did I ever say so?.

The first of these was "To the Finland Station" by Edmund Wilson. Really, I have never seen a book that was written on so low a level. The man has no understanding of history, philosophy, or any of the larger issues. His treatment of dialectical materialism was a joke. He never was able to rise above the level of petty personal gossip, some of it quite malicious. My impression from the book is that he is a "mean" man, in the sense that he is petty, pompous, and highly limited in his view of things (limited to his own petty personal concerns). I am surprised that you thought that this book was worth the time needed to read it.

Then come the books of Isaac Deutscher. While he gives some interesting facts, he analyses them on a childishly superficial level. For example, in "Russia, What Next", he admits that Russia, under Stalin, has made impressive achievements, and that in each crisis, Stalin had really no choice but to do as he did. But then he interprets this by assuming (without proof) that Stalin's intentions were to set up an "Oriental" dictatorship which would crush the Soviet people. In some strange way, he was then "defeated" by the very success of his own efforts, because the rise in technical level and the level of education made the people less willing to [word missing, perhaps: accept] this dictatorship. But Deutscher apparently did not consider the [missing] explanation that Stalin always had in mind that the [missing] level of the people would eventually make a stern regime less necessary. He started with the Russian people very backward and with the whole world out to destroy the new Russian society. Stern measures were necessary. But the whole theory of Communism is that the rise in the level resulting from the "dictatorship of the proletariat" will eventually make this dictatorship unnecessary. This is an elementary example of dialectics, namely, of how a given condition may contain in it the seeds of its own negation. But Deutscher transforms it into a mysterious and unexpected failure of some type of "Stalinism" that exists only in the mind of Deutscher.

Moreover, I cannot help feeling that Deutscher is hypocritical when he ascribes conditions in Russia to "Oriental" traits, implying thereby that we in the "Occident" are more civilised, and would never get into such troubles. He seems to forget how "Occidental" Germany murdered 12 million people, how "Occidental" Britain, Belgium, and France make life miserable for colonial peoples, how "Occidental" USA treats its negroes. Cruelty and barbarism are not an "Oriental" invention. It is true that much of the cruelty of the Russian revolution can be traced to the backward & barbarous state of Russian society at the time, but much more is due to the "Occidental" efforts to destroy socialism in Russia. And I can easily see how "Occidental" Brazil would have to have a much crueller revolution, if ever they are ever to get anywhere. What conditions are needed for a revolution in "Occidental" Germany?

Some of the other books that you sent me are even more superficial, e.g. "Community Power Structure" by Hunter. Most of what this man says is so obvious that I am surprised that anybody prints it. He might as well write a book on "The Structure of the Number 4 as a Sum of 2 plus 2".

I must apologise for accusing you of being dragged along in your opinions by the "Weisskopf crowd". Now, I think that what is most likely is that you are just absorbing the superficiality which [missing] in the American environment. It is characteristic that immigrants are more subject to the virus of "Americanism" in its [missing] than are many of the [missing]

[ends here]

Chapter 29
Letters to Miriam Yevick, 1954, Part 1

Letter 111. Folder C123, dated: January, 1954.

Universidade de São Paulo
Faculdade de Filosofia, Ciências e Letras
Caixa Postal: 8105
São Paulo

Dearest Miriam,

I am glad that we have cleared up some of the misunderstandings that arose in our correspondence. One thing you must realize; that lately you have been presenting in your letters a rather consistent political point of view; which is roughly what Time magazine calls the "anti-Communist" left, including that of such men as Koestler, et al. It is true that this is only a part of your point of view, as became evident in your later letters. But your letters by themselves already give a logically consistent set of ideas. How was I to know that you also held a logically opposed set of views, which you did not feel it necessary to repeat, because these views we shared in common? Now that I realize that you do not feel it necessary to maintain logical consistency, I shall be more at ease when you repeat ideas that might have come from Koestler or Orwell. Perhaps you wish to take advantage of your being a woman, (especially feminine, since you are now in the process of producing a child) to excuse your inconsistency?

More seriously, though, I think that you must make more efforts at being consistent. It is true that you may have to be a bit inconsistent when you want to experiment with new ideas. But the danger of such inconsistency is that you will gradually find yourself swept into positions that you would never originally have contemplated, and eventually you may find yourself throwing out your older ideas to remain consistent with your newer ones. This is the path that has actually been taken by many leftists, who first started out by only wanting to criticise the Soviet Union, and later began to think that perhaps "Western democracy" isn't so bad after all, and finally end up by

C. Talbot (ed.), *David Bohm: Causality and Chance, Letters to Three Women*,
DOI 10.1007/978-3-319-55492-1_29

doubting that Socialism is really possible without a violent and repressive revolution that destroys its very basis. Such people then become typical "Socialists" of which there are so many in Western Europe. They admit the desirability of Socialism in principle, but in practice, they contemplate an infinitely gradual process of transition that will disturb nobody, especially not the bourgeoisie, so that no violent reactions on the part of the latter could be provoked, which could lead to the need for violent suppression. Thus, whenever any concrete step is to be taken, such Socialists always tend to feel that it is too violent or abrupt. They are so much against the Soviet Union that they are blind to their own interests and cooperate with the US even in the rearmament of Germany.

You may think that none of this applies to you, but actually, the line of your letters in the past has implicitly lead in just this general direction.

Now, as for me, I agree with you that a relatively democratic and peaceful transition to Socialism is strongly to be desired, if it is attainable. Indeed, my own temperament is against either the exercise or the acceptance of dictatorial authority. If I ever have to exercise power, I could do so only if it were democratically done, after full consultation with all members of the group concerned. Unfortunately, in a critical situation this cannot be done,[1] especially considering the rather poorly developed state of most humanity at present. People are really caught in a terrible mess, because the existing society stupefies them, brutalizes them, blinds them, causes them to hate each other and to tear at each other's entrails in a vicious and meaningless struggle for survival and advancement of each individual against all others.[2] There are good currents in our society, to be sure, but these are very weak compared with the destructive currents arising from ignorance, narrowness, hate, etc. Eventually, in a different society, these very same people will change in a radical way, and will rise to a level at which really democratic gov't is possible. Indeed, I believe that people will eventually go beyond democratic gov't, and will reach decisions by mutual agreement of all concerned, without even the need for a formal vote. This already happens in many simple societies of savages, and it is possible there because: (a) Every person is assured of his basic needs of food, shelter, clothing, etc. (b) The problems are such that almost everybody has a more or less adequate understanding of them. When production and education in our society make the same conditions possible, (but at a higher level of course), then it will become possible to evolve a society in which there is no need for a formal government. All decisions will be by mutual consent (this is the famous "withering away of the State").

At present, however, only a small percentage of the people (especially in the US) has a conception of the need for a basic transformation of society. Every effort must be made to expand this number, but we must remember that at best, under existing conditions of control of press, education, communications, etc., the number of people who will really have an adequate understanding of these problems, must remain relatively

[1]For this reason, I would not wish to accept a position of great authority under present conditions. I should be satisfied to be a professor with a few good students, in an environment of intellectual stimulation, where I could perhaps contribute by writing a few books on theory of various things.

[2]This is true of leftists almost as much as of the people in general.

small, until several generations after the society has been altered in a fundamental way. The direction of social transformation will have to be determined largely by the small fraction that has some theoretical understanding.[3] The power for the transformation itself can however, come only from the people as a whole. There are two conditions for a social change: (1) By means of military, economic, or other crises, the control of the ruling class is greatly weakened. (2) There exists a deep and widespread dissatisfaction with the existing state of things among wide classes and among the majority of the people. When these conditions are satisfied, a small minority with a clear understanding of what it wants can (a) Infuse its goals into an appreciable fraction of the people as a possible solution for the crisis (b) Take the power. The more condition (1) is satisfied, the less violence will be involved in taking the power. Thus, in the Russian Revolution, comparatively little violence was involved, because the opposition was weak. (Read for example, John Reed's "Ten Days that Shook the World"). Violence developed largely as a result of opposition, first by the bourgeoisie of Russia, and later, by that of the rest of the world.

Even if Socialists could take the power peacefully under conditions in which the bourgeoisie affect no violent opposition, however (as is conceivable perhaps in England) complete democracy would not work. For the people as a whole, educated in the old system, do not really understand the need for reaching the goal of a radical transformation of society, even if this process is very costly. Now you may say that it is unjust after a socialist gov't takes power to make the people work harder than ever for 10 or 20 years. But this is just what is necessary in most countries of the world (including England) where the level of the economy is too low for Socialism. Now perhaps 20 to 30 percent of the people will understand this need deeply, and be willing to sacrifice what is needed. Another 30% will perhaps understand it shallowly. They will eventually begin to look on socialism as a desirable thing, but will not see the necessity of it, so that they won't make sacrifices. In other words, they won't understand that if the level of production is not raised, the socialist gov't will fail; there will be a reaction; and that this reaction will perhaps be much more violent than any steps needed by the socialist gov't to insure the success of its program. The remaining 40% of the people are just blockheads who understand nothing except the need for themselves to be comfortable and to get ahead. If the country happens to have many peasants, the fraction of such blockheads will be unusually large. For the peasants live a very brutish, superstitious, limited life, in which the only goal is to get hold of a piece of land, while they would like to ignore what happens to everyone else. In the early stages, the desire of peasants for land may make them allies of the Socialist revolution, but eventually the two are bound to come in conflict. In the long run, the peasants represent a two-fold danger. (1) They will control the food supply, and thus have the cities, the workers, and the gov't at their mercy. (2) The inefficient individual production on small farms absorbs so many people that an adequate development of industry is impossible. Now in Russia, the peasant problem was dealt with in part by turning the poor peasants against the rich ones (the Kulaks)

[3]Even if the understanding of many of these people is highly incomplete, it is necessary that it exists, for without it, no long range progress is possible.

but this was inadequate. Eventually, it was solved by Stalin in a very brutal way in 1932, but the question is whether, under the conditions that existed then, any other way was possible. Here we are confronted by a difficult problem. The backward countries are the most likely to have deep crises which could result in the loss of control of power by the ruling class but their backward industrial condition and large peasant population make the construction of socialism difficult. In the more advanced industrial countries, however, where it would be easier to construct socialism, the ruling class has a firmer hand, partly because the stronger economy is more resistant to crises, and partly because the more developed means of education communication and transportation help keep the people in line.

In the US of course, the conditions of industry and agriculture would make possible a rapid and easy transformation to socialism, say in 5 to 10 years. The relative absence of peasants and small farms is a very favorable factor, perhaps as important as the highly developed state of industry. I agree with you that the technical organization of society would not have to change much at first. But the educational process, communications, etc. would have to change fundamentally and rapidly. I would not leave Princeton University unchanged. For it is now organised to educate a minute group of arrogant, stupid snobs who look forward to ruling the country. It would have to be reorganized so as to become part of a broad programme aimed at educating wide classes of the people, but in a very different spirit from that which now exists. To do this, one might have to seek new professors, administrators, etc. Meanwhile, there could be a gradual solution of the technical problems, which would include a national integration of industries, expansion of productive capacity, decrease of hours of work, construction of comfortable housing for all (perhaps the gradual tearing down of excessively large cities like N.Y. and the building of new ones elsewhere) improvement of public transportation, an immense programme of scientific research, especially in medicine and biology, aimed at lengthening life and eliminating diseases like cancer, arteriosclerosis, heart failure, etc., etc. All of this could only be developed in time. One should adopt a point of view of 20 years as the unit of time in which fundamental technical changes can be carved out. It would take perhaps two generations before the majority of the people would change so much that socialism would be an integral part of their personalities, so that a really democratic gov't was possible.

In this regard, I think that what you say about the Socialist youth of Israel is a bit shallow. It may be true that they had a very modern and extensive education, with much discussion, etc. But the most important point is that they had no opportunity to apply Socialistic theories to anything except the impoverished life of the Kibbutzim. This life required only the most elementary ideas and concepts; indeed, to be concerned too much with general ideas would be a disadvantage to a person working in a Kibbutz. Thus, the environment itself served to turn most of the teachings into just words. To be affected deeply by an idea, a person must have the feeling that he can either apply it or contribute to its growth in an active way. This was one of the things which, for example, drove me to study physics very intensively. And a very small fraction of the bourgeois can have that feeling today, because it has the possibility of working in an intellectual profession. Needless to say, this fraction is now decreasing in Western Europe because of impoverishment and in the US because of the development of Fascism and preparation for war. But at best, those few of the bourgeois who have the opportunity to indulge in intellectual work, do so on a very unhealthy basis, which leads to tremendous unhappiness, neurosis, and a great waste of effort and energy. In Russia today, the opportunity for real intellectual development is growing, but it is still somewhat limited, first because what is needed most is still the dull straightforward technical work of building up the industry and agriculture of the country, and secondly, because the country is so threatened by the capitalist powers that there is not yet sufficient freedom for full intellectual development. I think that it may be a long time before a really deep, critical and yet constructive and creative culture grows in the Soviet Union. Its first manifestation will be in the sciences, where they have already shown very impressive achievements. This is, of course, because the sciences play an important and ever increasing role in the development of the country. Later, as the technical level rises, and there is more leisure and security from invasion, other branches will develop. Even now, a broad mass base is being laid, through the extension of education, and the diffusion of an unheard of number of books at very low prices. Considering the rather low initial level of the people, a period is probably needed, which may be called a sort of consolidation of cultural gains. Such a period is characterised less by brilliant new ideas than by a very wide and thorough diffusion of some of the older ideas, which seem almost trivial to the more advanced intellectuals, but which are a necessary base that most people lack, both in the "West" and in the "East". I believe that when a large fraction of the people (say 50%) have a good general foundation and an avid interest in scientific and cultural developments, this will create the soil in which new developments of hitherto unseen brilliance will become possible. This is not only because more gifted individuals will be available to contribute, but also because the whole environment will cooperate to stimulate them and to criticise them, thus forcing them to continually higher levels of achievement.[4]

Now to come to the question of a possible attitude of a Socialist party that aims at a peaceful transition to socialism in a country like England. The first problem in the world of today is what their attitude should be toward the Soviet Union. I believe that they should regard themselves as independent, but sympathetic. While admitting and abhorring the excesses of repression and cruelty which have taken place in the Soviet Union (and which may well continue for a while) they cannot place themselves in opposition. For in spite of all these things, the Sov. Union is a socialist country; it is moving toward increasing socialism and the solution of its problems. All efforts to remove the unpleasant aspects of the Sov. Union by overthrowing their gov't would surely only create the bloodiest reaction imaginable. Our <u>only</u> possibility is then that the Sov. Union will continue to move toward greater <u>security</u> and wealth, thus permitting a solution of its internal problems. Moreover, one must admit that because of the difficult Russian situation it seems doubtful that socialism could have been achieved there in any other way.

---

[4]Witness the large fraction of Jewish intellectuals, partly due to an environment where intell. activity is appreciated, partly due to greater leisure & wealth.

But a pious neutrality (such as that of Bevan) is not enough. For it is a fact that England cannot be independent. Its industry and agriculture do not permit it. In the long run, it must supplement its economy either by a connection with the US or with the "East". But the US now demands as a price for aid that Britain cooperate in the anti-Soviet foreign policy, aimed at the eventual overthrow of the Soviet Union. If socialists cooperate in this, then their Socialism must become corrupt, and they must eventually partake of the same decadence that is overcoming the US. This indeed seems to be happening, as British politics is now showing a lack of spirit, a dullness which has [unreadable] weariness and loss of hope. Witness the fact that there is so little British opposition to the armament of Germany. It seems to me an axiom that no Socialist party can remain true to Socialism if it follows a policy aimed at the ultimate destruction of Socialism in Russia. Thus, no Socialist party can accept an alliance with the US. Since some kind of alliance is needed, the only choice is a policy of at least friendliness towards Russia, involving trade relations, etc. This party can stand for its own method of achieving Socialism in its own country (e.g. Britain). But if only out of simple realism, it must be prepared to admit that in France and Italy, and other such places, the most probable means will be through the Communist Party, because there, they are already the principal socialist group.

Until the left wing of Laborites make the above critical step, they will be trapped in a hopeless contradiction, and will not be able to even formulate a practicable policy. Neutralism is no longer possible. They must make their choice on one side or the other. They should take the attitude that perhaps the critical Russian situation made the repression, purges, etc. inevitable there, but that the different situation in England may make possible a different kind of change. In other words, there should be no mechanical transference of Russian experience to new situations.

Next, they must have the courage to admit that a change to Socialism should not necessarily be accompanied by an immediate improvement in the standard of living of all. Indeed, for all but the lowest classes, there may be some drop in standard of living, for the first 10 years, while British industry, agriculture and commerce are being reorganised, nationalized, and modernized. But the people must be led to see that this is absolutely necessary because there is no alternative. To give people a small rise in the standard of living without solving the basic problem of production will only intensify the crisis and perhaps lead later to a more violent solution of the problem. Thus the British really have no choice. They are now faced with a crisis whose only resolution is an economic reorganization that cannot be carried out without a rather thorough transformation to Socialism. To continue "muddling through" may lead to a much deeper crisis or perhaps to a war in which Britain may be annihilated.

As for the US, I have no very clear ideas. I'm inclined to agree with you that the immediate problem is to prevent Fascism and war, by some move toward a New Deal. The situation may perhaps improve later, but I am inclined to believe that, at best, the next 25 years in the US will be very cruel years. Perhaps if I were in the US, I could see for myself whether any real new currents of opposition are developing. From here, one does not see any such thing.

Now I would like to discuss one more point. Much of our disagreement stems from a philosophical point, about which we have already argued for many years. You are always objecting that you do not want things put in "white and black", but want to consider all possible "intermediate shades of gray" which are closer to reality. You have the feeling that some of the brethren of the left see things much too simply. Now I would like to say a few words about this problem.

It is true that to see things in "white & black" (i.e. in terms of 2 simple + mutually exclusive alternatives) is a rather primitive way of thinking, unfortunately still indulged in by a large number of people of all shades of political opinion (especially by most Americans, who regard Communists as a sort of utterly evil devil). It is true that reality is neither white nor black, but as you say, gray. In other words, things are usually quite mixed up when you look at them in the real world. It is an important step forward to understand this. But this is not enough. To stop here would be superficial. For this mixed up "gray" state of affairs is usually only in the immediate superficial aspect of the situation. When you look more deeply at the processes that are taking place, you discover that there are usually opposing trends or currents. In the long run, these currents will carry you in either one direction or another. Thus, beneath the surface complexity, one finds the deeper laws which enable one to understand where the complex gray mixture comes from and where it is going. These laws do not display themselves in a superficial glance but by enough study, one finds them. And experience with a wide variety of situations teaches us that we cannot permanently stand in the middle gray region. Our choice will cause us to follow one current or the other, and in the long run, we cannot avoid this. Even if we think we are making no choice, the currents still drag us, whether we like it or not. I would compare your desire to choose an intermediate shade of gray and maintain it as analogous to what the artist does when he paints a picture. The artist can do this, because his materials are pigments, which remain static for a long time. But in most political situations, this cannot be done, because the materials are in a very violent state of flux, especially in times of crisis like the present. And I believe that the currents of today are such that in the long run, you will either be for socialism or against socialism, because these are the only two possible attitudes that can be maintained in the face of today's problems. And if you oppose socialism in Russia you will find yourself inevitably dragged to the side of capitalism, whether you like it or not. Witness what happens to practically all former Communists who recant. The pressures are too great for them to remain in an intermediate situation. The utmost limit of choice possible (perhaps in some countries like England) is that the socialist movement (in England) shall choose its own way of bringing socialism. Whether this is still possible in the US, I would not know, but I am a bit doubtful. But this can only be done in the framework of a friendly attitude to socialism in the USSR.

As to my own personal problems, I fear that they can be solved only by my leaving Brazil. You suggest that I get myself a girl to have an affair with. You do not realize that with the mid-Victorian standards that prevail in S. Paulo, only two sources exist for such girls (1) The street (inadvisable because syphilis is widespread) (2) The foreign colony (inadvisable because it consists mostly of American secretaries or post-war refugees from Central Europe, with neither of whom I could have much sympathy).

More seriously though, you seem to have an attitude towards sex which is quite stereotyped, but which you do not apply to yourself at all. You seem to think that one needs only to pick up some "hot babe" (as you called her) and have an affair with her. Actually you once told me that you had intercourse with some man whom you didn't like very much, and that it seemed a dull mechanical thing. The same thing seems to be true for me, and as far as I can tell, for a great many other men. There seems to be a widespread fable (partly due to American advertising, and partly due to the low position of women in our society, especially in its historical development) that true masculinity demands that man shall have an irresistible urge to jump on any creature in skirts, as long as she is reasonably attractively built. You seem to share this idea, as does George. However, I have never observed George do any more than talk about this urge. Even when you left him, he did not actually go around picking up arbitrary women and having affairs with them. Thus, neither you nor George practice what you preach. In this regard George is quite funny at times. Influenced, as he himself admits by American traditions, he sees nothing in women but their asses. Whenever he passes a girl, he must comment on her ass. Eventually, it appeared that George regards women as essentially consisting of asses, provided with legs to carry them around, mouths and stomachs to feed them, and heads to direct them. But all of this evidently takes place only in a very superficial level of his mind. I believe that as a sort of defence against being regarded as odd, George has adopted this superficial veneer of stereotyped masculinity, so common in America (and probably elsewhere too). In his actual relation with women, he forgets all this nonsense. The same is true about his politics, in which he has absorbed an absurd veneration for the sheer magnitude of American industry (mile after mile of steel mills!). As Mario Bunge said "It is a marvel that a man with such bad ideas can be so good". And there is no question that basically George is an extremely good person. The reason that this is possible is that he doesn't really take some of his crazy ideas very seriously himself.[5]

To come back to the subject, I fear that I have practically no desire to have intercourse with a woman who does not already attract me in other ways. I just can't help it, but if you take any old "hot babe", the whole business seems to be just a sort of mechanical pumping action, not too interesting in itself, and certainly not worth the long preparation needed before the girl is ready to go to bed. Perhaps it is because I am getting old, but perhaps it is because the current idea of sex is too simple and mechanical, being regarded as the mere satisfaction of a desire, like "taking a glass of water when you are thirsty". The sexual urge is given a concrete and independent existence, and the rest of the body is supposed to exist only to help satisfy this urge (as is the case with women's asses in George's point of view). At least, the whole behaviour of a person is supposed to be directed mechanically to satisfying this so-called drive, willy-nilly with whomever happens to come along. I believe that this is

[5]Some of George's ideas are individually quite good, but the aggregate taken as a system doesn't make much sense to me.

a distorted and exaggerated idea of sex; at least, it does not seem to apply to me. So, I am afraid that it will not be so easy to follow your advice, especially in Brazil.[6]

As for the book by [unreadable], please send it, also the MR books. You might also renew my subscription to MR.

I guess that we can still remain friends despite our disagreements, eh? It's perfectly OK for you to "chat", as you say it, but as I said before, what bothered me was that your "chatter" made such a logically consistent picture, that in my naive way, I thought you had abandoned your old ideas.

With regard to nationalizing industries, I too am doubtful that this is an adequate solution. What is needed is

(a) Socialisation of blocks of industries. Blocks because of interdependence (e.g. steel, coal, transport, power, automobiles, building, etc.). Socialization means more worker participation. In Britain, one needs practical confiscation because the people do not have the resources to pay for the industries. This could be done legally by a suitable policy of taxation.

(b) A more rational reorganization of industry to avoid waste and duplication. This is especially necessary in the distributive end. Commerce must be socialized. In the US, for example, the cost of "distribution" is usually much more than the cost of "production". Giant, public super-markets of all kinds could perhaps be established, very efficient and without profits. Automobiles could be ordered directly from the factory – a dealer is not needed here. Each make could be placed on display in a few convenient places, advertising would be reduced to perhaps 10% of its present level (merely needed to tell people what is available and where). Enormous savings are possible in this way.

(c) A tremendous modernisation of industry is needed in England. This requires a huge capital investment that can only come from a lower standard of living for a number of years.

(d) A correlated educational program to train workers to do the new kinds of jobs that will be needed – also to help raise their general cultural level so that they will be able to make good use of their leisure that will come ultimately.

I don't think our disagreement is by now really so deep, at least I don't disagree with the part of your point of view that you didn't express in your (earlier) letters.

I am sorry that my letter that seemed so cold to you. Perhaps it came not so much from the South Pole as from São Paulo, which is an extremely cold city (emotionally speaking). Next to Rio, it is the most uncomfortable city I have ever known. Dirty, noisy, full of smoke, chaotic traffic, poverty everywhere, it has practically no place where one feels comfortable. Every city I have known thus far has had a place where one could feel somewhat soothed by some aspect of beauty or peacefulness, but no such place exists in S Paulo. Also, the food makes me sick most of the time. It must be some bacteria to which I am sensitive. (Vigier also got sick very often, but not George). Right now I have gone to a sort of resort in the mountains, and feel

[6] If only this girl who was a physics student hadn't gotten married so abruptly, dammit. She now looks prettier than ever, and seems very happy, while she is also demonstrating more and more intelligence. She is the only girl in Brazil who has thus far had such an effect on me.

much better, especially because the food here agrees with me. Within one day after arriving, the sickness disappears. In Rio, which is even dirtier than S Paulo, I have without fail, always gotten sick within a day after arriving. This sickness, combined with the general annoyance of having to be in such a horrible city as S. Paulo, makes me irritable, and makes it difficult to be friendly in an argument. Actually I am only now (four days after leaving S Paulo) beginning to have some life in me. If only it would stop this eternal raining here, I would be even happier.

As for Einstein Bose and Fermi-Dirac statistics, there is no trouble in the causal interpretation. You must remember that I assume that a system of electrons for example, consists of a Schrödinger wave field, $\psi(\vec{x}_1, \vec{x}_2 \ldots \vec{x}_n)$ and a set of particles located respectively at the points $\vec{\xi}_1, \vec{\xi}_2 \ldots \vec{\xi}_n$. The probability density for particles lying between $\vec{x}_1$ and $\vec{x}_1 + d\vec{x}_1$, $\vec{x}_2$ and $\vec{x}_2 + d\vec{x}_2$, etc. is $P d\vec{x}_1 \ldots d\vec{x}_n = |\psi|^2 d\vec{x}_1 \ldots d\vec{x}_n$. Now a set of particles obeying Bose statistics has a symmetric wave function so that interchange of 2 particles leaves ψ unchanged; with F-D statistics, we have an antisymmetric wave function, so that interchange of two particles changes the sign of ψ.

Now, in the usual interpretation, the ψ function is regarded as a complete description of reality. In other words, once the ψ function is [word missing] the physical state of the system is completely determined. Thus, when the coordinates of 2 particles are interchanged, ψ, and therefore the physical state of the system, remain unchanged, when the wave function is symmetric. If the wave function is antisymmetric, ψ is changed to $-\psi$, but this wave function represents the same physical state as does $+\psi$. From this, one concludes that particles are "indistinguishable", because no physical change can possibly result from their interchange. One also goes further (and in my opinion unjustifiably) says that the B.E. or F.D. statistics are a consequence of this indistinguishability. But all that can actually be concluded is that both indistinguishability and B.E. or F.D. statistics follow from symmetry or antisymmetry of the wave function. In other words, there is no direct line of reasoning, from which one can prove that BE or FD statistics necessarily implies indistinguishability or vice versa. The basic element here is the symmetry or antisymmetry of the wave function.

Now in the causal interpretation, the wave field, $\psi(\vec{x}_1, \vec{x}_2 \ldots \vec{x}_n)$, does not provide a complete description of the system. One also needs the actual locations, $\vec{\xi}_1, \vec{\xi}_2 \ldots \vec{\xi}_n$ of the particles. When we interchange 2 particles, say $\vec{\xi}_1$ and $\vec{\xi}_2$, we do lead to an in principle new physical state of the system. But because the particles are equivalent in their physical properties, this new state is physically equivalent to the old state.

Now, we have a statistical distribution of particle positions with a probability density, $P = |\psi|^2$. I have shown in my papers that such a distribution leads to physical results that are identical with those predicted by the usual interpretation. This happens whether the wave function is symmetric, anti-symmetric, or has arbitrary symmetry properties. Since the BE or FD statistics follow uniquely from the symmetry or antisymmetry of the wave function combined with the statistical distribution implied by $P = |\psi|^2$, it is clear that the causal interpretation must lead to the same BE or FD statistics as does the usual interpretation.

However at a level at which the present form of the qu. theory may break down (e.g., 10^{-13} cm) it would be possible for the theory to modify in such a way that the individual particles could be distinguished. At the usual level, however, the identity of the particle is of no importance, since each particle always produces effects equivalent to those of all the others. Thus, we may say that there is a "relative indistinguishability" of particles, which applies only at the atomic level. At deeper levels, they may be distinguishable. One has the well-known saying "At night, all cats are gray" – In other words they are indistinguishable. Under other conditions they can be distinguished. Similarly perhaps with electrons. The proponents of the usual interpretation assume however (without proof), that the electrons must remain eternally indistinguishable, just because we have not yet found a means of distinguishing them at the atomic level.

Love
Dave

PS. With regard to socialisation in Britain, it just occurred to me that an essential step would be to give up all colonies and the policy of colonialism in general. To maintain colonies is just too big an inconsistency. It can only be done by cooperating with the most reactionary elements and thus corrupting the socialist gov't. Money from colonial exploitation also corrupts the workers.

More concretely, it seems to me that all colonies in Asia must be given up: (Malaya, Burma, etc.). Egypt must be given complete control of the Suez canal. England should stop interfering in Iran on behalf of the Anglo-Iranian oil co. In Africa, they should perhaps maintain their gov't for a while, because the people there need protection from the other colonial powers and from the South Africans. They are still too primitive to be safe. But a socialist policy should be followed there based mainly on confiscation of estates by heavy taxation, distribution of land to the people, improvement in roads, sanitation, etc., prohibition of exploitation, etc., expansion of education.

You can see how such a policy would conflict with the present US foreign policy. It could only be carried out by abandoning the alliance with the US. If not carried out, Socialism in Britain would almost . . . [text ends here]

Letter 112. Folder 126, dated: 1953.

[Just 1953 on original but seems to follow "long" letter 111]

Universidade de São Paulo
Faculdade de Filosofia, Ciências e Letras
Caixa Postal: 8105
São Paulo

Dearest Miriam

I hope that by now you will have gotten my long letter, in which it will be clear that I am not angry at you. I delayed answering; first because you asked me to delay, secondly, because I wanted to write the letter at leisure on my vacation, and finally, because there was no air mail at the place where I spent my vacation, so I waited to get back to S Paulo before mailing the letter.

You should realise that I am not really made angry by an occasional personal insult, especially when it is accompanied by a joke. Your last few letters made me laugh heartily at some points, as you made some good jokes, and even a few telling points in answering me. This is the sort of thing I like. It is a sort of relief from the eternal grimness of S. Paulo (so my sense of humor really exists, if you only know where to look for it!). However, there are other kinds of letters which may tend to get me angry, such as the one in which you made fun of the ∞ of levels, in a way which seemed to indicate to me that you had rejected it for very superficial reasons, and then made a joke to excuse yourself from thinking more deeply (I believe it had something to do with a comparison between accepting the ∞ of levels and accepting God). Sometimes you use your sense of humor in this way, and then your sense of humor makes me angry. Please remember that a sense of humor can (like everything) else be a double-edged weapon, so be careful how you lay around yourself with it, as you may some day [missing] cutting off your own neck by a misplaced thrust.

Your long series of letters written in the tone of Koestler and Orwell, combined with the sole diet of such books that you had chosen to send me, made me a bit worried about you; and for this reason I wanted to express my reaction in no uncertain terms. Perhaps I overdid it. But you should also remember that you (like me and everyone else) are not immune to the possibility of what I may call "corruption by an adiabatic shift in point of view". Indeed, so many leftists & liberals have travelled this route that you would have to be a paragon of iron-willed constancy (such as does not exist in reality) to avoid gradually drifting over into the "anti-Communist left" if you had continued to think in the vein of some of your letters of a few months ago, without some criticism. So I criticised you, partly to give you a chance to be exposed to another point of view, which apparently, you have lately had little chance to hear. In accusing you of absorbing some of the corrupt points of view prevalent in the US I meant nothing personal. Everyone is bound to do this to some extent, when flooded with ideas coming from only one direction, and that direction a corrupt one. Just because I am in a different environment I can see more easily how superficial are some of the ideas that you were expressing, just as perhaps you were able to see so easily how my own environment here might tend to move me further toward the left than you were prepared to go. However, although I was a bit angry at you for letting yourself be dragged along by the tide of superficiality and rottenness in the US, this didn't mean that I no longer regarded you as my friend. Just as you didn't feel it necessary to repeat the "80% of ideas that we had in common", I didn't feel it necessary to repeat what I took for granted – that we remained friends. Thus, much confusion arose on both sides.

Now that the confusion is, we hope, over, let me say that I think you are partly right. The Russians, by their brutal and ruthless behaviour, have made Socialism so unattractive to so many people that the problem of spreading it in the West has been made more difficult. But, as even Deutscher admits, most of this tendency of the Russians was a reaction to the way they were treated by the West. And in this treatment Western Socialists & Social Dems share much of the guilt, for they rarely made strong pressure on their gov'ts to secure a fair treatment for Russia. Of course, the reaction of the Russians was exaggerated, but one can hardly blame people who are faced with a super-human task for reacting with an exaggerated desire to protect the fruits of their labors and their dreams for the future. In any case, the question of blame is by now academic. Even people like Deutscher agree that on the whole, the Russian balance is favorable, and that great possibilities exist for correcting the excessive brutality and ruthless disregard of individuals. But these possibilities can be realized only if the "West" stops getting ready to annihilate the Russians by hydrogen, cobalt, or other bombs, and only if it stops trying to strangle all movements toward socialism on the periphery of Russia[7] (Iran, India, China, Korea, etc.) If Western "leftists" could achieve such a change of policy, they would not only give the "moderates" a chance to gain power in Russia, but also would destroy the basis of the anti-Red hysteria, which now functions as a hopeless trap in which all Western political parties are caught.

Practically, this means that extreme care + discernment must be used in criticising the Russians, to avoid several erroneous conclusions that are widespread among the more superficial members of the left. These are

(a) That the Russians are so bad that little is to be expected of them – At best, it is to be hoped that they will not interfere too much with us.

(b) That "Western democracy" can continue to guarantee individual rights and still aim at the destruction of the Soviet Union (either by war or by a [missing]

(c) That Russian ruthlessness is due to some intrinsic "Asiatic" trait that we advanced peoples of the "West" do not have in so high a degree (this idea makes me really angry. Think of the Nazis, think of the American Southerners, the South Africans, etc., and then think of the Chinese who are infinitely milder + more sensible than they are more fanatical Western counterparts).

So much for politics. Perhaps it would be wise for a time to call a halt to political discussions, especially since letters may be read in this time of movement towards fascism

As for your idea of writing an article for the Scientific American, this is worth trying. I would be interested in seeing the outline. Needless to say, there are serious difficulties, especially the lack of carefully thought out technical articles that could be used as a base. People will criticise this article, saying it is full of unproved assertions.

[ends]

---

[7] Suppose for example that Russia had organized a world alliance against the US, ringed US with A-bomb bases, and then threatened either destruction or peaceful "liberation" of the US. You can imagine the repressive measures that the US gov't would adopt.

Letter 113. Folder C124, dated: Feb 8, 1954.

Universidade de São Paulo
Faculdade de Filosofia, Ciências e Letras
Caixa Postal: 8105
São Paulo

Dearest Miriam

Since I have already written you two letters that haven't been answered yet, I must assume that you have become angry at me. I tried to explain in my previous letters that I was just disagreeing with your point of view. I don't think that it should have been regarded as an insult by you that I implied that you were being dragged along by the corrupt and superficial points of view prevalent in the US today. After all, when one is subjected to such a torrent of half-truths, it is inevitable that one will be influenced.

I very much hope that you will not stop writing. Your letters are stimulating and often quite amusing. In any case, I am always very glad to hear from you. I am quite lonely here, as the Schillers are the only people with whom I can really talk. Besides, my health is getting worse, and I am feeling discouraged about my work, because progress is so difficult. I am taking steps to get out of here, but it will take from 6 to 18 months. You don't realise how hard it is to live in this cold city, where there are so few people with whom one can even talk intelligently

I hope that the infant is progressing well. During the past six months, you have been spending your time quite profitably making this child grow. I wish that I had been able to spend my time equally profitably.

Love
Dave

Chapter 30
Letters to Miriam Yevick, 1954, Part 2

Letter 114. Folder C124, dated: Mar 1 1954.

Universidade de São Paulo
Faculdade de Filosofia, Ciências e Letras
Caixa Postal: 8105
São Paulo

Dearest Miriam

Thanks for your 2 letters. I shall answer them in parts. So this is the 1st instalment.

First of all, I have definite plans to leave here in a reasonable length of time (In fact several alternative plans). I don't know precisely when I can do it, but as I said before, in 6 to 18 months. Naturally I prefer that you don't discuss my plans with other people. I'll let you know when something more definite develops. I am waiting for various items of information now.

Secondly, it is not really practicable to have a "light-hearted affair" in São Paulo. You do not seem to realise the power of the medieval standards of morality here. This is typical of people who have never been seriously incommoded by such things. They imagine that such things do not exist. An affair with an unmarried woman is difficult, because the woman is literally too scared to try it. The consequences of discovery are just plain too serious. Married women are more likely to have affairs, but then it is important not to be discovered by the husband. For tradition demands that he murder or beat up the other man involved, – whether he wants to or not (More exactly, tradition demands that he take the classical attitude of the outraged husband, or else he may "lose face"). To have an affair with a married student however, as you suggest would be very complicated, even in the US, while here it would be impossible to carry off, because the thing could not be hidden, and it would create a tremendous scandal if it were known.

In any case, "light-hearted affairs" do not interest me much now, perhaps precisely because I am too depressed to be diverted much by them. It would take something

C. Talbot (ed.), *David Bohm: Causality and Chance, Letters to Three Women*,
DOI 10.1007/978-3-319-55492-1_30

more powerful to get me out of the miasmatic and tired state, into which life in São Paulo has dragged me. Also, my health is not too good, and perhaps because of my advancing years, I have a genuine decrease of interest in sex as such.[1] I am sure that the right sort of woman could re-awaken it but very few of the women here are of that type. Most of them would just plain be too dull. There would be nothing to talk about after the 1st hour. Perhaps this little physics student would be better, but the problems are too difficult. In any case, she is only about 20 years old, and has a point of view on life that seems almost like that of a child. So I wonder what kind of a relation we could have. The thing that attracts me about her is a certain manner, a way of talking and a look in the eye which suggests that she is capable of something more than the typical woman (especially Brazilians who have no chance to develop their intellectual capacities at all). Most women seem to be so passive with regard to ideas that I am sure they have long since lost the capacity to have an idea of their own. This little girl has the possibility, but I am sure she will lose it, since life in Brazil does not encourage people to think, and especially not women. The general pattern is to take it easy, to live only within the bosom of the family, to have only superficial relations with other people, to have only superficial ideas about things, etc. It is a sort of dream-like existence (at least that's the way it seems to me) in which each individual is basically isolated not only from other people but also from ideas, concepts, culture, etc. Nothing penetrates him deeply. He senses that something is missing, but his faculties are so dulled that this sensing comes only as a vague dissatisfaction, the origin of which he cannot pin down (perhaps an emptiness in life). Eventually, he loses the desire to have intimate friends and to share interests with people. When he does this, he feels that he has outgrown "childish" yearnings and has adjusted himself in a "mature" way to the "realities" of life.

Of course, all of this is true of the US middle class too, but in Brazil the middle class appears to be much more individualistic and isolated than even in the US. Brazilian culture is literally non-existent, except in connection with the Communist Party, and even that is not very high. All foreigners feel completely lost here, unless their only objective is to make money (which is quite easy).

Now I would like to say something about the probability problem. In addition to the need for having the relative frequency approach a given value independent of initial conditions (except for a set of measure zero) it is necessary that the distribution of events be "random". To illustrate this need, I may give an example due to von Mises.

Consider a set of posts on a road, with one large post every mile and small posts every tenth of a mile. The rel. frequency of large posts approaches $\frac{1}{10}$. Yet, the concept of probability does not apply here. For if we choose every <u>tenth</u> post, then the relative frequency approaches either 0 or 1, depending on whether the initial post chosen is large or small. Here, we have a failure of randomness, since there is a regular (periodic) distribution of posts. The applicability of probability theory requires, however, that such regularity be absent, or in the common way of speaking, that the distribution be "random".

[1] Kinsey seems to be right on this point.

The next question is "What does the word "random" mean". None of the existing probability schools have succeeded in giving it a clear meaning. Von Mises' first attempt was to state the following definition:

"A random distribution is one in which the relative frequency taken in an arbitrary subsequence of events approaches that in the main sequence". (In other words, it does not contain regular sub-sequences, of the type described above in connection with the distribution of posts on a road).

It soon became evident that the above definition doesn't work. For example, consider a series of presumably random throws of a coin. By choosing precisely the sub-sequence consisting of all heads, we can make the relative frequency of heads approach unity. (Indeed, it is clear that with a proper choice it could be made to approach anything at all.)

More generally, the difficulty is that any distribution, however irregular, can always be re-arranged into a regular distribution by a suitable order of choice of terms. Thus, no definition of randomness is possible which is completely invariant to the means of taking out sub-sequences.

From the above it is clear that randomness cannot be an absolute property. A distribution can be random only relative to some order. The order to which we wish to compare it will be decided by the physical conditions of the problem, so that in different problems, we may have random orders of things in time, in space, in temperature, or what have you.

Various followers of von Mises have tried to improve von Mises' definition. Von Mises wished to regard a game of chance as the prototype of all probability theory. He proposed the following idea:

"A truly fair game is one in which it is impossible to find a "system" of playing which will enable one to win in the long run". If the game is truly "fair", then the distribution of plays will be said to be "random". A "system" of playing corresponds to some means of choosing a subsequence of plays which will lead you to win.

Wald, a follower of von Mises, then considered various systems of rules of play, in which the choice of each play was based in a definite way on the results of all previous plays. Wald has shown that it is possible to define a wide class of rules of choice which could lead to the same relative frequency in a denumerable infinity of subsequences as in the main sequence. Wald then defines a random distribution as one which satisfies the above criterion for a particular rule of choice (or a denumerable infinity of rules).

Wald's definition can be criticised as follows:

(1) It cannot be applied in practice, since we must test with an infinity of rules of choice; with no reason given to prefer any one over the others.
(2) Its application to other fields (e.g. physics) is very clumsy and it is not too clear that the satisfaction of Wald's criterion would be enough to make all of the applications of probability in such fields valid.
(3) In any finite number of plays there always exists some rule of choice of sub-sequence which can lead to victory. Thus there is no unambiguous meaning to

randomness [unreadable] real [unreadable] consisting of a finite number of plays [unreadable sentences].

Nevertheless I believe that Wald has made a contribution, e.g. to call attention to the fact that randomness must be relative to some order.

Let us now come to the attitude of the axiomatic school of Kolmogorov. This school does not believe that the question of randomness belongs to the theory of probability. They merely want to define various probabilities, p, as entities which are not further analysed at the level of mathematics. Various rules of operation with these "p's" are then defined, and the deduction of the further consequences of these rules constitute all that there is supposed to be of the mathematical theory of probability. The question of where this theory works is to be decided empirically in each specific field.

Now naturally we know that the theory of probability won't work unless the distribution is really "random", where we use the word "random" in its intuitive sense of "irregular". But the axiomatic school says that this question of randomness cannot possibly concern the mathematical part of the theory. The mere fact that the theory works in a particular field should be regarded as a-posteriori evidence that the distribution is random enough.

More generally, the question of the applicability of the theory is supposed to be the concern of the scientist, the statisticians, common sense, etc. but never a question of mathematics. The mathematicians have furnished us with a theory of probability, whose foundations are regarded as completely finished and never subject to further growth, expansion or [missing word] in response to the needs to deal with new problems concerning [missing word]. The scientist, statistician, etc. will try to use this theory, and will find that it works, they will continue to use it, and where it does not work, they will not use it.

The axiomatic school can be criticised as follows:

(1) It is certainly true that the axiomatic school have furnished us with a self consistent and logically complete mathematical theory, which has been successfully applied in many domains. But this does not prove that it is the only possible mathematical theory, and that it must [word missing] impossible in all mathematical theories of probability to deal with the question of the applicability of the theory, or of randomness [word missing].

(2) Since it is clear intuitively that the theory of probability requires a random distribution for its validity, it seems likely that it might be possible to [word missing] mathematical theory of randomness which should permit [word missing] of the scope of the theory of probability. In this regard, the [word missing] Von Mises is right in trying to study the problem of randomness but they have not yet succeeded in giving a good definition [word missing] much less a theory of randomness.

(3) There is a wide domain of problems which the present theory of probability does not treat and which still contain statistical elements. These may be described as problems lying part way between regular distributions and random distributions. The following are some of them.

(a) Consider the theory of errors of measurements. We have a systematic reproducible error, and a fluctuating error that is assumed to be random. By taking N measurements, the fluctuating error can be reduced to $\frac{1}{\sqrt{N}}$ times the mean fluctuating error of each measurement. Thus, if we took enough measurements, the fluctuating error should be [word missing] as small as we please.

The above reasoning is, however, wrong. For in a real measuring apparatus, there exist errors which are not strictly reproducible, and which are not yet irregular enough to be strictly random. For example there are fluctuations in temperature, small irreversible changes in machinery (such as breaking minute bits off an axle, etc.). This intermediate type of error cannot be reduced indefinitely by taking more measurements. Thus we really need here a theory taking into account the possibility of all types of distributions, lying between completely random and completely regular.

(b) Consider the problem of social, economic, or industrial statistics. If we draw a graph, we often see a general trend, in which irregular fluctuations take place. The application of the theory of probability implies that the fluctuations about the mean trend have randomness[2] Yet, in reality, part of them may just be irregular without being totally random.

(c) In real games of chance (e.g., roulette wheel), it seems doubtful that the concept of randomness is perfectly valid. For again there may appear irregular changes in the machinery (as in theory of measurements) which are not reproducible and yet not totally random. One could try to treat the problem by defining an irregularly changing probability ($P(t)$) which changed, for example, when a small bit broke off the axle of the wheel. But even this becomes difficult to define, since the axle may maintain its state for only a limited number of plays, so that we cannot reproduce conditions for an infinite (or very large) number needed to give probability interpretation as the limit of a relative frequency. Here again we see the [word missing] introducing distributions which are neither perfectly random nor perfectly regular.

(4) Non-equilibrium problems, such as those arising in statistical mechanics. We may start with a very regular array of molecules, and after enough collisions, the distribution will be so irregular that the theory of probability applies. Clearly, the distribution must pass through some intermediate stage, where it is becoming random. To treat this, we need to be able to define such intermediate distributions.

To sum up, we see that there is a wide range of real problems, where statistical distributions arise which are neither regular nor random. To treat these, we must be able to define such distributions, and particularly to define the degree of randomness. In other words, we must stop making a sharp division between random distributions to which the theory of probability applies and regular distributions to which it does not apply. Instead, we must consider a wider class in which we approach conditions in which the theory of probability applies. In this way, we can

(a) Treat new types of real problems.

[2]Specifically it implies that there exists a probability, P, of any given kind of fluctuation whose interpretation is in terms of a suitable randomness in the process of fluctuation.

(b) Define the limits of applicability of the present theory of probability, as the limits in which randomness is being approached. We can also get a better estimate of the error in using the theory in a real problem.

(c) Treat the problem of how a random distribution develops from a more ordered one or vice versa. Thus we do not have to imagine randomness is an eternal, undefinable and not further analysable property of distributions, but we can show how and when it comes into being, how and when it can be destroyed, and we can deduce some of the characteristic properties of random distributions on the basis of a deeper understanding of the causal laws. In this regard, we see that randomness can arise out of causal laws, can be reflected in causal laws, just as it can lead to new causal laws at another level (e.g. law of large numbers – averages become practically determinate because fluctuations become so small). Thus randomness arises out of regular laws and regular laws arise out of randomness. Each is equally real, and both are inter-dependent and interconnected.

Thus, we aim to answer the axiomatic school by developing a general mathematical theory of randomness, which shows how to give a mathematical treatment of just those problems which the axiomatic school claims belong only to the scientist, the statistician, common sense, etc.

On the other hand, we will do this by going beyond the sharp distinction made in the von Mises school between those cases where probability can be applied and those where it cannot.

I wish to add here another important example of processes between random and regular. This is in hydrodynamics. In turbulent motion, we have some very highly irregular motion. Yet it is not totally random; for if we look at the fluid at any instant, we can see a complicated but irregular pattern of vortices. As we consider smaller and smaller vortices, the concept of random motion gets better. But at the large scale, it just isn't totally random. Now, a theory has been developed that works in terms of correlated motions at neighbouring points. But this theory cannot cover the problem completely, since it would be quite possible for a fluid in a non-equilibrium state to have irregularly changing vortices that were not reproducible, so that the theory of probability would not apply and which would still not be totally random. In other words, if we tried to do the same experiment again, we would be sure to change conditions somewhat. But these changes would not fluctuate at random.

Now let me outline my ideas on what should be done about this problem. First of all comes the problem of a definition of the degree of randomness. To do this, I shall use the concept of correlations. Let us consider for example a series of N coin throws with equal probability of heads and tails (where N is large but finite). We are interested not only in the over-all relative frequency of heads (approximately 1/2) but also in the question of whether or not there is some regularity in their distribution. For example, there might be many cases for pairs of heads to be produced in succession (e.g. $++--++----++-\ldots$) Such a correlation would imply a lack of randomness. To test for such correlations, we could evaluate various correlation functions. To define our correlation functions, we shall let $x = 1$ for heads, -1 for tails. Then we may consider functions such as:

$$\sum_n (x_n x_{n+1} - \bar{x}_n \bar{x}_{n+1}) \sum_n (x_n x_{n+2} - \bar{x}_n \bar{x}_{n+2}) \sum_n (x_n x_{n+3} - \bar{x}_n \bar{x}_{n+3}), \text{ etc.}$$

These are the two-fold correlations
Then we have the three fold correlations

$$\sum_n (x_n - \bar{x}_n)(x_{n+\Omega} - \bar{x}_{n+\Omega})(x_{n+s} - \bar{x}_{n+s}), \text{ etc.}$$

etc.
The above correlation functions are defined over the actual finite distribution (whatever it may be) and not as in the usual theory of probability over a hypothetical infinite distribution.

Now, a random distribution might be defined as one in which all correlation functions vanished (or were very small). It is clear that if all correlation functions vanished then all of the consequences of the theory of probability could be applied; for we should have complete statistical independence in the evaluation of the average of an arbitrary function of x. Thus, a random distribution would permit the theory of probability to be applied and by estimating the various correlation functions, we could estimate the error to be made in applying the theory in the evaluation of any given function of the x.

But now we come to another problem. There is an infinity of correlation functions. In any finite distribution, it would be impossible for all of them to vanish (or to be small). But most of these are higher order correlation functions representing very complex types of correlations. In any practical problem, we are usually interested only in lower order correlations. The higher order correlations represent special types of order which are however so complex and irregular that for practical purposes, they produce the same results as if such order did not exist. We may give as an example the series of posts suggested by von Mises. Suppose that the large posts were distributed, not uniformly, but by means of some very complicated rule. As the distribution grew more and more complicated and irregular the practical results of the correlations will grow less and less important, so that the use of the abstraction of a completely random distribution (and with it the usual theory of probability) would become a better and better approximation in a wider and wider set of applications.

Now you may ask "what determines which correlation functions are important?". The answer is – "the specific conditions of the problem". For example, in statistical mechanics, complicated correlations of particle positions which take place in a very small distance are of no importance in determining typical large scale averages (such as pressure, temperature, etc.) Thus, only a limited number of correlation functions is important, and the distribution is effectively random if these correlation functions remain small. Similarly, each problem presents its own conditions for defining what is a good enough approximation to randomness.

You can see from the above that there is no such thing as absolute and perfect randomness. There is only an approximate randomness relative to some order (for example relative to space or time). The proper order is determined by the conditions of the problem. It is that order in which only the lower order correlation functions are of practical importance.

It is clear that the correlation functions can be regarded as a sort of infinite dimensional Hilbert space. A re-arrangement of order is equivalent to a transformation in this space. Now, any order, however irregular, can by a suitable transformation be made regular (or vice versa). Thus there is no absolute way to specify the degree of randomness or of regularity. But we can specify the correlation coordinates in this Hilbert space. The proper axes in this space are determined by the physical conditions of the problem and an approximately random distribution is defined as one in which the physically significant correlation coordinates are small.

Now I shall illustrate these ideas in the example in which the angle is multiplied by K where K is a large integer ($\theta_n = K\theta_{n-1}$).

We have already shown that for all initial conditions except a set of measure zero, θ_n becomes equally distributed as $N \to \infty$. Now we want to show the statistical independence of successive steps. To do this, it is necessary to show that the mean value of arbitrary products such as

$$f(\theta_n)g(\theta_{n+m}) \quad f(\theta_n)g(\theta_{n+m})h(\theta_{n+s}), \quad \text{etc., etc.}$$

is just the product of their mean values. Proving the above would show that there are no correlations and that there is "statistical independence" of successive steps. Now an arbitrary function can be expressed as a Fourier series

$$f(\theta_n) = \sum_s e^{is\theta_n} f_s$$

But here is where we come to the crux of the argument. We are going to consider problems in which the distribution in a very small range of angles is of no physical importance. For example, we shall suppose that we are concerned only with problems in which an accuracy greater than $\Delta\theta = \dfrac{2\pi}{A}$ is of no physical significance. This means that we are concerned only with functions in which Fourier components with $s < A$ occur; for higher Fourier components can affect the distribution only in a region smaller than $2\pi/A$. Thus we write

$$f(\theta_n) = \sum_{s=-A}^{A} f_s e^{is\theta_n} \quad g(\theta_n) = \sum_{s=-A}^{A} g_s e^{is\theta_n} \quad \text{etc.}$$

The product $f(\theta_n)g(\theta_{n+s})$ can be written as

$$W = \sum_{s=-A}^{A}\sum_{t=-A}^{A} f_s g_t e^{is\theta_n} e^{it\theta_{n+m}} = \sum_{s=-A}^{A}\sum_{t=-A}^{A} f_s g_t e^{i(s+t)\theta_n} e^{itK^m\theta_n}$$

Now because of the equidistribution, $\overline{e^{iB\theta_n}} = 0$ where B is any integer.

The only non-zero contributions to W will therefore come from $s + t + (K)^m t = 0$. One contribution occurs from $s = t = 0$. This corresponds to the product of the mean of the functions, $f(\theta_n)$ and $g(\theta_n)$. Remembering that m is at least unity, we see that no other contribution can occur unless

$$s + t = Kt \quad \text{or} \quad s = (K - 1)t$$

The smallest value of s that will lead to an additional non-zero contribution is therefore $s = K - 1$ which occurs with $t = 1$.

If we choose $A < K - 1$, it is then clear that no two-fold correlation occurs, since we have

$$\overline{f(\theta_n)g(\theta_{n+m})} = \overline{f(\theta_n)}\ \overline{g(\theta_m)}$$

provided that we are not concerned with functions f and g that vary significantly in a region $\Delta\theta = 2\pi/K$

Three-way correlations are treated similarly. We have

$$\overline{f(\theta_m)g(\theta_{m+n})h(\theta_{m+p})} = \sum_{\Omega=-A}^{A}\sum_{s=-A}^{A}\sum_{t=-A}^{A} f_\Omega g_s h_t e^{i(\Omega+s+t)\theta_m} e^{i(sK^n+tK^m)\theta_n)}$$

We get a contribution from $n = s = t = 0$
Noting that n is at least 1 and m at least 2 (by definition), we see that the most favorable condition for another contribution is

$$\Omega + s + t = sK + tK^2 \text{ or } \Omega = s(K - 1) + t(K^2 - 1) = (K - 1)(s + t(K + 1))$$

If $|\Omega| < K - 1$, $|s| < K - 1$, $|t| < K - 1$, this condition cannot be satisfied.
(The most favourable cases are $t = 0$, or $t = -1$, $s = K - 1$, neither of which can be satisfied for $|\Omega| < K - 1$).
A similar proof holds with higher order correlations.
Thus we have proved that in taking products of arbitrary functions of successive angles, there is no correlation, so that the use of the theory of probability with the assumption of statistical independence of successive angles would lead to correct predictions in all applications which were not sensitive to a precision of angle higher than $\Delta\theta = 2\pi/K$. Nevertheless, there could still exist correlations involving smaller angles than this.

The above result shows that a perfectly determinate order of angles can reproduce to a high degree of approximation all of the properties that we are intuitively led to associate with a "random" distribution. It is merely necessary that the order be so irregular that no appreciable large scale residues of regularity persist.

It is clear that now that we have a definition of randomness, we can treat the problem of estimating the error in the theory of probability (By estimating correlation coefficients). We can also discuss the transformation of regular order to a random order and vice-versa, in terms of the changes in the correlation coefficients.

One more point arises, the fact that by means of Markov processes, correlations can be treated even in the usual theory of probability. But the treatment is different. Thus, the usual theory of probability would work with a probability $P(\theta_n) = \int F(\theta_n, \theta_{n-1})P(\theta_{n-1})\, d\theta_{n-1}$
Some correlation would exist between θ_n and θ_{n-1} (depending on the form of F), but after enough steps, this would disappear. But the question of correlation would be meaningless for a single chain. It could only be given meaning in terms of a statistical ensemble of chains, in which it would be clear that every time θ_n took a certain value, a large fraction of the chains would show a value of θ_{n+1} close to θ_n. But for θ_{n+s}, this correlation would disappear, as s became large.

We could get Markov-like chains in our determinate process by making the number K, very small. Thus it would take many steps before the correlation was destroyed. What would happen is this: The more steps we considered in $f(\theta_m)g(\theta_{m+s})$, the smaller the region of $\Delta\theta$ into which the correlations would retreat.

The important difference in our method is this: in a single chain we can show how all initial conditions (except a set of measure zero) must create a sequence in which all kinds of correlations gradually become smaller and smaller as the sequence grows longer and longer. If we take a small value of K, we can even retain local correlations (as in the case of a Markov chain) but we can see these local correlations gradually disappearing as the chain grows longer. Thus, we see with a single chain how the theory of probability gets to be a better and better approximation. Before the chain has grown too long, we have a case in which the distribution may be somewhat irregular without being effectively random.

I would appreciate it very much if you will send me your comments on these ideas. You do not have to study them carefully, as I should be interested even if you only give your general impressions as to clarity, etc.

Since this letter is getting so long, I shall close for the time and discuss the various other points raised in your letters within the next few days. I must apologise if I have been too cold to you. Again I say that this lonely existence here tends to freeze me up a bit, but I expect to get over it when I get out of here, which may be fairly soon.

Meanwhile I hope that the infant isn't bothering you too much. I have heard about a process called painless childbirth without anaesthetics. Have you heard of it? The idea is based on the theory that most of the pain is a conditioned reflex, which can be deconditioned by suitable training. I have heard of a number of people who successfully had children in this way.

Until a few days
Love
Dave

Letter 115. Folder C125, dated: March 15 1954.

Universidade de São Paulo
Faculdade de Filosofia, Ciências e Letras

Caixa Postal: 8105
São Paulo

Dearest Miriam

I am once again in the mountains for a few days rest, and with a more leisurely mood, I'll try to tell you a few things.

As I told you before, I am definitely planning to leave. Israel is one possibility. I wish you would send me the names of people to look up in case I should go there (naturally you must not mention my plans to anyone). Later on, perhaps when my plans are more definite, you might also write to a few of your friends there, to let them know that I am coming (But not till I tell you).

Right now, my health appears to be poor, for some reason. I really don't know what it is, but I have noticed in the past few months a very large drop in sexual desire. It may be purely psychological, of course, as I feel by now that I shall never bother with any women in Brazil. But I am afraid that there is a partly physical basis, since I have noticed lately that the muscles of the abdomen and the legs are very much strained, so that sometimes it hurts a lot when I ride in a bumpy car. I have always had a slight hernia, and perhaps it's getting worse. In any case, it's sort of annoying, because one feels that something that was always there is now missing. Otherwise I do not appear to be unusually run down; and my general physical strength is about normal, so I don't know what to make of it. I'll go to a doctor when I get back but I don't have too much confidence in these Brazilian doctors.

But I guess what is a lot worse is the steadily growing loneliness, which literally numbs the emotions, by a process of desensitization that is necessary for survival in this cold atmosphere where life is lived at such a low level (emotionally and intellectually). To save my life, I'll just have to get out of here, where there is really nobody to talk to except the Schillers (and I have by now pretty well exhausted what I have to say to the Schillers, too). To have a few isolated foreigners who cannot fit into the society is really very bad. Eventually, it stultifies even the relationships of these foreigners, because they form too small a group to maintain a vital interest for a long time. One needs a large group, with new ideas, problems, events, etc. always surging up, or else one will grow stagnant and dried out.

So at the moment I am living only in hope of leaving. What I would do if I did not have this hope, I don't know. Brazil seems such a desert as far as possible human relations are concerned that life would hardly seem worthwhile if I had to stay indefinitely.

Please forgive me for being depressed. At this moment, I am feeling tired and lonely, and also worried by this continual pain in the muscles near the crotch. I am sure that I shall feel better later, perhaps even tomorrow.

Having a baby must be a tremendous experience, especially the first time. The kid will doubtless require a lot of care, much of which is boring. But it is surely worth whatever trouble that may be involved. When I see a happy young child, I feel a sort of "tug" inside. Somehow, children seem to have the genuineness of feeling which adults lose in the struggle against each other. Merely to be in contact with such genuineness of feeling awakens in me emotions which I had thought were dead. I may have told you that when I was a child, I felt that there is a division line of the order of 14 years. Beyond this line, people ceased to be genuine, but had their attitude determined mainly by the current prejudices of society (which are mostly false). Therefore I felt that adults cannot be trusted. And since growing up, this feeling has become stronger in me. Perhaps this is one reason that George is appealing – he has some of the intensity and whole-heartedness which people usually lose when they grow up. For this reason, also, George should get along well with children (Unfortunately, he also retains some of his childishness in his thinking too).

I want to say here that I value your letters not so much because they are (sometimes) amusing, but mainly because I need your friendship. I think that you are, in most ways, the person with whom I get along best. Usually, with most other people, I exhaust all topics of conversation in periods varying from 15 minutes to a few hours, but with you, this did not seem to happen. Also, there is a feeling of peacefulness, a lack of tension, that I seldom have with other people. I guess this question of love is a difficult business to analyze. With me, in my present frozen state, I am afraid that it's a thing that must develop very slowly, like a hard and somewhat damp log that will burn only in a heavy fire. When we were together in Washington for a few days, I felt a bit cold at first. But gradually I began to get over it, and by the last day (Sunday) I felt a warmth and ease of comradeship that I can still remember. But with my very encased temperament, it is very difficult for me to start something, because I begin by feeling so cold and doubtful that it hardly seems to be worth trying to get acquainted with another person. With my natural shyness in these things, I therefore can never get started. And with you, things were complicated by the fact that it would have been necessary to make a very difficult decision before one felt sure that the thing would work out right. If you had not been married to George already, I probably would have felt justified in making a try of it, but as things were, it seemed too big a step. Perhaps it would have been wise, to take a chance. Who knows? By now I am probably too cold a fish (as you say) to be able to marry anyone. This life in Brazil has certainly contributed to extinguishing my emotions. Also, the way the situation has been developing throughout the world contributes. For my youthful enthusiasm about the possibilities of people is cooling off. I still believe in these possibilities, but only in the long run. People raised in the present society are so corrupted, stultified, and twisted in general that it is difficult for me to take most of them seriously, much less to love them. I feel that people in general fall into three classes: (1) Those that are too stupid to know what human beings can and should become (2) Those that are intelligent enough, but who have had the idea knocked out of them, or smothered by a flood of misinformation picked up in education, reading, or just in common life. (3) Those that are intelligent and who do understand these things, but who lack the courage to try to bring their ideals into reality. Most of these people will have to betray what is most fundamental in themselves, in order to make their peace with society. Then there are a few people who go beyond these categories, but still, so to speak, have their feet in the mud, even though their heads are part way out. Most of these people share almost all of the defects of the people in the other 3 classes. Those who have courage frequently lack intelligence or have only a superficial understanding, so that their actions are ill-conceived and do not gain success. So that a solution of the world's problems seems a long way off. And meanwhile, perfectly good children will continue to be transformed into beings who are more or less like the adults of today (if not worse). And those who resist will be trapped in bitterness, in the ferocity and apparent hopelessness of the struggle, so that their personalities too will be distorted, dried up, rigidified, etc. I really think that society always has been a sort of inferno, and that modern capitalistic society is even more so. In Brazil you can see it more clearly, as there is here a sort of caricature of capitalism, which brings out clearly its true nature without the softening haze of culture and social conscience that has been allowed to develop in other countries.

As for me, I seem to have only one strong emotion left – and that is hatred for the forces that have destroyed so many human beings (including myself). For relative to what I could have been, I regard myself as destroyed and the same is true of so many other people. I am not particularly bitter that they took away my job, and would probably like to send me to prison. Once you grant the premise that the "American Way of Life" is the right one, then what they did to me is quite logical, and indeed their treatment of me has been irrationally mild. What I cannot forgive them, however, is just precisely the "American Way of Life". It is a way of life that glorifies selfishness, cold calculation, conformity, insincerity ("Dale Carnegieism"), a way of life that stultifies all ideas except those narrowly connected with profit, that kills all real comradeship and trust between people, that turns human beings either into dull work horses or into unprincipled exploiters (often into both at the same time), that makes people lose their humanity, so that they can no longer really be friends, that works the life out of people and then threatens them with the insecurity of unemployment due to depressions, that throws them on the ash heaps when they are old, if it has not already killed or maimed them in wars aimed at solving "economic difficulties". It is a way of life that drugs their minds with sex, cruelty, and mass sports. It turns love itself into a commodity, compelling a large fraction of the women to prostitute themselves, and requiring larger numbers to marry for economic security.

I could go on like this for hours. I have hated the American Way of Life from the moment in which I was conscious of the need to take care of myself against others, from the moment in which, as a child, I had to hide books, so that people would not think I was "abnormal" or a "sissy". When I was younger, I didn't see that all of these things I hated sprang from features intrinsic to capitalism, and therefore deeply rooted in America as the country where the people as a whole have most thoroughly absorbed the ideology of capitalism in its most virulent form. And it is just for doing this that I find it hard to forgive the American people. For by my own experience I know that this kind of life has effectively negated most of what I could have been, and what most other people could have been. The American people have most thoroughly betrayed what is best in themselves. I would like to help them if I could, but the only thing that could help them (Socialism) is precisely what they are convinced is the worst evil in the world. They are in a trap so complete that I fear only some tremendous external shock will get them out of it.

To repeat, my loss of a job has never made me very angry, because I can understand that they want to get rid of me if I don't fit into their way of doing things. Likewise, the difficulty, as you put it, for a "Jewish woman" to do some intellectual work is quite natural, since the Jews have never quite fitted into their way of life either, so that one can understand why they mistrust, fear, and hate Jews. If their way of life were the right one, all could be understood and even approved of to a limited extent, since they would have the right to protect themselves from people who are untrustworthy from their point of view. But what is wrong is precisely that their way of life is a bad one, a way that looks backward instead of forward, that looks at human beings cynically instead of seeing their unlimited possibilities, that crushes and stultifies what is best in people, and brings out mostly what is vicious, cruel, stupid, selfish and cowardly. I cannot forgive them for creating conditions in which I could have so few real friends, and in which I lost the opportunity to work together in a comradely way with other people toward a common goal that was worthwhile. I cannot forgive them for making it necessary for me to live a life of loneliness and futility, if I wanted to go beyond the inane and petty interests that occupy most people. I cannot forgive them for turning most people into the dull, limited creatures that they are, and for breaking or corrupting most of the few who tried to resist these processes. I cannot forgive them for turning even their own lives into parodies of what life should be, where the futility and loneliness is reflected in the faces even of the most prosperous people, and much more in the young men in front of soda fountains and pool rooms in a typical American city, with nowhere to go, nothing to do, with only emptiness and futility awaiting them for the rest of their lives. So you see, I still have some emotion left, even if it is not a very good kind.

Another thing that disturbs me is the continual evidence of corruption of the left-wing movement throughout the world. For example, the declaration of the British Labour Party in favour of rearmament of Germany. How can they seriously maintain that all of this fear of a revival of Nazism is just something stirred up by the Communists. Apparently to remember that the Nazis have killed 12,000,000 people is nothing but Communist propaganda nowadays. The Germans have now all become good democrats, and must be allowed to control Europe to protect us all from Communism (just as Hitler himself said 10 years ago. Perhaps it was all a mistake to have fought the Nazis. Maybe we should have allowed them to destroy the Russians so that the world could then have gone on to Democratic Socialism instead of Despotic Communism). Really though, such a cynical betrayal is disgusting, if not surprising. If the left wing of the Labor Party does not soon dissociate itself from such an insane policy, it will be ruined. How can it claim to be Socialistic, if it follows the foreign policy of Dulles, aimed at crushing Socialism in the East? To imagine that they can be reactionary abroad and progressive at home is absurd. For the money spent on the war is alone enough to prevent social progress. More important, if they fought for the line that anti-Communism is the most important thing, then they will inevitably lose to the Conservatives, who are clearly better at anti-Communism than are the Labour Party (just as happened in the US with McCarthy). In any case, the cynical dishonesty of such a policy will eat into their desire to establish Socialism. The whole movement will rot away if they don't make a strong stand within a year or two (at most).

What seems most likely is that 90% of all socialist movements will, in fact, rot away, as is almost inevitable, since they represent only the vague desire of the middle class for a better life without making a fundamental change in the social order. Since this is impossible, they will continually be confronted with crises, in which they must choose between their desire for a better life and their desire not to change the social order in a fundamental way. Most of them will choose the latter and thus betray their Socialistic ideals. The only party that will have the courage to face the real issues will, in the long run, be the Communists. Whatever one may not like about their methods it seems to me that in the long run, they will take over just by default of all the other parties. Even though they are unpopular for the moment in Britain, one must remember that things change. Parties that are small at one time may be large at another. And in the long run, it is hard to see how any other party will have the courage to carry out the stern measures, that are already needed much less the even sterner measures that will be needed later if a solution to the crisis is not found soon. The only way that this could be avoided is that the left wing of the Labour Party should split and come out soon in favor of a clear policy of over-all Socialisation, withdrawal from the war, trade with Russia, etc. This is still a possibility, but every month of delay makes it less likely. It is still a possibility because France may take the lead in rejecting EDC and thus starting on a path that must carry it ever further away from the Dulles line. This would create a pressure on the left wing of the British Labour Party, just as the right wing of the British Labour Party is now putting pressure on French Socialists to back EDC and the rearmament of Germany.

All in all I am afraid that there are difficult times ahead. There is a real, but small probability of a world war, which would be a horrible thing. Then there may be economic crises, with much disorder and suffering which will eventually lead to an improvement. But I am afraid that progress will be slow and hard. It is doubtful if that, in most places, a democratic solution can be found, because even in the best countries, such as England, the Socialist parties are pretty corrupt. In any case, the possibility of a relatively non-violent solution depends on the left-wing movements getting out of the self-created trap of blind anti-Communism, which is the principal impediment to a solution of social problems today. This will require as much courage and clear-sightedness, perhaps more than exists among these people.

To change the subject let us return to the probability problem. I do not agree with your statement that the most important part of our programme is to go beyond the proof of quasi-ergodic theorems, in the sense that we can specify whether or not a given initial condition, θ_0, leads to the correct average of an arbitrary function $\lim_{N\to\infty} \frac{f(\theta_n)}{N}$. This is, to be sure, an interesting question, but not the principal question for [line missing] the following:

(1) It is possible to define the degree of randomness in any given distribution, $\theta_1 \ldots \theta_N$ (see my previous letter), whether this distribution arises from a causally determined sequence (for example, $\theta_n = f(\theta_{n-1})$), whether it has been obtained in an empirical way by observation, or in any other way conceivable. In other words, given a specific sequence, regardless of its origin, we can define its degree of randomness and thus estimate the error in using the theory of probability for calculating the

average values of various functions of the θ_n, or of combinations of the θ_n, such as $f(\theta_n, \theta_{n+1}, \theta_{n+2} \ldots)$. In this way, we do not have to make a sharp separation between sequences to which the theory of probability applies and sequences to which it does not.

(2) To the extent that we know the law governing a particular sequence, we can deduce whether or not it will ultimately become random.

(3) We can follow the process of the development of randomness.

(4) We can consider sequences in which some or all of the functions such as $f(\theta_n)g(\theta_{n+s})h(\theta_{n+t}) \ldots$ do not approach a definite limit. In such sequences, the theory of probability does not apply. But the failure to approach a definite limit may itself be limited. In other words, within a certain degree of precision, such functions may approach a definite limit, but not to a higher degree (this happens with $\theta_n = K\theta_{n-1}$). Thus, the limits of applicability of the theory of probability can be deduced in such cases.

(5) We have a conceptual framework which permits us to develop new forms of mathematics treating statistical distributions that are only partially random. For example, in the problem of errors in the theory of measurements, we may postulate an irregular but not random error which has however a component approaching randomness. Thus, within a certain degree of precision we have randomness, but to a higher degree, the correlation coefficients do not vanish, but have values that change in an irregular way that does not necessarily approach a limit.

As you can see, the question of whether the distribution of θ_n approaches a uniform limit is only one point. The problem of correlations is equally important.

Now the theory of probability is useful precisely in those conditions where it enables us to make predictions independently of initial conditions. For usually we use it either where initial conditions are unknown to us, or where the calculations would be too complicated for us to utilise a knowledge of such conditions. But such a use is correct only if the mean values of various functions are really independent of the initial conditions. If we postulate that the set of measure zero leading to non-typical averages will never occur in reality, then we can safely conclude that the real averages will ultimately approach those given by the theory of probability. But we do not know precisely when or how, since these things depend on the initial conditions. We may have to wait a very long time, but we may calculate a mean time, near which most of the initial conditions lead to an average close to that given by the theory of probability. But without a knowledge of initial conditions there is no way to tell how or when the averages will approach their ultimate value.

Thus, the conditions for applicability of the theory of probability imply both an objective and a subjective side. The objective side is this:

The theory is applicable only when the motions are sufficiently "random". When they are random, then any initial conditions that can occur will ultimately give the same averages as those of the theory of probability, whether we know these initial conditions or not.

The subjective side is this:

The theory of probability is usually used when we don't know all the initial conditions or when we cannot calculate the relation between these conditions and the final results. Our lack of knowledge is however, reflected in an uncertainty on our part as to precisely when the real averages will approach the predicted values. This is not an objective uncertainty, since this time is in principle determined by the actual initial conditions and the causal laws.

Here, we must make a postulate, verified by experience. It will be rare that the real average approaches the predicted average in a time much longer than the mean determined by averaging over all possible initial conditions.

This is a highly qualitative postulate. It cannot be proved mathematically, but it can be made reasonable, if we remember that every system is in reality part of a world that is infinite in extent and has an infinite number of levels and that all levels have some element of random motion. Thus random fluctuations outside the system or at other levels tend to cause a statistical distribution over possible initial conditions, making it rare to pick on any particular initial condition. Since conditions that lead to an unusually long time for the average to approach the theoretical value have a small measure it is reasonable to assume that such conditions will in fact appear rarely (that is, it is consistent with our general picture of the world).

Our postulate can be stated in another form.

Initial conditions of very small measure will appear rarely. This postulate is not invariant to a change of measure. But it doesn't have to be. Let us recall (see my previous letter) that a distribution cannot be absolutely random, but is random relative to a particular order, that is determined by the physical conditions of the problem. In most problems the order is space and time; in statistical mechanics, we also have momentum, to give phase space. Thus, the natural measure which is significant is already determined for us. It is the measure in which distribution is approximately random (evidenced by the variation of the physically important correlation functions).

So much for the probability problem. You asked about George's expenses. He was paid 20,000 cruzeiros, at that time worth \$500, for 2 months expenses. His total expenses for the two months were just about \$500, so that the two cancel.

You mentioned in your previous letters that many of my friends are no longer reliable. It is important for me to know who these are and why. Actually, aside from you and George and Melba, and the Grosses, and a few other people, I never had any close friends. I still trust Hanna and her parents, but of course, we are in strong disagreement politically. I must confess that I am a bit worried about Eugene, not only from what you have said, but also because a person who withdraws from all social problems is in danger of a transformation of his personality that will ultimately corrupt him. In order to feel secure, it is not enough that he voice no dangerous opinions, it is not even enough that he voice the current opinions insincerely. He is always under strong temptation to believe the opinions that he thinks will make him safer. Also Eugene's convictions are in my opinion 90% intellectual, so that they are not as securely rooted as they might be. In time, an intellectually based set of convictions can transform a person's emotions, when combined with the experiences of life. But I fear that Eugene has had neither the time nor the opportunity for this to have happened. Of course my fears may be all groundless, since I have no real contact with him at this level at present.

With regard to Hanna, I believe it was wise that we didn't get married. I really liked Hanna a lot and still do. But our temperaments, opinions, and aims were so different that we would surely have gotten into trouble especially in the very difficult situation that we would have met in Brazil, where she would have found no friends and very little to occupy herself. I think that for a fish as cold as I am supposed to be, it would not be wise to want to "fall in love" before getting married. Strong affection plus agreement in temperament and aims should be enough to stand on – the love will have to develop later, if at all. Anyway, 90% of cases of love die out after the first year of marriage, so if it doesn't develop, I would be no worse off than most people are. But with Hanna, trouble would have developed, in fact was already developing. Also, her emotional instability would have made it harder to take things if they hadn't turned out well. She has a tendency to get depressed, almost to the point of a stupor, which is a sort of exaggerated form of my own tendencies. (Perhaps if I had to live in the hopeless situation that faces most women who want to do some real work, then I would become hopelessly depressed too). She is basically, however, a very warm unselfish person who likes to be concerned with people and things outside herself. Under socialism she would do very well and be very happy. Her ideas are often stupid, but part of this is due to her education under a strongly Catholic teacher whom she admired very much, at a very impressionable age. I think that she has real talents in the artistic direction, but lack of stamina and courage has prevented her from realizing them. This lack of stamina is probably due to serious emotional conflicts, at least in part. In a better society, she could resolve them easily; indeed in a better society, they would probably never have arisen in the first place. But under present conditions, I feel that things will be very tough for her.

A few days later –

Well, I feel a lot better now that I have had a few more days of rest, plenty of hiking, good food, etc. I've been working very efficiently and am surprised at how clearly I can think now. What pain in the muscles near the abdomen has mostly gone away, though it returns sometimes. Really, it's quite wonderful to get away from São Paulo. It is a much worse city even than New York – hotter, noisier, smokier, more chaotic, worse traffic, and with absolutely nothing to do. It really wears you down living there, especially because I cannot find food there that does not give me diarrhoea.

My only complaint here is still that I have nobody to talk to. But with good health and work to do, I don't mind it so much. It's bad though when one is too tired to work, or in health that is not so good.

I am afraid I'll have to end this letter now as I am very tired from a long hike, and I must return to São Paulo tomorrow, were I shall mail this letter

Love
Dave

P.S. About your sense of humor, I agree that you should be a bit more careful with it hereafter. Sometimes (quite often in fact) you say things that are really funny, but very often your humor tends to hurt or embarrass people. Unless you really wish to do this, I would advise going easy on that kind of humor. As for me, I don't mind it if you occasionally try to show how clever you are – provided that you don't do it by trying to show how stupid I am. But something that really offends me is when you try to use your humor as a weapon; for example, when you use it to justify a superficial attitude to questions that you would prefer not to enter into more deeply. I am willing to take a reasonable number of jokes at my expense, but not at what I regard as the expense of a correct understanding of a problem or a situation.

Don't let me dissuade you from writing jokes. I really enjoy most of them and would miss them if you stopped (Some of these at my expense are the funniest, and it would be more of a loss not to have them, than to avoid the slight annoyance which some of them may cause). But jokes like those aimed at the infinity of levels, I wish you would try to avoid.

Letter 116. Folder C125, dated: April 54.

Universidade de São Paulo
Faculdade de Filosofia, Ciências e Letras
Caixa Postal: 8105
São Paulo

Dearest Miriam

I hope that you have received my long letter of a few days ago. I wish to write now merely to let you know that Routledge and Kegan Paul in England have agreed to publish my book on causality, provided that I make it a bit longer and treat my own new ideas in more detail.

I had already been making extensive revisions of the mss independently in just this direction. But it would help me if you were to send me your detailed criticisms, as soon as your health and the infant give you time to do this. No hurry. Please let me know all about the infant when it arrives.

Love
Dave

Chapter 31
Letters to Miriam Yevick, 1954, Part 3

Letter 117. Folder C126, dated: Spring 1954.

Universidade de São Paulo
Faculdade de Filosofia, Ciências e Letras
Caixa Postal: 8105
São Paulo

Dearest Miriam

Thanks a lot for your three letters. I must apologize for not answering sooner as I have been busy with a host of things–rewriting the causality paper, writing 2 new papers on causal interp (hydrodynamical model, and spin theory). Thus I haven't had time yet to think too much about probability, but now I shall.

You must be about ready to have your baby. I am sure that it will be a fine bright healthy child, which you will certainly be very glad to have brought into the world. It's hard to believe, but there it will be.

I don't want to bother you too much at this time with technical things but I'll answer your letters now, and you can read it later at your leisure if its too much for you now.

As for me, I am in good health. The weather is cooler and I have never had so many ideas in so many subjects. Sometimes I feel that I will run myself to pieces unless I can slow down the pace at which the ideas come. It seems that the long hot period when I didn't seem to be making progress was a period of germination (or gestation) in which all these ideas were slowly getting ready to be formed. So you are not the only one who does such things! Anyway, it appears that my natural rhythm of thinking has to consist of a long slow period of vague and painful germination, followed by a torrent of ideas, then a new period of germination, etc.

Now about the probability problem, you are 80% right in what you said in your letters. I disagree only in that you underestimate the importance of the concept of randomness. Let me discuss the other problems first.

C. Talbot (ed.), *David Bohm: Causality and Chance, Letters to Three Women*,
DOI 10.1007/978-3-319-55492-1_31

(1) I would appreciate all translations and better yet, digests, of articles on the subject that you can get, arrange, or make.

(2) I would appreciate a digest on your part on what v.Neumann and others have done on ergodic theory, including a definition of what they mean by "non-uniform" integral of the motion, etc.

Now let me summarize the situation in this regard. By setting up a definite causally defined sequence in which the relative frequency approaches a certain value, except for a set of measure zero, we solve a problem not even faced in the usual theory of probability. For the usual theory says that rel. frequency approaches probability, except for a set of trajectories of measure zero. But what are these special trajectories? Not only is no means given for defining them or recognizing them, but worse yet, they can be defined only after the series has been given. In other words, given a series of events up to a certain number, n. Is there any physical or other property in the system which in principle at least defines whether we are dealing with one of those special sets of measure zero? The answer is, no! The only way that we can tell is to go thru an ∞ of events, and then we will see whether the rel. frequency approaches the probability or not. Thus, the probability becomes a mysterious property of an ∞ of events which does not come from any physical or other property of the real systems, but which just simply emerges as "a thing in itself", after an ∞ of events, with no particular origin. Hence, there is no way by which the origin and conditions of applicability of the theory can become intelligible. It either applies or it doesn't, and to tell, we can only wait and see. Of course, if it has applied in similar systems in the past, we can postulate that it will, "almost always" do so in the future, and thus use the theory in practice. But we have no intelligible basis for understanding the origin of the exceptions, or the conditions which permit them to arise.

Looked at from the other point of view, however, we see that whether it will be an exception or a regular case is already contained in principle in the physical conditions of the system. Thus, in the case of a set of angles multiplied by a constant, it depends on θ_0. It then becomes a definite mathematical problem, difficult perhaps, but always in principle solvable, to decide whether a given θ_0 will be an exception or a regular case.

We must still make the postulate that the exceptions will in fact occur rarely, but now we can see what this postulate implies physically and can see that, considering the real nature of the world, (many, perhaps an ∞ of irregularly fluctuating levels of motion) it is a very plausible postulate indeed, and that any other postulate would in fact be extremely implausible. Of course, we are not usually interested in practice in deciding whether a particular case is regular or not in this way, but it is important in principle, for an understanding of how the theory of probability comes to be applicable, what are the conditions for the applicability and what one can do in the cases where it is not applicable.

Now let us come to the problem of randomness. I agree with you that there were some errors in my presentation of this problem. The principal error was in the attempt to define randomness in a finite sequence. Actually, like probability itself, it can only be a property of the sequence, conceived of as unlimited.

But you are wrong in supposing randomness to be of no great importance, either conceptually or practically. For it is necessary condition for the applicability of the theory of probability. Consider, for example a series of coin throws. If we always got the regular sequence H T H T H T . . . etc., then we would have a relative frequency of $\frac{1}{2}$ for H, $\frac{1}{2}$ for T in the infinite sequence. Yet, it would make little sense here to talk of a probability of $\frac{1}{2}$, for several reasons:

(1) All of the rules of combining probabilities would not work (in particular the product rule) as we shall see presently.

(2) Many essential aspects of the theory (especially Bernoulli's theorem) would not follow.

(3) In practice, this is important, for the existence of correlations greatly changes many consequences of the theory. Thus, in sampling problems, it would no longer be safe to infer the properties of the whole from a small sample drawn in an arbitrary way. (Consider for example the problem of predicting elections where unknown correlations and regularities exist in political opinions of people).

(4) In physics, correlations are important, because many properties (e.g. the mean energy) depend on them.

It is true that statisticians have, in practice, often been able to deal with this problem without understanding it very well. This, however, is a fortunate circumstance arising from the very nature of a random distribution – namely, that if it really exists, then its properties are preserved almost independently of how it is re-organized; or of how selections are made within it. But this doesn't make the existence of randomness less important, but rather it makes it more important – as practical statisticians such as Gallup discover to their sorrow when they must deal with distributions that are not entirely random.

Now, in order for the theory of probability to apply, there must be suitable irregularity or "randomness" in the distribution of individual elements combined with a certain regular trend in the over-all mean behavior of an aggregate containing a large number of elements.

But how can we define this irregularity? This is a problem that has thus far not been attacked seriously, but the lack of an existing criterion for such irregularity does not make the irregularity still less objective and does not prevent it from existing, without our knowing as yet how to describe it precisely.

I am now preparing a possible attack on this problem. This is to use the aspect of predictability as a criterion for regularity. Let me explain myself.

(1) The theory of probability applies only when the specification of the state of the system is, for one reason or another, less than that needed to determine the system. (Think this over a little and you will see that it is obvious – for with a complete specification, there would be no room for probability to apply). In this connection, I am not asserting that a more complete specification is impossible, but only that, at a certain level, a complete specification is not actually given (e.g., at the macroscopic level, the significant physical variables, pressure, temperature, etc., leave most of the details of the atomic motion unspecified. At the social level, certain over-all economic data leave the individual economic conditions practically unspecified, etc.)

(2) But an incomplete specification by itself does not make the theory of probability applicable – it only makes room for it to apply. Thus, if there were some objectively regular behavior at the deeper level, in terms of which a more complete specification was possible, then the theory of probability would still not apply.

Let me illustrate this idea in terms of the coin throws. Now a specification of the results of a series of coin throws (e.g. H T H H T . . .) is not enough to determine what the next result will be, because additional causal factors are involved (state of muscle, bone, nerves, brain of man who throws it). But this lack of specification alone is not enough for applicability of the concept of probability. The actual process of throwing it has to have a certain objective irregularity.

If, for example, the man controlled his hands so well that he could reproduce initial conditions of the coin at will, he could establish a causal connection between the results of previous throws and those of later throws, such as H T H T H T . . . which would yield a predictable series (predictable once the man told us which series he intended to establish) and to which conclusions of the theory of probability concerning statistical fluctuations and Bernoulli's theorem would not apply. In reality, of course, the man cannot do this, because the movements of his hands depend on irregularly changing factors in his nerves, muscles, etc. that cannot be controlled by him sufficiently to permit the establishment of such a regular series. Thus, one requirement for applicability of the theory of probability is that no significant direct causal connections exist between the various results, which are strong enough to establish some regular order in the events or things in the distribution.

But a regular order could exist even without such a direct causal connection, if the underlying levels were undergoing some regular process. Consider, for example, a coin throwing machine, which threw the first ten throws according to some random process, but which thereafter reproduced perfectly the results of the first ten throws periodically. By observing the regular sequence of coin throws, we could then learn the rule, and predict the entire series on the basis of any ten throws, even though we knew nothing about what was happening inside the machine, so that we could not give a complete specification of the state of the system.

In order that all of the consequences of the theory of probability apply, it is necessary that there exist no such underlying regularity, which would make possible a predictability on the basis of an incomplete specification of the system.

Now, let us look into the question of predictability more precisely. First of all, we have determinate predictability. But this is a special case of a more general statistical trend toward predictability. In other words, if a given sequence, such as H H T H T tends to appear unusually often, then we can predict a statistical trend toward producing this sequence.

Now, let us consider a very long sequence of heads and tails (with N throws where N is very large). In a typical case, the number, N_A, of heads, will be close to that, N_B, of tails. Let us now, however, go through our sequence and pick all cases that are heads.

We now define N_{AA}, the number of cases in which heads are followed by heads and N_{AB}, the number in which they are followed by tails. Clearly

$$N_A = N_{AA} + N_{AB}$$

Similarly, we can define N_{BA} and N_{BB}, the number of cases in which tails are followed respectively by tails and by heads. Clearly

$$N_B = N_{BA} + N_{BB}$$

We then define the corresponding relative frequencies

$$f_{AA} = \frac{N_{AA}}{N_A} \qquad f_{AB} = \frac{N_{AB}}{N_A} \qquad f_{AA} + f_{AB} = 1$$
$$f_{BA} = \frac{N_{BA}}{N_B} \qquad f_{BB} = \frac{N_{BB}}{N_B} \qquad f_{BA} + f_{BB} = 1$$

Now, if there are no predictable trends in the long run, the relative frequencies of heads and tails should be [the] same in that subsequence of throws following heads as in the one following tails, as in the main sequence. Thus, we should have

$$f_{AB} \cong f_{BB} \cong f_B$$
$$f_{BA} \cong f_{AA} \cong f_A$$

This equality will not hold precisely in any finite sequence, but for an absence of a trend toward regularity, we require that it hold in the long run in an infinite sequence, in the same sense that the rel. freqn. tends toward the probability.

If the above relations do not hold, we have a two-fold correlation of nearest neighbors. Thus, the rel. freqn. of heads may be higher in the throws that follow tails than in the throws that follow heads. A special case of this is a determinate correlation such as $f_{AB} = f_{BA} = 1$; $f_{AA} = f_{BB} = 0$. This leads necessarily to the sequence H T H T H T H T . . .

We may also define correlations of more distant neighbors in the same way. Thus, for a completely irregular distribution

$$f_{AB}^{(\Omega)} \cong f_{BB}^{(\Omega)} \cong f_B$$
$$f_{BA}^{(\Omega)} \cong f_{AA}^{(\Omega)} \cong f_A$$

This relation must hold for an arbitrary separation, Ω, but Ω must be a great deal less than N, the total number of throws. For fixing Ω, we then let $N \to \infty$.

Then we may define the multifold correlations. Thus, if heads are obtained in the nth throw and tails in the $(n + \Omega_1)^{th}$, the rel. frequency of heads in the $(n + \Omega_1 + \Omega_2)$ throw may be higher than in the main sequence.

An irregular distribution must have no such correlations. Thus,

$$f_{A,A,A}^{(\Omega_1,\Omega_2)} \cong f_{A,B,A}^{(\Omega_1,\Omega_2)} \cong f_{B,A,A}^{(\Omega_1,\Omega_2)} \cong f_{B,B,A}^{(\Omega_1,\Omega_2)} \cong f_A$$
$$f_{A,A,B}^{(\Omega_1,\Omega_2)} \cong f_{A,B,B}^{(\Omega_1,\Omega_2)} \cong f_{B,A,B}^{(\Omega_1,\Omega_2)} \cong f_{B,B,B}^{(\Omega_1,\Omega_2)} \cong f_B$$

The same must be true for multifold variations of arbitrary order, but $\Omega_1 + \Omega_2 + \Omega_3 + \ldots \ll$ total number of throws. In other words, $N \to \infty$ faster than $(\Omega_1 + \Omega_2 + \Omega_3 + \ldots)$

Thus far we have discussed correlations consisting of fixed patterns of events. This is already enough to cover most of those of importance in almost all problems thus far treated in practice by theory of probability. Thus, if there existed an unusual tendency for heads to appear after the sequence H T T T H, a gambler who knew of this could refrain from betting until he saw this combination, then bet on heads and win systematically. In physics the recurrence of certain patterns of events in time (for regular but fixed arrangements of things in space) is usually all that is important, as is true in most social, economic, medical statistics, etc.

But the concept of irregularity cannot be exhausted by testing only for fixed patterns of events. For example, there may exist a trend toward the production of a pattern of events that is changing in a definite way. To test for this, we could consider multi-fold correlations involving a changing pattern of events. Thus, in the above discussion, Ω_1, Ω_2, etc. could be specified functions of the time, as could the order A, B, B, A, etc. But since the number of possible changing patterns is inexhaustible, we cannot give an explicit definition that covers all possible types of regularity. Instead, we give an implicit definition.

A completely irregular or "random" distribution is one for which [missing words] have statistical independence. Now a random distribution has complete statistical independence for all possible combinations of events. But distributions are possible in which there is, for example, 2-fold statistical independence but not 3-fold, and vice-versa. Or there may be statistical independence for all fixed patterns of events, but not for changing patterns. Or some of the f_{AB} may approach no precise limit at all, but may nevertheless remain within certain bounds, permitting the concept of probability to be used to some approximation. Thus there are all kinds and degrees of approximate and partial randomness. We may postulate complete randomness in a given distribution, and then use this postulate as long as we do not obtain contradictions with the more elaborate predictions of the theory of probability (e.g. probabilities of various kinds of fluctuations, Bernoulli's theorem, etc. But we must always be ready to discover that what seems at first sight to be complete randomness is in reality partial and approximate randomness.

Also, given any sequence (such as $K^n\theta_0$) we can test it for randomness up to any given order of correlations, time dependent or independent, and thus see whether, in a given context it will act effectively like a random distribution.

Finally, at the philosophical level and using the ∞ of levels, we see that the irregular or random side of things is a basic thread which must be woven together with the regular side to come to a more nearly complete concept of reality. For not only do all causal laws follow eventually from the regular trend in random motion arising at other levels, but also we can never reach the infinite depth and breadth and height needed to exhaustively trace all random fluctuations to causal origins. Thus the statistical hypothesis concerning irregularity or randomness reflect objective irregular aspects of nature, which unite within themselves at any given level the effects of an inexhaustible number of causes, some known, some unknown, some due to laws which are unknown and which may never be known, and all tangled up into a mess that could not possibly be analyzed completely at that level in view of the fact that many of the essential elements needed in this analysis exist other unknown levels, most of which are working beyond the frontiers of existing knowledge. Thus statistical laws have an objective validity at their own right, not only within the framework of their own level, but also they make an essential contribution to complete our understanding of other levels (because all levels are reciprocally connected). Thus the dream of analyzing all nature into a single most fundamental level does not reflect the real structure of the world. The real objective structure is in reciprocally interconnected levels, and because of this each level makes its own contribution to the totality of all causal laws. Each aspect (regular vs irregular, motion vs [unreadable] thing, being vs becoming) also makes its own contribution which cannot be reduced completely to the other aspect. Because everything is interconnected, each aspect, thing, or side of nature is reflected in other aspects, sides or things. More than this, it is mapped in them. More than this, they exert to some extent a controlling causal influence on it (and vice-versa). But the naive point of view of most scientists is that we can eliminate things in terms of other things which reflect or map them. The ∞ of levels shows the impossibility of this. Thus we avoid a tangle of logical paradoxes, into which those who want to reduce everything to one final secure basis have always fallen.

Love
Dave

Letter 118. Folder C124, dated: May 54.

Universidade de São Paulo
Faculdade de Filosofia, Ciências e Letras
Caixa Postal: 8105
São Paulo

Dearest Miriam

Congratulations on the infant. I hope you got my telegram and the two letters that I sent before sending the telegram. Please let me know all about him. After all, I am a sort of god-father, since he has my name. I would be in favor of your nursing him in the regular way. A few months of this will not hurt you, and may even help you, while it might harm the child to wean him too soon.

As for me, I am very busy, and also quite tired by now from so much work. I need a rest, which I shall take in a few weeks.

Not having heard anything definite from the Promised Land yet, my plans are to ask for a new contract here, and stay at least another year, obtaining a citizenship perhaps if the US moves toward Fascism too far. I am making so much progress in my ideas here that I am reluctant to move for a while. Discussions with Schonberg on dialectics and on the notion of a turbulent ether explanation of the quantum theory have led to ideas which look very promising, and I hesitate to break the train of thought until the ideas have matured a bit.

The US appears to be losing very steadily on the diplomatic front. I think that the first faint flush of dawn is visible. And once the foreign policy of the US fails abroad, they'll have to change at home. The US gov't seems to be extremely stupid. All of the intelligent people are perhaps so scared by McCarthyism that no sensible policy can be followed (not even sensible from the point of view of the businessman).

The US appears to have let itself be outflanked on the H bomb. It seems certain that the first US H-bomb designed to work on tritium, was probably a dud. Then the Russians exploded a lithium bomb, which worked; and from this the US scientists learned how to make a super-bomb. At least, this is the story that one pieces together reading Time, Newsweek, and the N.Y. Times along with the rumors carried by European scientists. But the US made such a big explosion that the net effect was to scare the British + the French, for it became plain that they would be wiped out before the war could get very far. Even the US military position is bad; for the Eastern cities are sitting ducks for such bombs. And according to the NY Times, the Russians have got bombers capable of delivering them. Thus, the atomic situation has now reversed; and as time goes on, it will very definitely favour the Russian side. Even now, to go to war would be to invite annihilation. I doubt very much that anyone will be anxious to do it. I don't see how they could possibly drag the British people into certain annihilation. They may like capitalism, but they don't like it that much.

My impression is that American civilisation is slowly disintegrating. No intelligent policy can be formulated anywhere, because of fear of McCarthyism. And the methods of propaganda and education keep on spreading pure stupidity and demoralisation. There is a limit to how far this can go, before it begins to affect the efficiency of science + industry. Also, if you look at who is running the US, you can see that they are just a bunch of businessman who probably haven't the faintest idea of the deep historical currents that they are trying to buck. And whoever has such an idea is careful to keep it to himself. Under these conditions, it is not surprising that the US makes diplomatic error after error, and that its science is falling behind that, not only of Russia, but probably even that of Britain. The big businessman were so confident of their power that they thought they could just order things around arbitrarily, without thinking of the consequences, as if they were gods. It now appears that they were wrong; and the destruction of American reserves of people who had a deeper insight into things (which was always extraordinarily small) is now either non-existent, or ineffective. The consequences of this policy have already begun to manifest themselves, and in the future, they will do so more and more. Already, in Time magazine, one no longer detects that sense of smug superiority and security, based on their consciousness of thousands of atomic bombs and "mile after mile" of steel mills. They are now discussing an "agonising reappraisal" of their policy. This American dream of limitless power will soon be shattered; partly because it was always exaggerated, and partly because they are "killing the goose that laid the golden eggs" with all this hysteria.

So much for this subject. Give my regards to George, and to the infant

Love
Dave

Letter 119. Folder C124, dated: July 1, 1954.

Dearest Miriam

Thanks a lot for your letter. I am sorry for delaying my answer, but I have been very busy writing up an enormous number of things. Also, I have just been away on a vocation.

Well, it's good to hear that you are becoming adjusted to your baby quite well. Naturally, I shall, whenever possible, discharge my duties as god-father, rich uncle and whatever else may be expected of me (teacher of philosophy + physics, and perhaps even a little politics in time, who knows?) I don't see why you didn't nurse him, and why George insisted on his being bottle fed, however.

As for my plans, I am still not decided. There are several causes of uncertainty. (1) the people at Haifa haven't yet finished the formal steps of offering me a job. While there is no doubt that this will be done soon, the offer will be of no use unless they get me a special p.p. As yet, they haven't discussed this aspect of the case. (2) In order to protect myself, I am now applying for a new contract here in São Paulo. (3) I am now working very efficiently, and have many ideas. It will take about a year to finish the phase of the work which occupies me right now. It is particularly convenient to remain here (a) Because Schonberg is working on this problem too and he knows a great deal of the math involved (connected with Pfaffian forms) (b) Because I have finally found where I can get good food that doesn't make me sick. If I moved, it would take 6 months to a year to make similar adjustments, and also I cannot expect to find anyone there who could collaborate with me on these problems. Thus, all in all, I am inclined to delay any moves for about a year, but not 100% certain. I agree with you however that the political [missing words] and that I may be able to return home in 1957. [missing words] reason, too, I must avoid any move that might involve loss of US citizenship.

If one is in good health, it really doesn't make too much difference where one is. It would be nice to have companionship, but I do not feel that the level is so high in Israel that it would make a <u>qualitative</u> difference to me, at least to the extent of making it worthwhile for me to slow down on this phase of the work. By the way, things are going very well in my work on the causal interpretation. The Pauli spin theory is now under 100% control – a paper is being written on it. The Dirac equation is practically ready, and now the problem of 2nd quantisation looks ready to crack open. Once this program is finished, the causal interp. will do everything that the usual interp. does and one or two things besides (such as explain Fermi-Dirac + Bose statistics). It will also suggest a new sub quantum-mechanical level. The picture that seems to be developing is as follows:

Space is full of an "ether". This ether is constituted of small spinning bodies. But the spins are not independent, – they are related to the mass motion, through the fact that the main rate of spin at a certain point is equal to the mean rate at which the flow is turning there $\left[\vec{w} = \frac{(\nabla \times \vec{v})}{2}\right]$ The resulting interconnection between spin and mass motion leads to a new property – that of "coordinated motion". In other words, while there is a random aspect to the motions of the ether, (as in ordinary fluids) this is not able to determine the characteristic properties of the ether. Instead, the coordination is so strong that particles do not collide with each other, but get out of each other's way, adjust to each other, etc. The analogy would be to a ballet dance rather than to a crowd of people. The coordination is not mysterious, but is a stable form of motion that is consistent with the basic law of motion. Random fluctuations take place around the stable coordinated pattern of motion, but they are not powerful enough to destroy the pattern. This coordination is characteristic of non-linear equations in many already well known fields.

Now, in this ether, there can exist stable modes of motion, involving the trapping of extra ether particles in a given region of space (or the trapping of a deficiency of ether particles). It is this deviation from the uniform background that constitutes "elementary particles" as we have known them. Thus, "elementary" particles are not elementary. They are higher-level entities built on the basis of a moving structure in the ether. This explains why they can be "created", "destroyed" and transformed into each other. Indeed, one can already see that at short distances, current quantum theory must break down, and that qualitatively new laws will apply in the processes by which such "particles" are "created", "destroyed" and transformed into each other; laws which in principle determine which structures and modes of motion are stable, metastable, etc. and which determine the laws of transformation. Of course, only the general outlines of the new theory are now visible, but the completion of the program that I have in mind will open the way for studying these new theories. Even by itself, it should convince many physicists that the causal interp. has richer possibilities than the usual interp. For this reason, I should like to finish this phase of the theory before moving to new fields (both physically & mentally). (Incidentally, you may tell George that the general picture that we developed when Vigier was here is beginning to take shape).

How about the probability problem. I shall try to express my ideas on this subject here. First of all, let me say that I still hold to the notion that real distributions have only an approximate randomness. But the point is to explain why randomness has to be defined as a limiting property of an infinite (or at least very long) sequence. More precisely, the forms of regular order that are used in any criterion for randomness should involve sequences having many less terms in them than the sequence as a whole. Otherwise there will be no way to discuss the statistical tendencies toward regularity. If one has for example a sequence of 1000 members, and one wishes to discuss whether there is a tendency to establish a certain regular pattern that requires 1000 members to be defined, one cannot discuss this problem statistically. The problem either exists or does not. But in a sequence containing 1,000,000 members, we could raise the question as to whether this pattern occurs frequently or not, and of how frequently it occurs. Thus, statistical methods may be applied here. Statistical methods may be defined as methods that could apply with complete precision only to the abstraction of a sequence having an unlimited number of members, but which will provide a good approximation when the number is sufficiently large (like replacing a continuous variable by a very dense set of discrete points).

One other point. I still hold to the idea that regularity must be defined relative to some order, an order defined by the physical conditions of the problem. But then we come to the problem of defining the various kinds of regularity that can exist relative to a given order. Then if all the physically significant kinds of regularity are absent in a given problem, we may treat the problem in terms of the simplifying abstraction of a "random" distribution, i.e., one having no regular order at all (Just as small finite bodies may be treated by the abstraction of mathematical points, in certain applications).

Our problem is now to define what we mean by randomness. To do this, I want first to give a review of my current general ideas on this problem. First of all, we have the notion of chance or accidental events. These are events that are contingent on causal factors existing outside the data specified by the physical conditions of the problem. The contingent factors may exist outside the system of interest or at another level. But to obtain true chance, mere contingency is not enough. There must exist irregular fluctuations in these contingencies (i.e. irregular relative to the data specified in the problem). For example, if we specify the macroscopic properties of a system, the microscopic behaviour of a given atom is not determined, but remains contingent on factors existing at lower levels (e.g., positions and velocities of other atoms, quantum fluctuations originating at a deeper level, etc.). These contingent factors show an irregular or "random" fluctuation. One of the most important properties of truly random fluctuations is that one can count on their mutual cancellation in the long run, when many objects or events are involved, to permit the development of regular mean trends, which are treated by the theory of probability. This cancellation is necessary to make many things what they are (e.g. the macroscopic level). Thus chance or "accidental" fluctuations dependent on contingencies are a basic aspect of the world in which we live, which help determine the modes of being of things.

Now how can we define randomness more precisely? First of all, we may use the n-fold correlations that I described previously. Surely, a necessary condition for randomness is that these correlations vanish. But it is not sufficient. As you pointed out, many other forms of regular order could exist (for example, one could have a sequence in which a time-dependent pattern of regularity existed which would not show up as a correlation of n-fold fixed patterns of events, but which would still constitute a regular order). Indeed, in an infinite sequence, the possible types of regular order are limitless

[missing pages, the remainder numbered as p 10]

agree with me. All I hope is that you will not accuse me of following some hypothetical standard line. At least, perhaps you will admit that I do some thinking of my own. Also, you must remember that I do say many silly things, which can be ascribed to personal depression, too much self-pity, etc. I am trying to avoid this sort of thing from now on.

As for the infant, I shall be very glad + proud if you call it David, (provided of course that it is a boy). I am glad to hear that the both of you are doing well.

Can you explain to me why Eugene doesn't answer my two letters?

Give my regards to George

Love
Dave

Letter 120. Folder C124, dated: Aug 1, 1954.

Dearest Miriam

Thanks for your two letters. I hope that you got my last letter of a few weeks ago. I am still awaiting an answer to it.

I certainly wish that you would keep writing. I am interested in the infant and would like a photograph of him. I realize that it is a big experience for you (and probably for George too). Of course I am far off here where I can have little touch with all that is going on there. But your letters are important for me. I have a certain tendency now to feel that there isn't much to be hoped for from people as one meets them here – they are so limited in possibilities and have long since accepted their limitations without really struggling to get out of them. It is important for me to keep in contact with people who are more alive intellectually. Indeed, I feel that if it can be arranged, I may go to Israel anyway soon, even if I will almost certainly not have as much opportunity to work on my ideas there as here (Class hours will be longer, and there will be more paper work, etc.). I have to get out of the dullness of feeling into which I am falling, or else I shall get old. Intellectually I feel that I am sharper than ever, but this too will be lost eventually unless I can get into an atmosphere where people care about those things enough so that I can continue to get excited about what I am doing once in a while. But when people get set in their careers, they generally fall into a groove and don't bother about such things. In Brazil, it is particularly bad, because everybody is superficial, but I am not so sure that Israel is really a great deal better. That is one reason I don't make "heroic" efforts to get there. If it can be done without illegality I will try it, but I am not sure it is really worth more than this.

Now about the problem of probability, I hope that you got my last letter, where I answered many of your questions. I shall once again however summarise a few of the ideas, but this time in a somewhat different form.

[missing words] start from the fundamental problem of what is a causal law. A causal law is a necessary relationship that connects things at an earlier time with those a later time. A causal law may be an individual law; [if] it is, it predicts uniquely, in an individual case, that certain effects will follow. But to do this, it presupposes a specification of all the causal factors that are significant for the effects in question. Now, however, suppose that we specify less than this totality , which we evidently must in reality always do, since for perfect precision of prediction, an infinity of such factors is involved. How is it possible to make predictions anyway? The answer is that the individual causal laws do not exhaust all of causality. There are also statistical laws which make predictions even when all the significant causes are not specified. For an individual case, these statistical laws merely define a certain range of possible effects (e.g. a range of errors) but for a number of cases, the mean is predictable. As we specify more and more of the significant causal factors, the range of possible effects narrow down and we approach closer and closer to an individual causal law. To reach an individual law, we may however have to go to qualitatively new levels (e.g. to a molecule level, to atomic level, to sub-atomic level, etc.).

In those domains in which individual and statistical laws treat the same things they must agree. For example, in a certain aggregate of cases, if we can actually trace the causal laws in each case, all one by one, the predictions for the totality must agree with those coming from the statistical laws of the totality.

This does not mean, however, that the statistical laws do not contribute something new of their own. For because of the ∞ of the universe, there is always an ∞ of irregularly fluctuating things that lie outside the scope of a given theory. These fluctuations may be inessential in many cases; that is, by better isolation or by better construction of apparatus, they could in principle be avoided in a certain problem of interest. But in other cases, they are essential to the very being of the thing that we are working with. Thus, the pressure or the temperature of a gas could not even exist except in a context in which random fluctuations in molecular motions, coming from many levels, were practically cancelling each other's effects on the large scale. But eventually all causal laws will arise in this way out of irregular motions at lower levels.

Now we see that statistical laws do not represent ignorance, but rather, another kind of knowledge. They tell us that in an enormous number of connections, the irregular fluctuations in the ∞ of factors not taken into account in our laws, fluctuations which in many cases are essential for the very being of the qualities that appear in our laws, do have a certain random character, so that they will tend to cancel out, and to have a limited (more or less Gaussian) range of fluctuation. If this were not so, then our approximate individual causal laws could not even exist. Thus, statistical laws not only represent knowledge that is of interest in itself, but the fact that they are valid is what permits us to know something reliable about a given thing without knowing everything about it. In other words, the validity of statistical laws is what permits us to ignore, in many problems, the myriads of causal factors that are not accessible to us.

We see then that statistical laws and individual laws express different sides of nature. We may compare nature here to a many-sided object (really infinite sided). Any given shape in the object can be projected on one of its sides. In many cases, it doesn't matter much which side we use, so that we get a set of overlapping descriptions. All of these descriptions will be closely related, since they refer to the same thing. But each will generally portray some feature of the object better. Indeed, for shapes that are perpendicular to a given side, we have no choice but to project them on another side. Thus, even though different sides (individual and statistical) may yield overlapping treatments of certain phenomena, if we go far enough, each will be seen to do something that the other doesn't.

The overlap in the two points of view makes possible an approximate explanation of one in terms of the other. Thus, if we go deeply enough, we may find the individual causal factors that explain a certain statistical law, but then a still deeper view will show that the very being of these causal factors arises out of a statistical cancellation at still deeper levels. Thus, we start out by explaining a statistical law in terms of individual laws but we are led also to explain individual laws in terms of statistical laws. What really happens is that both types of laws are different sides of the same process, so that they will have to be related to each other, and this is what makes possible such approximate explanations. Thus, given a statistical law, we can transform it into an individual law by specifying more and more of the causal factors (which usually exist however at other levels). This transformation makes possible an explanation of the statistical law, because we see how in each case, individual laws are applying, while in the totality of the cases, we get an overall behavior which is just what the statistical law gives without specifying all these causal factors. On the other hand, we may go the other way. We may apply our statistical laws to larger and larger aggregates. The irregular fluctuations then cancel out better and better, so that their effects become less and less relevant and eventually we have transformed our statistical law into a new kind of individual law. Here, however, we have explained the individual law as a consequence of a statistical law. A still broader view however is that what is a statistical law in the context of an incomplete specification is also more nearly an individual law in the context of a specification that is more nearly adequate to a unique definition of a certain effect. Or in other words, the same process satisfies statistical laws with regard to an incomplete specification, and individual laws with regard to a more nearly complete specification.

Now the problem before us [is] to work out in a few simplified and idealised but nevertheless plausible cases how these transformations may take place. The problem of going from the statistical to the individual law is already well worked out (i.e., in the law of large numbers). The problem of showing how an individual law can be transformed into a statistical law is therefore what concerns us.

Here, it is evident that two levels of specification are involved. First there is the specification of the individual law; and then there is the less complete specification which enters into the statistical law. Our problem is, starting from the specification appropriate to the individual laws, to show how the form of the law is transformed as we decrease the completeness of specification, and to see where something new (if anything) comes into the statistical law as we do this. Thus, we will, so to speak, watch the statistical law come into being, as we reduce the degree of specification, just as we watch the individual law come into being as we increase the number of cases. In both cases, something new will have to be introduced, which will involve the replacement of "always" by "almost always", or vice-versa, and this something new will represent a certain physical hypothesis about the nature of things; This hypothesis is plausible, but it is nevertheless a hypothesis. It is a hypothesis that is at the root of all theorems that aim to explain how and why things are what they are; for if this hypothesis were not valid, then things just could not even be what they are.

Now let us consider the example of $\theta_n = K\theta_{n-1}$; The precise specification is that of the individual law, given above. We then introduce a specification of precision, $\Delta\theta$. Even if $\Delta\theta$ is small, after a number N of steps, $K^N \Delta\theta$ will exceed 2π. This is the first step in the transformation to a statistical law. For in terms of the specification to a precision, $\Delta\theta$, a unique prediction of any precision at all becomes impossible for $K^N \Delta\theta > 2\pi$. Thus, the individual law disappears. Then, as we consider a longer time, new regularities appear which make possible a statistical law. For many properties, such as the mean number of times the system enters a given element ($\Delta\theta$) becomes practically constant and independent of the angle, θ_0, except for a set of measure zero. We see then that for most trajectories, a new kind of law has come into being. On the other hand, it is not yet a statistical law. For we are sure that this law applies only if the system is not in one of the special states of measure zero. And the problem is more serious than this. For in a finite number of steps, then we are sure that our statistical law applies only if the angle is not in a set of θ_0 of very small but non-zero measure. To obtain the statistical law, we need an additional hypothesis, plausible it is true, but finally based only on experiment and observation; namely:

In an enormous range of real statistical problems, sets of very small measure do in fact occur rarely; and in a given statistical problem, we make the hypothesis, that the systems that we are interested in have a similar property.

This kind of hypothesis has very far-reaching significance; for as we have seen, it is what makes it possible for us to formulate individual laws of limited precision without knowing everything about everything and it also makes possible the use of statistical laws in their own right. It is really a hypothesis concerning the behaviour of the infinity of irregularly fluctuating things that are always left out of any real theory.

We see then that we do not actually deduce the statistical laws completely from the individual laws. Certain aspects of them can be deduced, but in general what we do is to show that the individual laws leave room for statistical laws and help determine their form, but that the statistical laws are more than just consequences of the individual laws.[1] Similarly, when we go to the limit of large numbers, we do not deduce the individual laws. For the statistical laws merely show that the overwhelming majority of all possible motions leads to laws that remain within certain limits that grow moreover as more objects and events are considered. We need also the postulate that those that do not do this occur so rarely that they can be ignored. In other words, we have merely shown that within a certain approximation, a definite individual law, whose form can be deduced approximately from the statistical law, is consistent with the statistical law. A further hypothesis, based on experience and observation, is needed to bring us to the fact that such an individual law is practically always satisfied.

But now it is not enough to prove that the over-all relative frequency almost always approach a certain value. We must do the same for all of the other main consequences of the theory of probability. Thus, we must show that the approach of the relative

[1]In a less idealized treatment, this conclusion would be even stronger, since the very being of the angle θ_n would depend on the cancellation of statistical fluctuations at a deeper level.

frequency to its mean value is such that the error is less than $K\left(\frac{\ln n}{n}\right)^{\frac{1}{2}}$ except for a set of θ_0 of measure zero. And as von Mises pointed out, if we choose a given sub-sequence of the θ_n, then the relative frequency must be the same as in the main sequence, except on a set of θ_0 of measure zero. The same will hold true for any number of subsequences of measure zero relative to the total number of subsequences in the main sequence. And as I have stressed in my previous letters, if one takes any pattern of K events, where $K \ll N$, (N is the total number of events) then if we choose any set of measure zero of the possible subsequences drawn out of the totality of such possible patterns, then the relative frequencies must be the same as those given by the theory of probability, except for a set of θ_0 of measure zero. Thus, we justify the multiplication rule of probabilities. The net effect of all this research is to prove that the sequence is irregular enough so that it corresponds to the practical way that we deal with "random" sequences. No sequence will be perfectly random; but a given sequence may be so with regard to a set of properties of measure zero relative to the totality of properties that can be defined in the sequence as a whole. If it is random relative to all properties that are significant in the problem of theoretical or practical interest, this is all that is needed.

If one proves all the above, then one has made the following achievement. One has shown that probability is not a mysterious property, totally unexplained which because of your mere lack of complete knowledge, permits us to make predictions. Instead, we have shown how:

(a) Statistical laws are continuously and naturally connected with individual laws so that one can grow out of the other.

(b) How even the form of statistical laws grows, at least in part, out of individual laws.

(c) How statistical laws do in fact represent something new relative to individual laws, but that the new feature is not an unfathomable mystery but is a definite physical fact about the nature of things; viz., that in many real problems, sets of small measure occur very rarely. Of course, it is not true in all problems that sets of small measure occur rarely but then, statistical laws do not apply to all aggregates containing a large number of things. The essential point, which we may regard as a general principle, is that they apply in a large number of kinds of such aggregates, so that it is worth our while to try out a statistical law every time that we meet an irregularly fluctuating aggregate.

Another point to consider is that statistical laws have a certain approximate autonomy relative to the individual laws. For the same statistical laws could be the result of a wide range of kinds of individual laws. Thus, in the case of an equal distribution over angles, θ, an enormous range of kinds of individual laws could lead to the same equi-distribution. But we can work with the statistical laws, up to a certain point, independently of the individual laws underlying them just as we can work with individual laws independently of the statistical laws underlying them. Indeed, if this were not so, these kinds of laws would have little or no value.

But our problem here has been to correct the current false impression that no individual causal law can lead to a statistical law, so that there is an eternal and unbridgeable gap between the two kinds of laws, rather than the possibility of a continuous and rationally understandable transformation from the one to the other.

The relative and approximate autonomy of the general form of statistical laws is important not only in practice but also as a matter of principle. For it means that even if there are a large number of things that we haven't taken into account correctly in our individual laws, the same statistical laws (more or less) would apply. Thus, the statistical laws are not sensitive to precisely what is going on at other levels or in other systems to which the system in question is connected. This helps give them a certain permanence and being in their own right. Similarly, the individual causal laws are not sensitive to the precise statistical laws that underly them. This helps make possible two clearly defined levels, the individual and the statistical.

In your previous letter, you asked for an example of a non-equilibrium problem. Any problem in macroscopic physics will do. Thus, in hydrodynamics, certain macroscopic equations are valid (the Navier Stokes equation). Nobody has yet deduced these from the microscopic equations in a logically consistent way without arbitrary hypotheses, which are probably not entirely true (e.g. Boltzmann's hypothesis of "molecular chaos" in collision processes). The only equations that have been so deduced are the equilibrium equations of thermodynamics (e.g., pressure as a function of volume, and temperature). In hydrodynamics, a serious new problem arises. The statistical distribution of molecular properties has to be a function of the changing macroscopic parameters (pressure, density, velocity, in a certain neighborhood), while the macroscopic parameters are in turn determined by the statistical properties. In the equilibrium case, one has a theorem, Louiville's theorem, which proves that the statistical hypothesis of uniform probability in phase space is at least consistent. No such theorem is available in a non-equilibrium problem. Indeed, no clear definition of probability is even available; for since the distribution of probabilities is itself changing with time, in terms of what will we interpret these probabilities? Here, I think that what is needed is a generalized idea of a statistical law. A statistical law is one, which on the basis of an incomplete specification of an individual case, permits, in a large number of such cases, a more precise prediction of some properties than would be possible from one's knowledge of the individual factors alone. Laws of probability are special cases of statistical laws. But one can have statistical laws even in very regular systems. Thus, in a crystal, the small vibrations of the atoms about equilibrium lead to sound waves. One can predict the behavior of one of these sound waves without specifying just what all the atoms are doing. Thus, the sound wave is a statistical property of the crystal. Statistical properties are possible whenever certain over-all properties have a behavior that is, for [piece missing]

We may also use an a-posteriori justification. For the empirical discovery (or theoretical proof) that a distribution is random relative to a certain measure then shows us that it will be useful to treat the distribution in question in terms of this measure.

As for the rest of the questions of developing a detailed math theory, I think it is pretty straightforward to take care of them, now that the main outlines have been sketched.

Finally, there is one more application that can quite easily be carried out, that of the multiply periodic system. Consider, for example, the motion $x = a\cos(\omega_1 t + \phi_1)$, $y = b\cos(\omega_2 t + \phi_2)$ where ω_1 and ω_2 are not commensurable.

It is easily shown that in time, every element of area, $dA = dxdy$ will be entered by the orbit and that the time spent in the element will be proportional to dA. Also, considering two elements dA_1 and dA_2 one easily shows that if we consider two times, t and $t + \tau$, where $\omega_1\tau$ and $\omega_2\tau$ are both much greater than 2π, there will be no correlation between the times dA_1 and dA_2 are occupied.

The case of multiply periodic motion is broad enough to cover a large number of applications. Thus, it may be shown that any system of atomic particles moving in a confined space, V, is multiply periodic. This means that the coordinate of the nth particle is given by the multiple Fourier series:

$$x_n(t) = \sum_{\Omega_1,\Omega_2,\ldots} e^{i(\omega_1\Omega_1+\omega_2\Omega_2+\ldots+\omega_s\Omega_s)t} a_n\,(\Omega_1, \Omega_2\ldots\Omega_s)$$

In this case, the Hamiltonian can be written as a function of N action variables $(J_1, \ldots J_N)$ which are constants of the motion.

$$\mathcal{H} = \mathcal{H}(J_1, \ldots, J_N)$$

The phase angles, θ_n, which are the coordinates canonically conjugate to the J_n satisfy the relations

$$\dot{\theta}_n = \frac{\partial\mathcal{H}}{\partial J_n} = \omega_n$$
$$\theta_n = \omega_n t + \phi_n$$

We then write

$$p_n = \sum_n b_n(J_1\ldots J_n)e^{i\sum_{\Omega_n}(\theta_n\Omega_n)}$$
$$\chi_n(k) = \sum_{\Omega_n} b_n(J_1\ldots J_n)e^{i\sum_{\Omega_n}(\theta_n\Omega_n)}$$

where Ω_n are integers.

Now, the J_n are equal to constants. The periods ω_n will, in general be incommensurable. Thus, in the long run, we will have a motion in which each region $dX_1\ldots dX_n dp_1\ldots dp_n$ is equally occupied (as can be shown by the methods of Weyl). Thus, we can prove the uniform distribution in phase space and even study the way in which it is approached, with the passage of time, as the various $\theta_n = \omega_n t$ get out of phase with each other.

With regard to the book, I have already written a brief (6 page) qualitative summary of these ideas on probability. I do not think that a quantitative discussion would be justified here. Also, I think it will take about another year before all these ideas develop to a really sharp quantitative form. Nevertheless to sharpen your own ideas (and more) as well as to prepare for an eventual article in perhaps the Scientific American, I suggest that you start to write something. In my opinion, such an article will have to be re-written 5 or 6 times, but it will be worth it. Send me copies when they are ready if you decide to do this.

Well so much for the present. I shall continue soon.

Love
Dave

P.S. Is there anything the baby could use from Brazil? I might as well begin my duties as a god-father as soon as possible.

Chapter 32
Letters to Miriam Yevick, 1954, Part 4

Letter 121. Folder C124, dated: Sept 10 1954.

Dearest Miriam

Well, it's a long time since we have written to each other. Thanks for your two letters and for the picture of the infant. He seems to be quite an active fellow, very interested in life. Perhaps I shall be able to see him one of these years. I suppose that you are now beginning to recover from giving birth to him, but he will take up a lot of your time from now on.

As for me, I am afraid that I am entering one of those depressed phases again. I have reached a new impasse in my work, and I seem to be losing my strength. Maybe I am getting old, who knows? I seem to have no sexual feelings or desire left whatsoever, in fact the whole of the region between the legs feels as if it were quite empty. Also, there is a general lassitude. Doubtless part of it is due to the lack of stimulation in this environment but I am afraid that I just don't have as much energy and adventurousness as I used to have. I don't get a kick out of new ideas (perhaps because there is nobody to discuss them with). The fact that I am so sick of this place may also contribute. We shall see later whether this is really the main cause.

I haven't made much progress in the probability problem since I last wrote to you. I would appreciate it if you would explain to me in great detail the method of getting a Gaussian distribution causally from a series of "independent functions". I can't get the reference to which you refer, and I didn't understand what you said about it.

About the article, I think you are making the scope too broad. The ∞ of levels alone will get you in a tangle of problems that would take several chapters to clarify. I would suggest that you stress instead the idea that no matter how deeply you go in your analysis, experience thus far has shown that something has been left out. This could be other levels, or the effects of things that are spatially separated from the system of

C. Talbot (ed.), *David Bohm: Causality and Chance, Letters to Three Women*,
DOI 10.1007/978-3-319-55492-1_32

interest.[1] All theoretical treatments of real problems must operate in terms of some limited cross-section of the universe. But what has been left out has an influence. How then is it possible for us to make predictions (approximately) even though significant effects have been left out of account? The answer is that in a wide variety of cases, the effects left out have an irregular distribution and motion, which tends to cancel out, and which tends to remain within narrow limits (given by the Gaussian or Poisson distribution). Thus, to a first approx., we treat them by the theory of probability, which is valid in a wide variety of cases, practically independently of just what the individual causal laws are, as long as these lead to an irregular distribution. The essential point is that certain statistical or average properties of distributions are very insensitive to the precise details of the irregularity. A wide variety of kinds of irregularity lead to statistical properties that can be treated approximately in terms of the theory of probability. Hence, the theory of probability can, at a certain level, simply be postulated to apply, and we can verify it by direct observation. At a deeper level, however, we must show that the individual causal laws are consistent with the probabilistic statistical laws, and even more, that the irregular individual motions make the statistical probabilistic laws follow necessarily. Vice versa, statistical laws can lead to determinate individual laws for a large aggregate considered as a new super-individual (or higher level individual). Thus individual and statistical laws work together. Indeed, the validity of statistical laws is just as much needed to make the world what it is as is the validity of individual laws. Thus, if statistical laws did not apply in measurements, for example, we could never know the accuracy of our measurement without knowing everything about the irregular details of functioning of the apparatus. Another example is that of a book on a table. We say that because of the law of gravitation (an individual law) the book will remain on the table. But this conclusion depends also on the cancellation of the irregular motions of the molecules in the book. If, by chance, they all began to move in the same direction, the book could rise. The statistical law which states that this will practically never happen is just as much at the basis of our predictions as the individual law of gravitation. In other words, to make predictions about the behaviour of the book without having to specify what each molecule is doing, we have recourse to the fact that the molecules really do obey statistical laws. More generally speaking, the world is so constituted that certain over-all or statistical properties can be determined (approximately) without a detailed specification of everything; and this is possible because these over-all properties are so insensitive to their detailed properties. In a wide range of cases, the theory of probability has proved to be an adequate mathematical instrument for predicting these over-all properties.

There remains the problem, however, of showing how statistical and individual laws are continuously connected. This is the central new problem that we are trying to solve. One way of solving it is to show how an individual law can be transformed into a statistical law if we simply seek the relationship between new aspects of the system (over-all average aspects). To do this properly, however, we must show how the property of randomness can develop continuously, either as we consider a larger

[1] Work out a few simple examples of this in your article.

and larger aggregate, or as we consider more and more time. We need an enlarged concept of distributions that are neither perfectly regular nor perfectly random, so that we can see how random distributions come into being, and so that we can treat new kinds of problems. Otherwise, we are left with two totally different theories, the theory of probability, and the individual laws, which are assumed to be eternally separate, each existing in itself, each given once and for all, as mysterious ultimate constituents of the universe, which themselves never come into being, and which are therefore not subject to rational explanation. Thus, one must suppose that in some miraculous way, man has received the theory of probability as an inspiration, a flash of genius, which permits him to know a certain absolute type of law that prevails in certain statistical phenomena, a law that has no relationship whatever to individual laws that prevail in other types of phenomena, which are conceived of as equally absolute and final.

The picture which I propose is this: The totality of causal laws includes both statistical and individual laws. We start with this totality as our basic reality. Then, we may take various views of this totality, some of which stress the individual aspect of the laws, and some of which stress the statistical aspect. But there is no such thing as a perfect individual law, because there are always fluctuations and errors coming from what has been left out. In many cases, these errors do not go beyond a certain range, so that we have approximate individual laws. On the other hand, there is no statistical law that does not say something about the individual. Thus, in the case of coin throws, an essential aspect of the law is first that in every case, the result must be either heads or tails (and not for example 1, 2, 3, etc. or birds and fishes). Thus a statistical law must begin by specifying the range of possible effects which the individual can suffer. Then it gives the probability of each effect. But when we come to the higher level individual consisting of the average of a large statistical aggregate, then we discover that this becomes predictable, within a certain range of error given by Bernoulli's theorem. Thus every statistical law has individual aspects, and every individual law has statistical aspects. The notion of a completely individual or of a completely statistical law is therefore an abstraction, like that of a mathematical point, essential for analysis, but not existing strictly in reality. And just as we do not consider the world to be made up of an infinity of mathematical points, but rather, we regard the mathematical points as abstractions that are taken out of the more complex reality, we also do not regard the totality of causal laws as a collection of individual laws (or of statistical laws) but rather regard individual and statistical laws as abstractions which are likewise taken out of a more complex reality that is neither entirely individual nor entirely statistical.

Let me explain this idea better. We start with the idea of a real world, which is in a continual process of change and development. We must now find means of analyzing this change and development. To begin, we seek those aspects that have a relative permanence. Over a short period of time, these aspects may be idealized and abstracted as having a being, conceived of as static. But like the mathematical point, the notion of a property or an aspect of things as having such a static and complete being is only a simplifying abstraction. In reality it does not have such static being, as is shown by the fact that it changes after some time. The fundamental reality is that of matter in being and in process of change, or of becoming, as it may more accurately be called. Thus, we avoid Zeno's paradox. Let us recall that Zeno could not understand motion, because he regarded it as the occupation by an object of a series of points, each of which he conceived to be fixed. But this turns the problem upside down. The fundamental reality is motion, and the series of fixed points is merely an abstraction, which we use to analyze some of the effects of motion. Similarly, when we define a derivative, the fundamental reality is the distance, Δx, covered in the time, Δt. If the ratio, $\Delta x/\Delta t$ approaches a limit, then we may form the new concept of a derivative existing at the point in question, but like the notion of the mathematical point itself, this is also an abstraction. But we avoid Zeno's paradox by starting with motion and regarding the point as a limit of motion as Δt approaches zero. Thus, while we use the concept of the point to analyze some of the properties of motion, we do not assume that motion can be synthesized as the sum of a series of points. For the motion has additional properties (e.g., continuity, inertia, etc.) which were always present, but which are simply lost when we go to the abstraction of a point. (Of course, these properties will find a partial reflection in over-all or local properties of the curve of points, e.g., continuity of curve, existence of derivative, etc.)

To sum up, then, our procedure in dealing with nature is to abstract out certain aspects, and to study the relationships between such aspects. In doing this, we inevitably leave other aspects out of account. We may then study these other aspects later in terms of other kinds of abstractions.

Because we are dealing with abstractions, which leave out parts of reality, our theories will eventually demonstrate an inadequacy to deal with some phases of reality. This inadequacy will usually manifest itself in some contradiction between the different kinds of abstractions that we use in dealing with different phases of reality (e.g., contradiction between concept of object in being at a point, and object that moves from one place to another). This contradiction must be resolved at a higher level (e.g. we introduce notion of a continuous curve, which has a point structure as the limit of division into smaller and smaller segments, but we always take the segment structure as basic, and not the point structure).

Now to apply these ideas to probability. We note that causal laws are relationships between various aspects of reality at different times. Depending on which aspects that we find are necessary, possible, or convenient to relate, we will have different kinds of causal laws, some more nearly statistical and some more nearly individual. But the essential point is that one and the same system simultaneously obeys individual and statistical laws. For in reality, this system is infinite in its qualitative and quantitative complexity and richness. Out of this infinity, we abstract certain aspects to be related and thus obtain different kinds of laws. All of the different kinds of laws must be consistent with each other, and related in rational ways. But because of the ∞, we cannot reduce the system to one set of aspects, or to one kind of law. Each kind makes an essential contribution to our understanding of what the system is. Thus, we do not regard the world as made of certain fixed eternal basic elements, satisfying corresponding laws. Rather, we say that out of the ∞ totality of reality, we abstract certain aspects satisfying certain approximate relationships. Different aspects correspond to different views and cross-sections of reality. The way in which all different views and cross-sections fit together consistently and rationally is then a demonstration that they do arise in the same unique objective reality, and are not figments of our imagination, or coincidental combinations of sense impressions.

Well, that's about all I have to say for the present. Will write more later.

Love
Dave

P.S. I shall make a few comments on your plan for the article.

1 It is not entirely right to say that it is "too difficult" to deal with the individual laws (see p. 5. of your letter). Because of the ∞ of nature, statistical laws may contain information not contained in the best individual laws known to date. Thus; statistical laws are not purely a matter of convenience and practicability. Moreover every level of individual law ultimately has some deeper statistical basis. A more accurate statement of the problem is thus:

Both for reasons of practical convenience and for reasons of principle, we study statistical aggregates in their own right. Our theoretical justification for this is that the structure of the world has so often demonstrated a characteristic insensitivity to individual details that it is quite reasonable to look for such an insensitivity again every time that we come to a new statistical aggregate. Such an insensitivity can be demonstrated directly without a detailed knowledge of the individual, provided that it exists. If we find such an insensitivity, then we have a statistical law. Such a law can often be treated in terms of the theory of probability (just as individual mechanical laws can often be treated with the aid of the differential calculus). This does not preclude the possibility that there may be new types of statistical laws requiring a different mathematical instrument, just as we cannot be sure that the differential calculus is the last mathematical instrument that will be needed for the treatment of individual laws. (New mathematical methods may perhaps be needed to treat problems where randomness is not complete, for example).

After we have a statistical law, then we may investigate the individual laws to see whether they lead to this statistical law, or whether they are at least consistent with the statistical laws. If we discover such consistency, well and good. If we do not, then either the individual or the statistical law or both, are inadequate; and a deeper study must be made. Thus, in physics, statistical laws have often served as clues to the correct individual laws, especially in connection with the atomic theory. The essential point here is to maintain the balanced position that both individual and statistical laws are laws valid in their own right. But the inadequacy of causal theories is in the separation that is maintained between them. We must not go so far, however, as to assert that the use of statistical laws is only a matter of practical convenience. It is true that given a certain individual law we may often deduce a statistical law (and vice-versa). [The next sentences is very blurred in the photocopy] But the [unreadable] when given that the statistical law is an [unreadable] the individual law but that a [unreadable] of individual laws can give the same statistical law. Thus, admitting the ∞ of nature, we have no justification for favoring the individual law as more fundamental, since underlying it are statistical laws, while a given statistical law may depend on unknown individual laws, in a way that is so insensitive to their details that it is synthesizing the effects of individual laws in its own new characteristic way. The best point of view is to regard these laws as different but related sides of nature.

Now that I have re-read the rest of your letter, I see that we are on the whole in agreement about the rest of the basic ideas. What must be stressed however is that individual and statistical laws are abstractions as limiting cases of laws in general, and that there remains before us the problem of formulating more general types of laws that could connect these two limiting cases in a continuous and rationally understandable way.

The fact that individual and statistical laws are abstractions does not, of course, detract from their reality. For the essential point is that they are correct abstractions. Just as in the case of geometry, where we know that to deal with reality, the correct abstractions are points, lines, surfaces, etc. we know here that in a wide range of problems the correct abstractions are individual and statistical laws (though in new problems, new kinds of abstractions may have to be made). In any given problem, the abstractions describe the essential properties within the context under investigation, leaving out an ∞ of others that may be important in other contexts.

One more point. The problem of deducing a statistical law from an individual law (such as $\theta_n = K\theta_{n-1}$) can be looked at like this: Such an individual law is an abstraction, if only because any real motion is made of matter having a constitution in terms of molecules in irregular motion. Nevertheless, even if we had a situation where such an abstraction was adequate, we would still get a statistical law. This shows that individual and statistical laws are tied together in a naturally understandable way. But now comes the second important point. Even if we take into account the myriads of factors (e.g. molecular motions) that have been left out, the statistical law is so insensitive to these details that it will still be true. In other words, certain kinds of statistical laws operate at certain levels, with a certain being in their own right, that has some autonomy relative to the underlying individual laws. Therefore, the object of showing the connection between [word missing –presumably: statistical] and individual laws is not to eliminate the former in terms of the latter, but rather, to show a typical example of how the one could generate the other, merely so that we can see in a specific case how both types of laws could apply in the same system. In reality of course, all of the randomness in the statistical behavior does not arise from the motion $\theta_n = K\theta_{n-1}$, but a part always arises in other kinds of randomness already existing elsewhere or at deeper levels. Thus, when in a real case we analyze the origin of a given kind of randomness, we discover that part of it comes from irregular motions and part from other kinds of randomness. This emphasizes the inter-woven character of statistical and individual laws. But by considering idealized problems, we can see how individual laws are always leading toward statistical laws and vice-versa. In the idealized case, we can consider an individual law which exists apart from the statistical laws to which it gives rise. In a real problem, considering the ∞ of nature this cannot be done, because the same fluctuations at deeper levels whose cancellation leads to individual laws also lead to part of the statistical fluctuations existing at higher levels. Thus one cannot in reality make this neat separation between individual laws and statistical fluctuations. For they are united inseparably, and we can separate them approximately only in concept, with the aid of simplifying abstractions [words missing] the abstractions serve as [missing] approximations, but we must be careful not to identify the whole of reality with abstractions that apply in some limited sphere.

Letter 122. Folder C125, dated: Sept 20, 1954.

Dearest Miriam

This is just a supplement to what I said in my previous letter about probability.

Now the essential points in a statistical theory are:

(1) Because of the ∞ of qualities and aspects in nature, we must, in a given problem, abstract those qualities and properties that are essential, and seek relationships between them.

(2) To some approximation, these relationships will be independent of the <u>detailed</u> behavior of the ∞ of qualities left out.

(3) This independence is not perfect. Therefore, the properties of interest will depend, to some extent, <u>contingently</u>, on the ∞ that has been left out of account.

(4) In most cases, these contingencies fluctuate irregularly, relative to the qualities and properties that have been take into account.

(5) Nevertheless, certain average properties will, in most cases, fluctuate within narrow bands that are essentially independent of the details of the ∞ of things left out. When this happens, we have a set of conditions in which the theory of probability can be applied.

(6) In this case, the individual events as objects fluctuate "by chance". In a broader context, in which more of the significant properties and qualities are taken into account, the domain of chance fluctuations may be reduced. This can also be done by going to a "super-individual" consisting of an average of a large number of individuals. Thus real causal relationships always have some chance fluctuation in them (∞ of nature guarantees this).

(7) Unique causal connections and perfect chance fluctuations are limiting cases which never occur in reality. Real laws correspond to a point of view of reality somewhere between these limits. Nevertheless, the two limits are essential to our own thinking (just as geometrical points and lines are). They are abstractions, but under certain conditions, they are abstractions that are able to take into account what is essential to the real physical process. They do this, not only when considered in isolation, but even more when considered in their interconnection. For real physical processes correspond to a transformation from one limit to the other. Thus, by increasing the number of elements, we transform chance into necessary causal connections (approximately). By following the irregular motions (causally determined) we transform necessity into chance (approximately). Thus, each pole helps us to understand the other. The transformations between the two poles reflect the fact that they are just opposite poles of what is basically a unity; viz, objective reality as a whole.

Now for a more technical presentation of these ideas. As is well known, a criterion for equidistribution is that the relation

$$\lim_{N\to\infty}\sum_{n=1}^{N}\frac{f(x_n)}{N}=\int f(x)\,dx \tag{1}$$

must hold for an arbitrary function. Now Eq. (1) is a necessary condition for randomness, but it is not sufficient. Another necessary criterion is

$$\lim_{N\to\infty}\sum_{m,n}\frac{g(x_m,x_n)}{N^2}=\int\int g(x_1,x_2)\,dx_1\,dx_2 \tag{2}$$

This criterion tests for correlations between x_1 and x_2. This criterion can be generalized to

$$\lim_{N\to\infty}\sum_{m,n,o,p,\ldots}\frac{g(x_m,x_n,x_o,x_p,\ldots)}{N^{\Omega}}=\int\int dx_1\,dx_2\,dx_3\,dx_4\ldots g(x_1,x_2,x_3,x_4\ldots) \tag{3}$$

where we go to an arbitrary number of variables.

I believe that the satisfaction of this criteria would be sufficient to give us an adequate definition of randomness.

But now, a few qualifications are needed.

First of all, in a real problem, N will be finite but large. In this case, the total number of functions, such as $f(x)$, $g(x_1,x_2)$, etc. with which we work must be small compared with N. Otherwise, we could not have a statistical theory. If we had as many functions as there were values of x_n, then a determination of the means of all the functions would determine all the x_n. In this case, it would not be possible for these means to have a considerable degree of independence of the precise values of the x_n, and hence no possibility would exist for a statistical theory.

You may ask, "What determines the functions that we will use?" The answer is "The actual conditions of the problem. "In a given problem, the mean of $\sin x_n$, or x_n^2 may be significant (for example).

In a collection of charged particles, the functions of interest are

(a) The potentials, $V(\vec{x})=e^2\sum_n\frac{1}{|\vec{x}-\vec{x}_n|}$ (the sum being taken over all the particles)

(b) The interaction energies, $V(x_1,x_0)=-\frac{e^2}{|x_1-x_0|}$

For (a) determines the effect of the distribution on an external charged particle, located at $\vec{x}$, while (b) determines the motions of the particles. To obtain a statistical theory, (b) should be averaged over a large number of particles, for example in a given region. In my paper on plasma oscillations with Pines, I used instead of $V(x_1,x_0)$, the Fourier component $e^{ik.(x_1)} \sum_j e^{-ik.x_j}$, which is averaged over all the particles. But we had to take a number of k that was small enough compared with the total number of particles.

The particular criteria I have given on Page 1 of this letter apply for an equidistribution. We may, however, consider more general distributions with a probability $P(\vec{x})$. Then we have for arbitrary functions

$$\frac{1}{N}\sum_n f(x_n) = \int P(x) f(x)\, dx$$

$$\frac{1}{N^2}\sum_{m\,n} g(x_m, x_n) = \int P(x)P(y)g(x, y)\, dx\, dy$$

$$etc, etc$$

Still more complex cases may be considered. Thus it may happen that there exists a conditional probability, $Q(x, y)$. Then we have

$$\frac{1}{N}\sum_n g(x_m, x_n) = \int Q(x_m, x) g(x_m, x)\, dx$$

etc., etc. The condition for complete randomness is that $Q(x, y) = P(y)$

In the subsequent work, however, we shall deal only with the case of a uniform distribution. The ideas will evidently be easily generalized to arbitrary distributions.

These ideas are easily illustrated. Thus, in the case of $\theta_n = K\theta_{n-1}$, the class of functions, $e^{i\alpha(\theta_n)}$ provide a good example, where α is an integer, such that $\alpha <$ K. Then, it is evident that $\frac{1}{n}\sum_n e^{i\alpha\theta_n} = 0$, for this follows from the theorem of equidistribution that has already been proved for this case.

Functions like

$g(\theta_m, \theta_n)$ can be built as a Fourier series

$e^{i\alpha\theta_m + \beta\theta_n}$ Writing $\theta_m = K^{\Omega}\theta_n$, where $\Omega = m - n$, we have

$e^{i(\alpha K^{\Omega} + \beta)\theta_n}$

This too is equidistributed, since $(\alpha K^{\Omega} + \beta)$ is just an integer

Similar results follow for $g(\theta_m, \theta_n, \theta_o, \ldots)$ etc.

Thus, for a fairly wide class of functions, the series $\theta_n = K\theta_{n-1}$ has been shown to act like a random distribution.

Now comes another example of much greater practical importance. This is the group of harmonic oscillators with incommensurable frequencies.

If we have N harmonic oscillators we may write for their coordinates and momenta

$$x_n = A_n e^{i\theta_n}$$
$$p_n = i\omega_n A_n e^{i\theta_n}$$

where A_n is the amplitude, θ_n the phase, and $\theta_n = \omega_n t + \theta_{0n}$.

This problem is important

(a) because many systems (e.g. collections of atoms oscillating about equilibrium points) may be approximated as harmonic oscillators

(b) Many field equations (electromagnetic, Scrödinger, etc.) have solutions consisting of a sum of normal modes, each of which oscillates harmonically

(c) Even when eqns. of motion are not linear, motion is often multiply periodic, so that it can be expressed a function of amplitudes A_n and phases θ_n of a set of new effective non-linear oscillators.

Our first problem will be to prove the development of random phases when we have a large number, N, of oscillators. To do this, let us consider the phases $\theta_1 = \omega_1 t + \theta_{01}$, $\theta_2 = \omega_2 t + \theta_{02}$, etc. Let us suppose that the time t is large. Now there is a theorem that can easily be proved:

(A) For a set of $(\omega_1, \omega_2 \dots \omega_n)$, then the numbers $\omega_1 t - [\omega_1 t]$, $\omega_2 t - [\omega_2 t] \dots$, etc. become equidistributed for large enough values of t, provided that $\omega_1, \omega_2, \dots$ etc. are incommensurable with each other. ($[\omega_n t]$ is the largest integer in $\omega_n t$).

This result will hold for all t greater than a certain, T, except for a set of measure zero.

A few precautions are needed here in expressing the theorem. We divide θ into N spaces of size $\Delta\theta = \dfrac{2\pi}{N}$. Then, we count the number of oscillators that have phases in the ith cell. It is this number that will become equi-distributed provided that t is large enough and that $\dfrac{N}{\pi} \to \infty$ (so that many oscillators may be in each cell).

Thus, we prove the tendency to develop random phases. From here on readily prove that

$$\sum_n \frac{f(\theta_n)}{N} = \int f(\theta)\, d\theta$$
$$\sum_{m,n} \frac{g(\theta_m, \theta_n)}{N^2} = \int g(\theta_1, \theta_2)\, d\theta_1\, d\theta_2$$
$$etc, etc$$

Hence, all the rules of the theory of probability can be used in calculating averages of typical functions.

This example justifies the use of random phases, which is characteristic of the application of statistical mechanics to such systems.

Well that's all for the present. Will write again soon

Love
Dave

There is one more point. We return to the averages $\sum_n \frac{f(x_n)}{N}$, $\sum_{m,n} \frac{g(x_m, x_n)}{N^2}$, etc., which are physically significant in a certain problem. Now, let us define the space of the $x_1 \ldots x_n \ldots x_N$. This, we call the Ω space, which is an N dimensional space, representing the range of possible values of x for each case. Now we give our characteristic averages, such as

$$A_s = \sum_n \frac{f_s(x_n)}{N} \quad B_t = \sum_{m,n} \frac{g_t(x_m, x_n)}{N^2} \text{ etc.}$$

Fixing the A_s, B_s, etc., we then determine a certain range of Ω space that leads to these values of A_s, B_s, etc. If N is large compared with the number of functions A_s, B_s, then a large number of values of $x_1 \ldots x_n$ is possible leading to A_s, B_s.
This is what makes statistics possible, for averages of a few functions are independent of a wide range of details of the individual events.

We now have an easy way of going from an individual to a statistical law. For as we increase N, we increase the range of Ω space that corresponds to the same values of A_s, B_s, etc. In order for a statistical law to exist, however, it is necessary that the system move in such a way that it stays within the region of Ω space corresponding to certain values of A_s, B_s. For if the system moves out of these regions, then the results will fluctuate without coming to some limit.

Letter 123. Folder C125, dated: Nov 5 1954.

Dearest Miriam

I was very glad to receive your long delayed letter. Really, I was beginning to wonder if you were angry at me or had just begun to lose interest in writing. The letter came at an opportune time, as it helped get me out of a steadily deepening depression into which I have been falling (at least it helped for the time being). I am really fed up with Brazil, with no one to talk to and nothing to do at all in my spare time. Also the climate is too warm, and my health generally seems to be [line of text missing] whom I could have something interesting to discuss. But I am afraid that part of it is due to the fact that I am not so very young any more. I am approaching a time in life when this depression and isolation is becoming more dangerous to my health – there no longer exist reserves of strength which help pull me out, but rather in order to maintain my strength, I need to have at least occasionally some stimulation in the form of satisfactory contacts with people. As it is, I am beginning to feel cold inside, a coldness based on a steady growing sense that I will never have the kind of life that I have wanted, a coldness that seems to foreshadow that of death. It can still be remedied, this state of mind, if I can manage to get out of this place, but I had better do it soon. Then my impatience to get out of here is also very hard on me, because my efforts to do so seem to get smothered in all types of bureaucratic delays. (Please do not discuss the fact that I am trying to leave with other people). I have an offer of a job in Israel, and my problem is to get travelling papers. I hope that Israel will be better. Sometimes I think however, that in the long run, one will be in the same rat race everywhere. Nevertheless, a change by itself is desirable, even if it does not lead to an improvement. However, there is a good chance that socially speaking at least, Israel will be better.

As for the US, I am inclined to give it up as a bad job. The general decadence and corruption among middle-class intellectuals that you describe is no news to me. It is just what I would have expected. I would never have expected these people to "stand up" for one of their own group. Indeed, I am surprised that I got as much support as I did in regard to my contract from the Princeton professors. You see, I know these people better than you do, because I come from the same background. If it had not been for various circumstances that brought me into this political conflict, and for various special traits of temperament and character that made it difficult for me to "adjust" to the stupidity and inanity of middle-class life, I would have been one of them. So I know how weak and vacillating they are, how terrified they are of risking their positions, because I know just how I would feel in their position. I know how weak are the bonds of friendship between them, because I know that each person must take care of his own career first, of his own family first, and only then can he consider his relationships to his colleagues, which are much too tenuous and ambiguous to permit the establishment of real solidarity of class or profession. Of course there are a few notable exceptions, but these are too few and too isolated to carry appreciable weight these days. In fact, I have always known that almost none of my fellow physicists was really to be trusted, and this was one of the things that kept me from being happy in Princeton or in California. And I have always sensed how thin are the ties between the people whose principal social contact is through cocktail parties, and how hypocritical are such social relations. Of course, I did find a few friends among those who were ready to place other things above the necessities of career and personal security and this is one of the things that I miss the most. However, I feel sure that there will be similar people in Israel.

As to the immediate future of the world, I am not very optimistic. There will be no general war, to be sure, because as you say, it is too dangerous. But all parties in the West except the Communists are still only too ready to sell out, if America will only pay the few billion dollars needed to buy them. In this regard, the election of the Democrats will delay a final solution, because they are not quite so stupid and stingy as are the Republicans. The thing they like to do most is to engage in small but expensive military adventures, but they are also willing to offer loans, etc. to other countries, and thus to buy out the bourgeoisie of these countries. The Republican failure to do this has helped a great deal towards splitting the West, as has their unwillingness to expand trade, and their very brutal and overbearing attitude toward other countries. The Democrats will be smoother and slicker, and will take more of the steps needed to delay a US crisis. In this regard, we must remember that with the enormous wealth of the US, they have great resources at their disposal for delaying a crisis for many years. Of course eventually problems will come up that they can't handle, but this may take 10 years perhaps as much as 20, and even then, the first effect of such a crisis will only be felt in Europe. Considering how indoctrinated and corrupt are the American people, I would put 30 or 40 years as the minimum time needed before there will be a socialist movement of serious proportions in the US (unless of course something totally unforeseen happens, such as a Martian invasion with flying saucers or the Second Coming of Christ). Meanwhile I think that at least this generation and the next generation in the US will not amount to much. The rottenness will steadily grow, but it may take a long time before it affects the technical capacity of the country (e.g. even in Nazi Germany, their technical capacity remained high until the end). Even in Europe, I expect that the middle-class part of the left wing movement will continue to rot away, as these people sell out on their basic principles, for example with regard to arming Germany and placing American atomic bomb bases in their countries. Socialist and Labour parties will share in this process of decay, because they are just not able to take the stern and difficult measures needed to solve the problems that they face.[2] In time, the Communists will be the ones to take over, because they are the only ones who have the steadiness of purpose and clarity of goal needed to keep them moving in the same direction over 10 or 20 years, in spite of the temptation to sell out, also they are the only ones who do not basically fear the working class rule more than they fear the bourgeoisie. Thus, when the moment of chaos and weakness of the existing society comes, they will be in a position to take advantage of it, as has already happened in Russia and in China.

Of course, there are many things wrong with the Communists; they are narrow, dogmatic, often callous to the point of stupidity. As an example, one may take the case of this American, Field, who was imprisoned in Poland. After 5 years, they finally admit that he was falsely imprisoned on the word of an American agent Swiatlo. But still they don't seem to realise that this arbitrary police power is often more dangerous to the Communists themselves, than to the Americans (as happened in Hungary too, where they now admit that many innocent people have been imprisoned). Also, I noted that Field is "recuperating" in a sanatorium before he returns. One would think that they would be more civilised than to treat a prisoner so badly that he needs to recuperate in a sanatorium before he can go home.

To be sure, things seem to be improving now in this regard. Thus Malenkov in Russia appears to be following a somewhat milder course, and also in Poland and Hungary, there is strong criticism of what has been done in the past. One must also take into account the strong pressures under which these people work. They are trying to build socialism under conditions, not only of powerful bourgeois opposition, but also of lack of understanding on the part of the peasantry and of a large fraction of the working class. As has happened throughout the whole world, the people raised in the old society have been conditioned in such a way that they cannot see the need for a change, or the advantage of such a change, particularly if it means that they

---

[2]How can the Labour Party propose socialist measures and re-armament at the same time? Rearmament means that further benefits to the people are impossible, so that their socialist program is empty.

must give up something, however little it was (such as a plot of land) which they once had under the old set up. Moreover, even the Communists themselves are drawn from a background that conditions them to be stupid, short-sighted, narrow, full of hatred, etc. The really dangerous position in which they find themselves when they have taken, but not yet consolidated, the power, intensifies all these weaknesses. For every Lenin who understands deeply what the problems are, there are 10,000 who have only the sketchiest and most superficial ideas, who must therefore accept their theory in neat and clearly labelled packages. All of this can be remedied only with time, perhaps 2 or 3 generations. In Russia, they are proposing the idea of the new "Soviet Man" who will develop under Socialist conditions. I have seen some of their recent literature, in which this idea is embodied in many forms. I believe that there is a sincere effort going on to bring into being a new type of person who will look at things in a better way, and who will have a deeper and less mechanical understanding of political problems. But in these days of international tension and fear, this is quite difficult. Nevertheless, with the rising standard of living, and with increasing security from war, there is a good chance that this may come about. In this regard, I hate to think of what the first American Socialist state will be like. Because of their extreme superficiality, Americans fall into thinking mechanically in terms of slogans much more easily than other peoples. I can already see the possibility of television "jingles", urging people to over-fulfil their quotas, while dialectical materialism is explained in three letter words in the Reader's Digest.[3]

In sum, then, I think that things will remain tough and discouraging for a long time. In fact, I doubt that I will live long enough even to see the beginnings of a trend toward Socialism in the US. Perhaps I will be able to see Europe go Socialist, but even this, I am not so sure. Meanwhile, I don't think that very much of importance will be taking place in the US. The people there are so unreceptive that little can be done in regard to spreading socialism there. Meanwhile I think that I can make some contribution by trying to clarify our thinking a little bit. In the long run (and the run is going to be plenty long, I am sure), most of our trouble comes from confused and wrong ideas that are very widespread. The material conditions have been ripe for socialism for a long time; many opportunities have arisen during which the bourgeoisie has been weak. If socialism hasn't come, it is because people are trapped in a point of view toward the world propagated in the existing society. And everybody is trapped: from the highest to the lowest. Everybody imagines that the state of society and the character of people has to be more or less as it is for all time. In other words, while we may expect changes in the direction of making the existing kinds of things "bigger and better" (i.e., big television sets, aeroplanes refrigerators, cars, etc.) no qualitatively new way of living is even to be conceived of, because the way people are is something that cannot be changed. But this whole concept that what exists is inevitable is closely tied up with a similar concept that tends to grow up in the sciences, namely, that the general outlines of the existing point of view will be with us permanently and will be able to take care of any conceivable problem that

[3]Perhaps a comic book entitled "dialectical Dick" will be more appropriate, since by the time America is ready to become socialist, most of the people will have become effectively illiterate.

could possibly arise. In other words, while we may discover new details in the future, they will have to fit into the present general scheme of things. Thus, in physics of the 19th century, it was felt that the general scheme of classical mechanics was the final one, while the 20th century physicists feel that it is that of quantum theory. In genetics, it is generally felt that except for details, the basic patterns of heredity are contained in the genes as currently conceived, so that mutations can only be chance mutations having no regular relationship to the cellular environment. In economics, the tacit assumption of the final and unquestionable character of a certain general framework of ideas is even more obvious. It is clear how such an attitude both grows out of and re-enforces the notion that the general form of the existing society is inevitable and in the nature of things, so that it is not to be questioned. This is also connected with the empiricist attitude which starts out from the facts as they are and will not go beyond these facts, so that one tends to assume implicitly that the basic setting out of which these facts arise is unchangeable.

Another closely related attitude that helps trap people is to suppose that while changes can occur, these are beyond rational human understanding and control, so that they must simply be left to occur by themselves and not to be the subject of attempts [missing word(s)] at conscious direction. You can see how closely this attitude is bound up with the a-causal point of view in the quantum theory. One may say that in the 19th century, the bourgeoisie felt that it had things under pretty good control, and that the future would consist of a continual expansion of the existing state of affairs by paths that were pretty well determinable. This general attitude found its parallel in science, in terms of the attitude that the whole world would have to fit into the narrow determinate scheme of classical mechanics, and that nothing qualitatively new would ever develop. By the twentieth century, however, the bourgeoisie began to feel itself at the mercy of wild changes that it could not understand.

More accurately, it did not want to understand these changes, because such understanding would inevitably have shown the need for socialism. Thus, it tried to comfort itself by supposing that there were irrational social forces which however, would always operate within certain limited forms, such that qualitatively new social relationships could not arise (e.g. the Nazis suggested mysterious racial factors, thinking with the blood, etc.) this trend toward mysticism and irrationality had already arisen during the late 19th and early 20th century among philosophers, sociologists, etc. And in science, the parallel attitude grew that changes at the atomic level took place in a way that is beyond rational human understanding. However, the assurance was given that "the more these things change, the more that they would remain the same". Thus, the electron could jump from being a wave to being a particle, but it could never become something qualitatively new that is neither wave nor particle (just as you may elect the Democrats or the Republicans or choose either Fascism or Democracy, but in the end, it comes to the same old story).

Now you may feel that the attitude toward science affects only a few people, but this is not really true. In the long run, it has a great effect when a certain general point of view is backed up by scientific authority. This is more evident in Europe[4] than in the US, where there are so few people who think that ideas do in fact have little importance. I feel convinced that it will have an important long-run effect to show that while the world is governed by rationally understandable causal relationships, these relationships do not imply the impossibility of qualitative changes, but rather, they imply instead the necessity of such changes under appropriate conditions. The knowledge that the world is built in this way will give people an impetus toward looking at society in a similar way. Thus, the effect of such a philosophical point of view will be important, not only in the sciences themselves, but also in the whole of human life in general. It will help people begin to regard each other in a new light, not as creatures which must operate within certain fixed and narrow limitations, but rather as beings that can develop in qualitatively new ways, without limit, especially over a period of many generations. But this development will be possible only under suitable conditions of society. And if the conditions of society hinder a development that has become necessary, then one still cannot keep things static. For if a people is ready for a qualitative change forward, and does not take this step, it still cannot keep itself from changing. But the change will now be backward toward decadence and rotting away (witness the American people). One cannot satisfy the nostalgic desire to keep things as they are, or to return to a previous state of affairs.

So you see I do not agree with the view that you once expressed, namely that with the coming of Socialism, all of these philosophical problems can be swept up quite easily, so that efforts made in this direction at present are of little use. On the contrary, I believe that an incorrect philosophical attitude to the world is at the root of our difficulties. And this is also true in the Soviet Union and in China; for there is a great deal of narrow limited thinking about socialism there, which has led to bad and costly mistakes.[5] It is true that the philosophy of science is only a part of our whole philosophical point of view, but it is a part that plays a crucial role because in the last analysis, our philosophy must be based on what we think the world is and on how we think it operates. But what we think we are and what we think we are capable of becoming is closely connected with what we think the world as a whole is, and what we think it can become. And this latter opinion is at the base of whether we think a qualitative social change is really possible, necessary, and desirable. Indeed, if more people had a clearer idea of what they could be in another society, and of how much better this is than what they are and have to be in this society, they would realise that to sell out in exchange for refrigerators, cars, and pleasant houses is like committing suicide. For selling out in return for such trifling rewards (rewards that are so trifling in comparison to what one loses) is like killing the best and most essential part of oneself. In this connection, it is unfortunate historically that the word materialism has come to have the connotation of a pig-like

[4]For example, I have heard from someone that in a debate on causality given in Paris, when our friend Vigier got up to defend causality, he was strongly cheered by the audience (which contained a great many students). I will guess that many of the younger people in Europe recognise that the question of causality has important implications in politics, economics, sociology, etc.

[5]This is also a proof that these problems are not so very easy to solve.

concentration on food, sex, and various commodities such as automobiles, which is actually characteristic of much of the bourgeoisie. In reality, materialism means just the opposite, since it stresses the fact that the human spirit, being a manifestation of matter organised to form human beings, is capable of the endless possibilities of qualitatively new development, that is shown by all matter, and that is described as the result of a dialectical process. On the other hand, most forms of idealism regard the human spirit more or less as a finished article, "full of sin", and incapable of real development, except by some external influence such as the Grace of God. Thus, they effectively condone and sanction the present state of society as the only one appropriate to the permanently evil human nature that shows itself in "materialistic" preoccupation with the material symbols of wealth power and security. But people tend to associate the materialist point of view of Communism with a brutal disregard for all the higher possibilities of mankind. And for this reason, perhaps some new word needs to be found, in order to permit the clear expression of the ideas involved.

Of course, the problem is not purely one of ideology, since obviously important practical considerations of political organisation, military strength, industrial organisation, etc. will enter into the problem. But my point is that at the moment, the principal bottleneck in the Western world is that people have a false idea of what the world is, and with it, a false idea of what people are, and of what are the possibilities in "human nature". If these things were understood more deeply by an appreciable fraction of the people, it would be impossible to maintain the present social order for another year. And even if they were understood by a few of the leaders, I think that the transition to socialism could not long be delayed. But as one looks at the leaders of the socialist movement, one sees no such understanding. At best, they want to give the workingmen the comforts of a bourgeois existence, not realising how inadequate such a solution is, not only from the point of view of providing the basis of a practical political movement, but also in relationship to what is both possible and necessary if stagnation and decadence are to be avoided. And I have the feeling that even most of the Communists perceive the end in view so poorly that they cannot make it very clear to the people where they are going. But I think that a better understanding of the infinite possibilities in nature will help people to see better for themselves and to get across to other people a better notion of the infinite possibilities of humanity. And I think that this notion will be of crucial importance in the coming transition to Socialism in Western Europe, for example. Moreover, a clear perception of these points in those countries that are already socialist will perhaps help reduce some of the inhumanity and blundering callousness that is characteristic of people that have only recently emerged from a very backward, ignorant, and cruel social system.

Well, this has been more or less a statement of my principles and aims, as they stand at present. My only worry is that in my present state of isolation, I tend to lose the feeling that what I do will have an effect, so that I lose interest. It is necessary to discuss these things once in a while, and this is difficult to do here.

About your baby, I think it only natural that you do not find that it fills your whole life. Perhaps when there were families of 8 or 9 people closely united, home children and family could have made an adequate career for a woman, especially if she was not excessively intelligent. But now, with one or two children, who will not be influenced by the home nearly as much as by other factors (school, other children, television, etc.), it is absurd to suppose that the same situation prevails.

In this connection, I have been thinking that it would be good for your children if you did not stay in the US with them too long. Considering the natural tendency of children to want to be like all the others, and the enormous pressure that comes from television, school, comic strips, etc., the chances are very strong that any child will grow up with all the narrowness, superficiality and emptiness that is characteristic of an American. If his parents are reactionaries, narrow minded and selfish, there is some hope for him. For in revolting against his parents, he may also be led to revolt against the society. But if his parents are progressive, this only increases the probability that he will revolt against his parents in order to conform to the society. Of course, if you could establish an "island" of progressives (as was possible until recently in a few places like NY and Los Angeles) then the child's environment would be better. But for you, the chances are against this. Your children will play with other kids who come from "normal" American families; and it is very likely that your children will be ashamed of their parents' "Communistic" tendencies. Of course, you may overcome this difficulty, but I don't see how you are going to do it. The chances are that you will be just as estranged from your children as your parents are from you.

With regard to the question of happiness, I think that this is just plain impossible in the present society, unless a person is so stupid and short-sighted that he is unable to overlook the mess in which we live. I have never known anyone who is happy. Everyone admits that life brings no satisfaction. Even in the papers today, I noticed a writer who said that at the age of 40, we learn that happiness is impossible. Only, he doesn't realise that this is because we live in a rotten society. In this regard, Brazil is even worse than the US. For most people, life here is utter misery; and even for the middle classes, it is a pretty dull and tiresome affair. The city of S Paulo is hell to live in, noise, smoke, heat, impossible traffic, electricity, etc. don't work, nothing to do, nowhere to go, people never see each other, but just sit at home, etc., etc. Thus, the amenities of life that help cover up the emptiness in the US are not present here. Even the [Paulistas – unclear] is dissatisfied, if he has any intelligence at all. Outsiders find this place impossible. As for me, if I do not manage to get out of here soon, I don't see how life can continue. There is no one to talk to, nothing to do, nowhere to go. I detest every square inch of this ugly city. I cannot even walk the streets, because the buses give off a dense black cloud of sulfurous smoke that has given me a persistent infection of the sinus that I can't get rid of (all of this to save a few pennies by burning the lowest grade of fuel oil so that the rich businessmen can import Cadillacs to burn up the high grade gasoline). And I can't travel anywhere. Wherever I go, I get a diarrhoea that weakens and sickens me, so that I have no energy to drag myself through the hot sunshine, and cannot sleep at nights from nausea, as well as the heat. Anyway, I am so sick of primitive conditions that I have no curiosity to see any more of these things. If I stay here much longer, I shall become old very quickly and lose all interest in life. Each day one has somehow to find a way of getting through so as to wait for the next. I don't suppose that it is a great deal worse than in a typical American small town, however. My real difficulty is that I have no one with whom I can talk (actually there are one or two people, but one very rarely has an opportunity to see them). The women that I have seen thus far are out of the question, because there would be nothing to talk about. As for sex, the only hope of reviving my flagging powers is that I should be in an environment that would be generally stimulating, so that I could get out of the miasmatic state in which I find myself. On the rare occasions on which I do manage to find people to talk to, I feel much better all around. I think that I could manage to live in São Paulo if there were a group of 5 or 6 people here interested in scientific and other questions. By the same token, I would be in the same difficulty if I returned to the US and had to take a job where I had no contact with such people, particularly if I had to work all day at some job which did not seem to have any value other than that it furnished some money. So it seems to me as if the only thing for me to do is to try all possible means of getting out, and trying Israel as the next step.

Well, so much for this depressing subject. Please let me hear from you soon.

Love
Dave

Chapter 33
Single Pages of Letters to Miriam Yevick and Letter to George Yevick

Folder 126, Page 1

[Same as the first page of Letter 119, Folder C124, dated: July 1, 1954].

Folder 126, Page 2

[Reference to Levinson suggests Letter 110, Folder C123, dated: Nov 10, 1953].

5 [written on photocopy]

I think that as bad as some of the groups on the left may be in some details, they are the only ones with a sincere enough determination to do some something and with a good enough theoretical and practical basis to create even a hope for resolving the problems. As for the others, I think that none of them will ever be determined enough or have clear enough ideas to carry things forward to a real conclusion. In addition, most of the remaining left & liberal groups have pretty well sold out on crucial issues (especially, the armament of Germany).

As for my own problems, I must make a final decision soon. One possibility is to return to the U.S. I realise that there is a good chance of my landing in jail in a year or two if I do this. But when I consider what is happening in other countries, I can see how I might fare even worse there. Besides, there is the fact that in another country I wouldn't have much influence of any kind except through my scientific work and frankly, I no longer feel that I shall be able to make any fundamental progress in this direction. The most I could do is to consolidate and defend my present position without basically altering the situation. So if I really feel that there was something I could do in the US, I might seriously consider returning.

C. Talbot (ed.), *David Bohm: Causality and Chance, Letters to Three Women*,
DOI 10.1007/978-3-319-55492-1_33

Give my regards to George, and to the Grosses, if you see them.

Love
Dave

P.S. In your previous letter, you mentioned some advice of Levinson to the effect that one can always find some way of doing something. I'm not sure that Levinson is an especially suitable person to be offering advice. After all, I understand that as a result of his testimony, several people are in danger of being fired, some of whom have been working for 25 years or more in their universities and have families, for which they are responsible. I am sure that if one acts as Levinson [ends here]

Folder 126, Page 3

["Throwers of monkey wrenches" suggests date April, 1952].

Dearest Miriam

I hope that my long delay in writing has not caused you to imagine that I am losing interest in you. Far from it, your letters always cause me much pleasure, and I look forward to them with excitement. I too get angry when you don't write me. Your last letter but one (the harsh one) went astray, as I haven't received it yet. I hope it didn't contain the proof that the set $\sum e^{ik\cdot x_n} \rightarrow \sqrt{NlnN}$ except for a set of measure zero. I am very anxious to see this proof, and also the references to the articles. I have made some progress on the general case in statistical mechanics and now see the general outlines of a method for showing that: Except for a set of trajectories of measure zero, the macroscopic averages of quantities such as energy and momentum are numerically equal to the averages of these quantities over phase space. I'll write you more about it later.

I have been impossibly busy with classes, rewriting notes in Portuguese, and having an endless number of meetings. At present, there are 4 profs here & two of them are professional throwers of monkey wrenches into everything. Thus, the score in physics dep't meetings is likely to be 2 to 2. However, the director of the School backs up my side. We are trying to take the steps that would permit us to expand the dep't, creating a few new professorships, and other such things. We got our point through, but what a waste of time to argue with these professional saboteurs. Also, I hope to bring 2 recent Ph.D's here next September. One is a former student of mine (Weinstein) and the other is a student of Peter Bergman's (Schiller). These people come in the category of special assistants to me. However, I hope we can invite some people here as professors in a year or so. Also, we have a good chance of getting a new professor in cosmic rays soon, paid by Unesco. I have been doing most of the correspondence on this, and I have to see the head of the dep't about 3 times before I am sure that he sends the letters he ought to send (a typical Brazilian).

You may ask why I get involved in all this stuff. The [ends here]

Folder 126, Page 4

[Also from about April, 1952]

. . . fight them (and I only one in the dep't [missing words] they will wreck everything and put me in danger too. You [missing words] how everything depends on immediate personal influence here, and how things can change very swiftly. There is no stabilising tradition, as the school was founded in 1934. However, all factors are now on my side, and I had better use them while the tide is in my favour, before things go the other way.

Now there are more complications developing. An assistant of mine (whom I inherited) appears to be a Fascist. He is very wealthy and has direct connections with the highest military authorities. He has just established a foundation for an "Institute of Theoretical Physics" having obtained about $200,000 from the state government and $800,000 from private sources. He is a strong Germanophile, and has invited Heisenberg + von Weissacker to head it. They will work alternately in shifts of 3 months. Already Weissacker has come with 2 German students, and with the flotsam + jetsam of Brazilians who didn't make the grade in physics they can have about 7 or 8 people. The situation is loaded with dynamite, as this assistant of mine, when he was still a student, went to the gov't and asked for the army to take over the physics dep't because there were too many reds in it. The move was stopped, but the fact that a mere student can do this shows how shaky things are here. My assistant & I are still on courteous terms, but much trouble could easily develop here. The entire dep't resents this "Institute" very much. They are a menace, with all this money, big names, and political + military connections. They probably will never do any work, but few people in South America are competent to judge them, so they can build themselves up as a sort of "Princeton Institute of Brazil".

A few other points. Pauli gave a seminar against my theory in Paris, and appears to have convinced de Broglie to give up the causal interpretation once again. I have read a resume of Pauli's arguments, and they are almost childish. I certainly hope that he publishes them; for if he does, I can tear him to pieces very neatly.

As for Pais + the rest of the "Princetitute", what these little farts think is of no consequence to me. In the past 6 years, [missing words] at that place, not even on the [ends here]

Folder 126, Page 5

[About June, 1952, (see Letter 38)].

basis of the currently accepted theories. I am convinced that I am on the right track, and have made much progress on the ether theory. I have also finished up the causal interpretation by showing that an arbitrary probability density, $P(x)$, ultimately changes into $|\psi(x)|^2$ after enough time. This I shall publish soon.

As the girls, I have had little time for them. As I said, I am attracted by a cute little blonde in the physics dep't who is quite leftish, but every time I look at her again, I feel as if she is a very little girl indeed (which she is). However I think that she has possibilities (which will develop in a few years). Tell George that I have too much to do to be interested in forming the characters of little girls. I would rather have worn with an already formed character of a kind that satisfies me, (more or less).

Well, I'll write you more later. Do you have any more news on how my theory is being taken. I have had many friendly letters. Incidentally I am told that Courant likes it too.

Love
Dave

P.S. No word on passport situation yet. I asked Massey to get me invited to England, so I could ask for p.p., but haven't yet heard from him in two months. It doesn't look too good.

Folder 126, Page 6

[Same as the first page of Letter 108, Folder C123, dated: Oct 24, 1953].

Folder 126, Page 7

[Reference to Navier-Stokes also made in Letter 120, Folder C124, dated: Aug 1, 1954].

10 [written on photocopy]

. . . one reason or another, insensitive to the precise behaviour of the individual members of an aggregate.

The sharp separation between laws of probability and other types of statistical laws is artificial. Thus, in a liquid, the sound waves depend on the irregular chaotic motions in the atoms, but these in turn depend on the state of the sound wave. A new point of view is therefore to look for those over-all properties of the motion that are insensitive to the details of molecular motion. These will include probabilistic properties and collective properties of other kinds. But such properties constitute a new level of properties. Thus, the problem of statistics is closely connected with the problem of the appearance of new levels of approximately autonomous properties.

Well, this is about as much as I want to say for the present. In the next letter, I shall comment on the Hungarian papers, and discuss other possible examples to work on

Love
Dave

P.S Eugene Gross is interested in the Navier-Stokes equ. problem. Why don't you talk to him about it?

Letter to George Yevick

Letter 124. Folder C115, not dated.

[Early 1952, referred to in Letter 65, Folder C117, dated: Jan 7, 1952].

[Circulation list:]

1. Dr Montroll
2. Dr Melba Phillips, Brooklyn College, N.Y.
3. Dr Eugene Gross, Lab for Ins. Res., M.I.T.
4. Mrs Yevick
5. Mr Yevick

Universidade de São Paulo
Faculdade de Filosofia, Ciências e Letras
Caixa Postal: 8105
São Paulo

Dear George

I was glad to hear from you, and to learn that the cold war fever is abating. I am rather busy nowadays, and don't feel too well, because of diarrhea. I can get rid of it all, and feel wonderful by taking sulfa drugs, but as soon as I stop, I get it back again. It's very much like what I had in the States very often, only harder to get rid of. I'm now being examined by a doctor for bacteria, and we'll see whether something can be done. The weather is hot and damp. Summer is a bad season here, as it rains a lot. By March, it will get drier and cooler, and then we can expect cool weather until November. I have seen some of the cool weather here, and it's really beautiful with a hot sun and a cool crisp breeze. We are at a fairly high altitude, which helps make for a better climate.

I was particularly interested in your questions about the ether. I agree with Dirac that there is much evidence for an ether, and I have even more arguments, as the case in favor of an ether is even stronger if you adopt the causal interpretation of the quantum theory. I wish to try to summarize these points here. I also wish that you would show this to Miriam, as it contains certain points connected with some points on which we have been arguing.

1. Historical Aspects of Concept of Ether.

In the 19th century, physicists sought to explain the transmission of forces between particles in terms of the concept of a material medium, or substratum, filling all space, which carries electromagnetic waves, gravitation, etc. Serious work was done in order to find a mechanical model that would explain Maxwell's equations. A correctly functioning model was found, but it required the assumption of three small mutually perpendicular gyroscopes at each point in space. This gave the right results for Maxwell's equations, but everyone recognized the implausibility of the model.

Further experimental and theoretical work were even more difficult to interpret in terms of an ether. Thus the ether was either dragged along by the earth or not. If it was not dragged along, the Michelson Morley experiment was expected to give a positive result. If it was dragged along, there should be no aberration of light from a distant star. The negative result of the Michelson Morley experiment, plus the existence of aberration showed that neither assumption could be true. Meanwhile, theoretical work by Lorentz led to further confusion. Considering that all matter is made up of charged particles, he was led to conclude that if there were an ether, and if the charged particles making up a measuring apparatus moved through this ether, then measuring rods would shrink and clocks change their speed in such a way that the measured velocity of light would still be constant, even if measured by a moving observer (His results were carried only to order $\frac{v^2}{c^2}$; it was too difficult for him to go further). Thus, at this stage of history, the concept of the ether became tangled up in impossible confusion. Einstein short-circuited the whole ether concept in the following way: If, both experimentally + theoretically we are led to the idea that the ether has the properties that prevent us from observing our velocity relative to it, it would be better just to consider the properties of space and time abstractly, asking what these properties must be in order to explain the observed facts. This led to special relativity, a great triumph. But unfortunately, the insight gained from this theory was clouded by a positivist distortion. The positivists said "If the ether cannot be observed, we may as well assume that it does not exist". This statement was just true enough to be convincing, but just false enough to be capable of leading physics up a blind alley. The fact is that there is no way of deducing that something that we do not know how to observe does not exist. (The same problem arises in the causal interpretation of the quantum theory). What does it mean that something exists that we cannot now observe? Merely that with the laws of physics as known at present, we cannot measure it. But the laws of physics at any stage are always approximations. It is fatal to assume that any specified law contains the final truth. Thus, if, as is almost certainly true, relativity is (like pre-relativistic mechanics) also an approximation to the correct laws, then ultimately we may be able to discover conditions under which the relativistic approx. breaks down, and these conditions might permit us to observe the velocity of an observer relative to the ether. Thus, the idea of an ether cannot be thrown out just because we do not now know how to observe the ether. It must always be kept in reserve as a possibility.

2. Theory of Relativity.

When Einstein progressed to the general theory, he began to move away from the positivist direction, for he assumed that each point in space was the locus of a field, $g^{\mu\nu}(\vec{X})$. This field was not, in general, zero anywhere. Far from "matter", it was small, and near the coordinate of the particle, it was large. But in a sense, the field associated with matter now covered all of space. Thus, in a very altered form, something like an "ether" began to come back into physics. In this point of view, space, time, and matter were all united, in such a way that each could affect the properties of the other. Thus, it was an advance over the Newtonian concept of space and time, whose definition was independent of the material located in this space and time.

Effects of Quantum Theory.

With quantum field theory, "empty" space began to be filled with all sorts of things. First, there are the "zero-point" fluctuations of the electromagnetic field, which are needed to understand "spontaneous" radiation, the Lamb shift, the gyromagnetic ratio of the electron. Then there are the infinite number of negative energy electrons (and protons and neutrons). In the usual interpretation of the quantum theory, the meaning of these things is a little vague, but if one goes to a causal interpretation, one is led necessarily to the conclusion that "empty" space contains real particles moving in every conceivable direction and really fluctuating electromagnetic fields. Moreover because these particles are spinning, there is in a sense a collection of gyroscopes at each point [missing words] old ether theorists needed to make a model of Maxwell's equations. Thus, we appear to have an "ether", but it differs from the old ether, in that its components must satisfy the laws of quantum mechanics, a requirement of which the old ether theorists could not have known. But the word "ether" is a bad word, because the density of matter in the vacuum is very high (in fact, infinite in present theories, and presumably finite and very large in a correct theory). It is much denser than any known form of matter, so it is hardly very "ethereal", a better word would be "substratum".

4. Suggested New Concept of the World.

Let me now picture the world as viewed from this point of view. All space is full of this very dense substratum. "Matter" as we know it now, consists of small localized deviations from uniformity in this substratum. The substratum can also carry electromagnetic and gravitational forces, waves, etc. The "annihilation" of matter as we know it is just the destruction of a localized inhomogeneity while its energy gives rise to a wave that spreads out. "Creation" is the inverse process. Two forms of energy exist; "localized" i.e., electrons, protons, neutrons, etc., and "diffusing" (i.e., light waves) which are inter-changeable. The transformation of "matter" into energy and vice-versa can now be visualized in terms of the motion of the more fundamental material "substratum", which is not "created" or "destroyed" in these processes, but just transformed.

When we look at the uniform background of substratum, we see nothing, because our means of detection are just waves and other inhomogeneities in the substratum itself, which move without being impeded in a uniform background. Thus space appears to be "empty" and all we see are the inhomogeneities, which scatter light, deflect other inhomogeneities (i.e., "matter" as we now know it, etc.). A similar phenomenon actually arises in a metal lattice, where electrons move without being scattered, through very dense matter, as long as the lattice is regular. Only irregularities scatter an electron. Thus, in a certain approximation, an observer noting that electrons move through a metal with little scattering might conclude that a metal is almost empty, except for a very thin haze of "matter" corresponding to the lattice vibrations. A similar conclusion is drawn by us when we look out at the sky at night. It is just as possible that the naive positivist point of view, which attributes reality only to what is directly visible, may be wrong in one case as in the other.

How can we discover this "substratum"? We already have some evidence suggesting it exists. We can get more direct evidence only if by using these concepts, we can correctly explain experiments that the present point of view cannot explain.

5. Relation of Plasma Theory to Substratum Theory.

There are many elements of similarity between the plasma problem and the substratum problem. In both problems, we have a dense background of strongly interacting particles, which because of their interaction, can have some of the properties of fields. The substratum particles have spin and other properties which complicate the problem, but which give rise to the possibility of vector fields, tensor fields, etc., whereas the simple plasma theory can lead only to a scalar field. The full problem is still not understood by me, but I will show you how a simpler problem leads to results that suggest where we should look for a solution.

Let us consider a scalar field, $q(\vec{x})$, and its Fourier components, $q_k = \int e^{-ik.\vec{x}} q(\vec{x})d\vec{x}$. If the frequency of oscillation of q_k is ω_k, the field acts like a collection of harmonic oscillators of frequency, ω_k. Hence the Lagrangian can be written as

$$\mathfrak{L} = \sum_k \frac{1}{2}(|\dot{q}_k^2| - \omega_k^2|q_k^2|) f_k \quad (q_k \text{ is complex}) \tag{1}$$

The function f_k is arbitrary, because no matter what f_k is, we obtain the same equations of motion, $\ddot{q}_k = -\omega_k^2 q_k$.

Let us now consider a collection of particles which are very dense, and see if we can invent interactions between them which will lead to the above equations of motion for a set of collective coordinates of these particles. As I showed in my plasma paper, a good collective coordinate is the particle density

$$\rho(\vec{x}) = \sum_i \delta(\vec{x} - \vec{x}_i) \tag{2}$$

where $\vec{x}_i$ is the coordinate of the ith particle, $\vec{x}$ that of an arbitrary point.

The Fourier component is

$$\rho_k = \sum e^{-i\vec{k}.\vec{x}_i} \tag{3}$$

This describes a wave-like perturbation of the particle density.

Now let us suppose that our field q_k is nothing more than the collective coordinate, ρ_k, of a very dense set of particles that fills all space. By inserting (3) into (1), with $\rho_k = q_k$, we get

$$\mathfrak{L} = \sum_k \sum_i \sum_j \{ f_k e^{ik.(x_i - x_j)} (k.v_i)(k.v_j) - f_k \omega_k^2 e^{ik.(x_i - x_j)} \}$$

Now in the above sum, we first remove the term with $i = j$. In treating the remaining terms, we write:

$$\sum_k f_k e^{ik.(x_i - x_j)} = U(x_i - x_j)$$

$$\sum_k f_k \omega_k^2 e^{ik.(x_i - x_j)} = V(x_i - x_j)$$

We get

$$\mathfrak{L} = \underbrace{\sum_{k,i} f_k (k.v_i)^2}_{A} - \underbrace{\sum_k f_k \omega_k^2}_{B} + \underbrace{\sum_{i,j} (v_i.\nabla)(v_j.\nabla) U(x_i - x_j)}_{C} + \underbrace{\sum_{i,j} V(x_i - x_j)}_{D}$$

Term "A" is analogous to a kinetic energy. [Summing over] all k, and assume that f_k is isotropic, we get

$$A = \left(\sum_k \frac{k^2}{3} f_k \right) \sum_i v_i^2$$

and if f_k approaches zero fast enough as $k \to \infty$, this sum will exist. Thus, the first term is just the kinetic energy of the particles.

Term B is constant, and therefore of no significance.

Term $\underline{D}$ corresponds to interaction of particles with potential $V(x_i - x_j)$

Term $\underline{C}$ is a velocity dependent interaction, analogous to the vector potential terms, which describe <u>magnetic</u> interactions of charged particles.

Conclusion.

With suitable but plausible velocity dependent interactions, as well as ordinary potentials between pairs of particles, a scalar field can be represented as <u>exactly</u> equivalent to a collective treatment of a set of background particles in a substratum, acting collectively. My idea is that when spin, etc., are taken into account, perhaps all fields, electromagnetic, gravitational, etc. can be described in this way. The "substratum" particles are perhaps the negative energy electrons and protons, or perhaps the latter are only a crude description of the substratum particles.

Sincerely,
Dave

P.S. When you & Miriam have read this letter, will you please send it to Eugene Gross, Lab. for Insulation Research, Mass, Inst. of Technology, Cambridge, Mass.

MIX
Papier aus verantwortungsvollen Quellen
Paper from responsible sources
FSC® C105338

If you have any concerns about our products,
you can contact us on
ProductSafety@springernature.com

In case Publisher is established outside the EU,
the EU authorized representative is:
Springer Nature Customer Service Center GmbH
Europaplatz 3, 69115 Heidelberg, Germany

Printed by Libri Plureos GmbH
in Hamburg, Germany